CLASSICAL PRESENCES

General Editors

Lorna Hardwick James I. Porter

CLASSICAL PRESENCES

Attempts to receive the texts, images, and material culture of ancient Greece and Rome inevitably run the risk of appropriating the past in order to authenticate the present. Exploring the ways in which the classical past has been mapped over the centuries allows us to trace the avowal and disavowal of values and identities, old and new. Classical Presences brings the latest scholarship to bear on the contexts, theory, and practice of such use, and abuse, of the classical past.

Sciences of Antiquity

Romantic Antiquarianism, Natural History, and Knowledge Work

NOAH HERINGMAN

OXFORD
UNIVERSITY PRESS

OXFORD
UNIVERSITY PRESS

Great Clarendon Street, Oxford, OX2 6DP,
United Kingdom

Oxford University Press is a department of the University of Oxford.
It furthers the University's objective of excellence in research, scholarship,
and education by publishing worldwide. Oxford is a registered trade mark of
Oxford University Press in the UK and in certain other countries

First Edition published in 2013

Impression: 1

British Library Cataloguing in Publication Data

Data available

ISBN 978–0–19–955691–5

Printed in Great Britain by
MPG Printgroup, UK

for eli

Acknowledgments

This is a book about knowledge work, so it is a special pleasure to thank all those who have shared in the labour of writing it. Thanks are due first of all to my extraordinary research assistants, Crystal Lake and Karen Laird. Crystal was my first co-pilot on the seas of antiquarianism; without her findings and our conversations, I could not have envisioned the initial shape of the book. Much later she read several chapters and fruitfully challenged my arguments. Karen worked her way through late drafts of many of the chapters with great care and insight, allowing me to recognize the book as finished. My esteemed colleague Ted Koditschek was the first to read a complete draft, and he did so with his inimitable generosity and scholarly clairvoyance. Mike Horton's scrupulous attention at the proof stage prevented numerous errors from seeing print.

I am grateful to the staff and numerous libraries and archives, all of whom generously provided scholarly and logistical support. I would particularly like to thank Alla Barabtarlo and the staff of Special Collections, Ellis Library, University of Missouri; Gail Gilbert, Margaret M. Bridwell Art Library, University of Louisville; staff in the Rara-Lesesaal, Staatsbibliothek Berlin; Bernard Nurse and Heather Rowland of the Society of Antiquaries of London; the Royal Society Centre for the History of Science; and Bruce Bradley, Linda Hall Library of Science, Technology, and Engineering. Thanks are due as well to the University of Missouri Research Board and Research Council, respectively, for funding my time at these archives, along with writing time at two important stages of the project. The Research Council also supported the publication of the colour images in this book.

I would like to thank the publishers of earlier versions of two chapters in this book for allowing the material to appear here in revised form. An earlier version of Chapter 4 appeared as 'Romantic Explorations of Antiquity: Customs and Manners in d'Hancarville's *Antiquités Etrusques, Grecques, et Romaines*,' in *Romantic Explorations: Papers from the Koblenz Conference*, ed. Michael Meyer (Trier, Germany: Wissenschaftlicher Verlag Trier, 2010), 149–71. I would like to thank my hosts in Koblenz and the lively original audience for that paper, especially Christoph Bode, Frederick Burwick, and Michael Meyer. An earlier version of Chapter 7 appeared as 'Picturesque Ruin and Geological Antiquity: Thomas Webster and Sir Henry Englefield on the Isle of Wight,' in *The Making of the Geological Society of London*, ed. Simon J. Knell and Cherry Lewis (London: Geological Society, 2009), 299–317. I am indebted to this paper's original Geological Society audience for improvements to the material and to Cherry Lewis for organizing a magnificent field trip and conference. I am very grateful as well to Angharad Hills and the modern Geological Society for permission to publish this version. An apology is due to the editor of the Routledge series *Sciences of Antiquity*, whose competing claim to the title only came to my attention when this book was already in press.

Many generous friends and colleagues have read drafts of chapters, including Sam Baker, Thora Brylowe, Elizabeth Chang, Jeremy Coote, Stefani Engelstein, Sean Franzel, Joshua Gonsalves, Jud Herrman, Scott Hess, Amy Mae King, Richard Lansdown, Devoney Looser, Mike O'Brien, Pat Okker, David Read, Jonah Siegel, and Jeffrey J. Williams. For help with references, translations, and general erudition, I would like to thank David Constantine, Rich Foley, Antony Griffiths, Johanna Hobratschk, Pippa Letsky, Anatole Mori, Mary Jo Muriatore, and Dan Sipe. My editors at Oxford University Press, Hilary O'Shea and Taryn Campbell, have been both professional and kind. I would like also like to thank my three anonymous readers for essential contributions, my series editors for their generosity and patience, and especially Jim Porter for encouraging me to write the book in the first place.

I am grateful to my family and friends for providing the day-to-day inspiration that I needed to keep writing. Among many informal contributions, I would like to acknowledge 'The Dark Abyss of Antiquarianism,' the alternate title for my book suggested by an early anonymous reviewer. I would like to thank Mike Barrett, Trudy Lewis, and Ray Ronci for their instructive insights on tri-unity and early man. My mother, Christiane Marks, discussed issues of translation with me and continues to inspire me with her creativity and dedication to languages. Above all, I would like to thank my wife, Elizabeth Hornbeck, and our son Eli, who has written at least fifty books since I began this one. This book is for Eli.

Contents

List of Figures

CHAPTER 1

CHAPTER 2

CHAPTER 3

CHAPTER 5

CHAPTER 6

CHAPTER 7

List of Plates

Abbreviations

The following works are cited parenthetically in the text.

AEGR	Hancarville, Pierre de [P. F. Hugues], *Collection of Etruscan, Greek, and Roman Antiquities from the Cabinet of Sir William Hamilton.* 4 vols (Naples: F. Morelli, 1766–1767 [1768–1776]).
Antichità	Griener, Pascal, *Le Antichità Etrusche Greche e Romane 1766–1776 di Pierre Hugues d'Hancarville. La pubblicazione delle ceramiche antiche della prima collezione Hamilton* (Rome: Edizione dell'Elefante, 1992).
Antiquaries	Sweet, Rosemary, *Antiquaries: The Discovery of the Past in Eighteenth-Century Britain* (London: Hambledon and London, 2004).
CP	Hamilton, William, *Campi Phlegraei: Observations on the Volcanos of the Two Sicilies*, 2 vols (Naples: Pietro Fabris, 1776). Vol. 1 (text) is cited by page number and vol. 2 (plates and explanations) by (Roman) plate number.
H & N	The Hamilton and Nelson Papers, Vol. 10–11 of *The Collection of Autograph Letters and Historical Documents Formed by Alfred Morrison*. (Second Series, 1882–1893).
ILH	Nichols, John and John Bowyer Nichols (eds), *Illustrations of the Literary History of the Eighteenth Century: Consisting of Authentic Memoirs and Original Letters of Eminent Persons; and Intended as a Sequel to the Literary Anecdotes*, 8 vols (London: Nichols & Bentley, 1817–58).
LA	Nichols, John (ed), *Literary Anecdotes of the Eighteenth Century: Comprising Biographical Memoirs of William Bowyer, Printer, F.S.A., and Many of His Learned Friends; an Incidental View of the Progress and Advancement of Literature in This Kingdom During the Last Century; and Biographical Anecdotes of a Considerable Number of Eminent Writers and Ingenious Artists*, 9 vols (London: The Author, 1812–16).
SM	Gough, Richard, *Sepulchral Monuments in Great Britain. Applied to Illustrate the History of Families, Manners, Habits, and Arts, and the Different Periods from the Norman Conquest to the Seventeenth Century*, 2 vols. in 5 (London: Printed by J. Nichols, for the author, 1786–96 [1799]). Citations include part numbers (e.g. II.i.9) in the case of the prefaces, which are separately paginated in Arabic numerals.
V & V	Jenkins, Ian, and Kim Sloan (eds), *Vases & Volcanoes: Sir William Hamilton and His Collection* (London: British Museum, 1996).

Introduction: Knowledge Work and the Proliferation of Antiquities

It is not here the statues of Polycletus that we admire, but the admirable monuments of the workmanship of nature, taken from the ruins of the great Herculaneum overwhelmed by the ocean. Robert Jameson, 'Advertisement' to Essay on the Theory of the Earth by M. Cuvier (1822)

I EXPLORATIONS OF ANTIQUITY

In its historically specific sense, 'antiquity' refers to 'the period before the middle ages, the time of the ancient Greeks and Romans'. Yet during the markedly neoclassical eighteenth century, two new forms of antiquity emerged as strong competitors for public and scholarly attention. 'I defy Homer, Virgil, or any of their Bardships to produce so great a Hero as Syr Chas. Bawdyn', exclaimed the poet Thomas Chatterton in 1768.[1] Echoing Chatterton's cultural nationalism a generation later, the self-styled 'public defender of our Antiquities', John Carter, observed that 'in styling our ancient gateways grand, I hold them in competition with the triumphant gateways or arches of Greece and Rome', and chided 'blind' Englishmen for their 'unlimited adoration' of 'armless Venuses'.[2] Along with these medievalists, many naturalists too laid claim to antiquity, altering the landscape of classical antiquity itself. William Smith implied a deeper geological antiquity in claiming that 'organized fossils are to the naturalist as coins to the antiquary; they are the antiquities of the earth'. Other naturalists, however, continued to pursue geological evidence for 'the great antiquity of arts and civilization', while colonial ethnographers

[1] 'Antiquity' II.4.b., *OED* online; Chatterton, *Complete Works* I.17n.

[2] Carter, 'Pursuits' XXXVI, p. 524; 'Pursuits' II, p. 825; 'Durham Cathedral' 231.

documented this 'antiquity'—the supposed origins of human civilization—among the living peoples of North America and the South Pacific.[3]

Sciences of Antiquity redefines antiquarianism by articulating the component parts suggested by these examples: neoclassicism, medievalism, and global natural history. These subjects flourished together, fostered by the disciplinary umbrella of antiquarianism and by an increased demand for, and division of, scholarly labour that preceded the 'second scientific revolution', the rapid proliferation of specialized disciplines after 1800.[4] Beginning in the 1760s, new accounts of human and natural history competed in public debates over aesthetically and politically charged objects (fossils, ancient or 'primitive' artefacts, decaying churches); in the stylistic pursuit of objective forms of writing and illustration, transcending the merely literary or romantic; and in the commercial pursuit of learned and popular readerships, which demanded technological innovation and successful collaboration among fieldworkers, authors, illustrators, and publishers. By the mid-nineteenth century antiquity in this expanded sense supported an unprecedented range of new and emerging disciplines. In addition to a booming print culture, the learned societies, especially the Society of Antiquaries of London, did much to promote this expansion. The expansion of the timescale, I will argue, depended on an expansion of the republic of letters traditionally defined by these elite learned societies: fieldworkers and illustrators were indispensable witnesses in any experiment on deep time, whether prehistoric or pre-human, and as they achieved wider recognition they helped to transform intellectual life.

The empirical sciences promoted by the European Enlightenment gathered and organized an unprecedented volume of natural and historical particulars, gradually revealing a 'dark abyss of time' beneath the orderly chronologies of geological, biological, and human antiquity. The story of this historical turn has been vividly and variously told by intellectual historians, philosophers of science, and historians of many particular disciplines including archaeology, anthropology, and geology.[5] This book takes up the literature of antiquarianism and natural history at an early stage of this massive specialization, identifying what might be called a 'prehistoric turn' made possible by these early modern disciplines. While attending to these revisions of the timescale, each of the chapters also explores particular social conflicts within antiquarianism and

[3] W. Smith, *Stratigraphical System* ix–x; Whitehurst, *Inquiry* 272; and see, among others, Lafitau, *Customs of the American Indians*. Smith's analogy between fossils and coins was not new, but he supplied this early modern commonplace with a newly scientific basis in stratigraphy.

[4] On the 'second scientific revolution', see Cunningham and Jardine (eds), *Romanticism and the Sciences* xix–xx, 6–8.

[5] On the proliferation of data, see Lepenies, *Ende der Naturgeschichte* 16–21. On the individual cases, see Schnapp, *Discovery of the Past*; Foucault, *Order of Things*; and especially Rossi, *Dark Abyss of Time*.

natural history—two allied 'sciences' often practised within the context of a single knowledge project and bringing together learned with self-educated, aristocratic with humbly born, and metropolitan with provincial or colonial practitioners. *Sciences of Antiquity* seeks to recover a shared enterprise of knowledge production that drove the system of aristocratic patronage toward an early form of professionalism, an enterprise that determined the pursuit of objects as disparate as Tahitian *marae*, Mount Vesuvius, Greek vases, and Salisbury Cathedral. Though now they seem to belong to the history of four distinct disciplines, these objects were initially claimed by antiquaries for one larger discipline that was the primary force behind the late Enlightenment proliferation of 'antiquities'—of specimens and of ancient pasts. Antiquarianism in its many forms is the source of many of the practices informing later, more specialized study of the prehistoric, classical, medieval, and even geological pasts. Recent scholarship has placed natural history squarely in the context of colonialism, and antiquarianism too was in some sense a 'conquest of the past'.[6]

The proliferation of antiquities holds rich possibilities both for the history of ideas and for social history. This book draws on both to account for a scientific turn in the study of the past as the product of old affinities between antiquarianism and natural history and new approaches to the division of labour. One of my goals is to reinsert labour into the literary history of scientific knowledge by highlighting the crucial role of the knowledge workers—fieldworkers, guides, illustrators, and other, often self-trained, practical scholars—who contributed so substantially to the research and publication projects associated with the learned societies and yet were often ineligible for membership in them. Chatterton, Carter, and Smith were among those who fuelled the operations of publishing and institutional knowledge production from the outside with objects—natural and artificial, ancient and primitive, real and fabricated. The particularity of these objects requires more detailed attention than it typically receives in scholarship on print culture and learned societies. The intellectual activity that produced the objects drew from both these publics, mediating between scholarship and the literary marketplace in a variety of ways too complex to be reduced to popularization.[7]

[6] *La conquête du passé* is the original title of Schnapp's *The Discovery of the Past*. In the British context, Lord Elgin's removal of the Parthenon sculptures promoted a new synthesis between antiquarian nationalism and imperial practices of collection. See my 'Stones So Wonderous Cheap' and cp. Jasanoff, *Edge of Empire* pt. 3, on Napoleon's extraction of antiquities from Egypt. Miller and Reill (eds), *Visions of Empire* provided the impetus for much subequent work on colonial natural history.

[7] For a complex analysis of popularization in the context of geological discipline formation, see O'Connor, *Earth on Show* 10–13. By contrast, institutional histories such as Herries Davies' *Whatever Is under the Earth* (within geology) or Evans' *History of the Society of Antiquaries* tend to emphasize the inner workings of the learned society. Sweet's *Antiquaries* succeeds admirably in addressing both popular culture and the learned societies. Detailed historical attention to

The individual tactics of knowledge workers, including discovery, recovery, classification, fabrication, and publication, also account for the proliferation of antiquities, of the objects and the historical domains that multiplied in such a characteristic way at the moment preceding the modern disciplines. The intensified scrutiny of Greco-Roman sites and objects in this larger context further unsettled traditional chronologies and helped to precipitate the fragmentation of antiquity into historical, prehistoric, and pre-human domains.

The careers of these knowledge workers—my central examples include Sydney Parkinson, Pietro Fabris, Pierre Hugues d'Hancarville, Jacob Schnebbelie, and Thomas Webster—all provide avenues for exploring the relations among antiquity, research-for-hire, and specialization. Cultural histories tend to credit their more famous employers—Sir Joseph Banks, Sir William Hamilton, Richard Gough, and Sir Henry Englefield—for many of their shared achievements, but these practical scholars produced a substantial portion of the relevant knowledge in the field. Neither the patrons nor their clients, or more properly collaborators, were motivated by an exclusive disciplinary programme. While many patrons of antiquarian research, including Banks, Hamilton, and Englefield, simultaneously pursued research in antiquities and in natural history, the fieldworkers, guides, and illustrators whom they employed—the latter in particular are a major focus of this study—developed independent expertise and empirical techniques that took them beyond the bounds of the client-patron relationship and enlarged the nexus between writing and fieldwork.[8] It is worth noting, too, the international makeup of this cadre of knowledge workers, which included (among many others) German contributors to British and imperial antiquarianism and natural history such as Rudolf Erich Raspe and John Reinhold Forster, who with his son George accompanied Cook's second voyage to the South Pacific (1772–75).

Knowledge work accounts for much of the cosmopolitan flavour of Enlightenment research. As Adriana Craciun has noted, Cook and his associates appear as 'explorers' only through the modern framework of the disciplines. Like that of Craciun and her collaborators on 'The Disorder of Things', my work aims to contribute to the emerging theory of 'predisciplinarity'.[9] Foucault's model of an 'epistemic shift' is undeniably important for this theory, as

objects of knowledge seems to thrive best in exhibition catalogues or multidisciplinary edited volumes such as *Things That Talk* (ed. Daston), where there is less space for institutional and/or social history.

[8] Clifford Siskin has emphasized the literary part of this equation, defining disciplinarity itself as a way to facilitate, by regulating, the boom in knowledge production (*Work of Writing* 20–1). The most prolific antiquarian writers, however, were often stigmatized as the least productive, especially in the absence of fieldwork. The proliferation of antiquities carried a very different set of connotations in the satirical tradition represented by Walter Scott's novel *The Antiquary* (1816) and George Cruikshank's cartoon *The Antiquarian Society* (1812). Here antiquities proliferate because of antiquarian credulity.

[9] Craciun, 'What Is an Explorer' 30–1; Calè and Craciun, 'Disorder of Things'; <http://ideasandsociety.ucr.edu/disorder_of_things/index.html>.

the term suggests. My response, however, is to recognize antiquarianism and natural history as disciplines in their own right, rather than defining disciplinarity itself as modern (as he does). Foucault's binary opposition between a static natural history and dynamic biology fails to capture the dialectical relationships between natural history and antiquarianism—which 'resist or escape current disciplinary mapping'—and their more specialized successors.[10]

II ROMANTIC ANTIQUARIANISM

Even today the term 'romantic antiquarianism' carries faintly pejorative connotations. The term is associated with a body of literature long consigned to the limbo of 'pre-romanticism' by literary scholars and set aside as empirically unsound by intellectual historians. According to Stuart Piggott's *Ruins in a Landscape* (1976), one of the groundbreaking studies in the history of antiquarianism, the study of British antiquity lost its way in the romantic period, exchanging a Baconian attention to physical remains for 'irrational beliefs of Gothick romance and a past full of Noble Savages and mystical Druidry'. This is a strong distinction between text and material culture, verging on the polemical, but for Piggott there is little danger of overstating its importance for the history of archaeology. Art historians such as Sam Smiles have followed Piggott in establishing visual documentation as the meeting place of art and material culture.[11] Piggott's analysis and more recent scholarship on scientific illustration by Smiles and others, in both antiquities and natural history, have influenced me in this study toward a consideration of plate books—scholarly descriptions illustrated with copperplate engravings—rather than literary antiquarianism as conventionally understood. My project, however, is to reincorporate these works with literary history—not to valorize them as nonliterary and therefore objective. To achieve a synthesis we must recontextualize visual documentation as knowledge work and examine its close affiliation with authorship. Literary antiquarianism in this sense refers to the production of new texts (rather than the study of old texts) that generated authority both for new kinds of antiquities and for a wider range of inquirers.

[10] Foucault, *Order of Things* 275 and *passim*; Calè and Craciun, 'Disorder of Things' 1. Craciun's own term 'predisciplinary comprehensiveness' seems to merit more strictly than Foucault the 'radical historicity' with which his model of 'classical' sciences is credited by Craciun and Calè. Their co-authored essay ('Disorder of Things') provides a spirited sketch of the historiography of the disciplines.

[11] Piggott, *Ruins in a Landscape* 120; Smiles, *Eye Witness*, esp. 27–9, 53, 71. In contrast to Smiles, some art historians tend to emphasize the discrepancy between representations and archaeological realities, as documented in Smiles and Moser (eds), *Envisioning the Past*. See further Chapter 6 below.

Parkinson, Fabris, Piaggio, d'Hancarville, Schnebbelie, and Webster were all aspiring and, to some degree, successful authors, and like their patrons in this respect. None of them rested content with fieldwork, illustration, and classification, fundamental as these contributions were, but took their technical knowledge—traditionally seen as menial—to be a qualification for authorship. From this point of view, scientific antiquarianism too may be seen as belonging to literary history and the world of the text, rather than a space apart.

To situate writing on antiquities and natural history in this way means revisiting familiar distinctions between romanticism and classicism, on the one hand, and literature and science, on the other. Print culture provides one common denominator for both pairings, albeit a very basic one. The technology and form of the plate book, and the issues it raises for commercial popularization, cut across all four domains and provides a starting point for discussion. The 'Gothick romance...and mystical Druidry' associated with literary antiquarianism were as neoclassical as they were romantic, though certainly they depended on manuscripts (when present) and philology more than artefacts or fieldwork, as suggested by poems of Ossian and the criticism of Thomas Warton, among others. The discourse on 'Noble Savages', however—here Piggott conflates literary antiquarianism with something quite different, conjectural history—had elements of scientific fieldwork as well as classicism and romanticism. Thus not only a shared marketplace for print but also a shared repertoire of concepts connected scientific antiquarianism with literary culture. In this light, the literary pursuit of Celtic and Gothic antiquities is less a falling away from the Baconian empiricism of seventeenth-century scholarship than a point of entry for authors of more empirical studies. Even when the content is poetic or 'mystical', a book is part of material culture itself and can influence perceptions of the material past.

The military garb and deportment of heroes in the poems of Ossian or *The Ballad of Chevy Chase*—revived by Thomas Percy's *Reliques of Ancient English Poetry* (1765)—surely provide a less reliable guide to ancient material culture than the uncannily realistic watercolour of a newly excavated Romano-British helmet commissioned by the Society of Antiquaries in 1798.[12] Nevertheless, these representations share a field of antiquarian possibilities. As demonstrated by Chatterton's dramatic career as a forger/scholar, authorship provides the common literary locus of both antiquarian speculation and invention, on the one hand, and empirical studies, on the other. Although my focus here is on the latter, this emphasis would not be possible without the recovery in recent decades of authors such as Chatterton, Warton, Percy, and Macpherson ('translator' of the 'ancient' Gaelic works of Ossian). Literary

[12] Gaimster et al. (eds), *Making History* 114–15. *Making History* also features this watercolour on its cover.

scholars including Katie Trumpener and Margaret Russett have explored the contributions of these poets and critics to the evolution of print culture and of British cultural nationalism. Their revival of vernacular literature in turn figures in the larger European project perhaps best summed up in the words of the distinguished Hellenist and pioneering romantic Friedrich Schlegel: 'our poetry lacks a center of the kind that mythology provided for that of the ancients. We have no mythology. But... it is time that we work together diligently to create one.'[13] Literary treatments of these sources, however, run the risk of reducing all antiquarian practice to literary mythmaking, and thus reimposing Piggott's distinction by other means. To recover the epistemology of antiquarianism, it is necessary to restore the connections between the literary and the empirical, to understand both as part of a larger field of possibilities presented by romantic antiquarianism.

The example of Piggott may serve to indicate the hazards of generalizing about these large bodies of early romantic thought—in this case literary primitivism, conjectural history, and neoclassical aesthetics, a third area that seems to me equally essential for understanding antiquarianism. Each of the chapters that follow takes up one or more of these areas to situate the empirical knowledge projects of Banks and Parkinson, Hamilton and d'Hancarville, and the others. In addition to the romantic revival, my argument depends on stadial conjectural histories such as those of Rousseau and Adam Ferguson and on German neoclassicism from Winckelmann through Lessing and Schiller to Schlegel. As Andreas Huyssen observes of the latter, his 'natural history of... art' from ancient Greece to the present makes any 'strict separation' of neoclassicism and romanticism untenable. Jonathan Sachs, insisting on the importance of ancient Rome for a romantic movement often assumed to have rejected Rome for an idealized Athens, offers a similar revision.[14] John Reinhold Forster, narrating his fieldwork on the Pacific Islanders during Cook's second voyage, habitually compares ancient Greeks with the noble (and ignoble) savages that Piggott associates with romanticism as opposed to empirical observation. As Jonathan Lamb points out, conjectural history and exploration narratives reciprocally informed each other, and so narratives like Forster's crossed the boundary between science and literature as well as the boundary between classicism and romanticism.[15] Rousseau's second

[13] Schlegel, *Kritische und theoretische Schriften* 190. On British literary antiquarianism, see Russett, *Fictions and Fakes*, and Trumpener, *Bardic Nationalism*, among others. McLane's work is also notable for making connections between the human sciences and the ballad revival, most recently in *Balladeering*.

[14] Afterword, in Schlegel, *Kritische und theoretische Schriften* 231; Sachs, *Romantic Antiquity* 30.

[15] It should go without saying that no such boundaries were in place at the time, but dismantling them is nonetheless productive for cultural history. Lamb suggests that the Forsters' 'theories of cultural difference and change' reflect, but in some cases also precede and determine,

Discourse—admittedly a product of the study rather than the field—stages the encounter between the savage and the ancient Greek as a competition in which 'the *ataraxia* [tranquility] of the Stoic does not approach to [the savage's] profound indifference to every other object' besides 'liberty and repose'.[16]

Jonathan Kramnick has argued that English critics such as Warton and Richard Hurd produced a hybrid antiquity, a concept that helps to make sense of Forster's and Rousseau's comparative projects as well. In both cases the recovery of a 'vernacular high culture' also serves the 'tendential suturing of antiquity to consumption' that, for Kramnick, defines print capitalism. When Percy suggests that 'rude' old ballads may 'throw light on our earlier classical poets', he makes England's 'ancient genius' coeval with classical antiquity, at least in the synchronic domain of conjectural history.[17] Warton similarly boasts of recovering the 'rude inventions preserved in 'ancient manuscript[s]', but also ventures into architectural history with his theory of 'absolute Gothic'—a contribution even recognized by Piggott for its originality. Macpherson, in his editorial guise, offers surprising parallels in Homer for certain passages in Ossian (including descriptions of military gear) and yet laments that 'quaint' Herculanean epigrams are preferred to 'natural' Celtic rhapsodies.[18]

Broadly speaking, this book incorporates two principles from recent scholarship on the literature and culture of the romantic era and on the history of science. First, it depends on the insight that 'literature' was narrowly redefined as imaginative literature in the first several decades of the nineteenth century. This redefinition, initially mooted by Jon Klancher among others, has since been extended to imperial cultural production by Tim Fulford, Peter Kitson, Debbie Lee, and others who productively dismantle the retrospective boundary between 'imaginative' and other kinds of literature.[19] The same restriction on imaginative literature also, until relatively recently, prompted a critical rejection of poets such as Chatterton and Macpherson, as both inauthentic

those of authors such as Adam Ferguson, Lord Kames, and Johann Gottfried Herder (*Preserving the Self* 77).

[16] I quote Adam Smith's translation of Rousseau (Smith, *Early Writings* 27).

[17] Kramnick, 'Making of the English Canon', 1090, 1099; Percy, *Reliques* I.ix, I.vi.

[18] Warton, *History of English Poetry* I.viii; Warton, *Observations on the Fairy Queen* II.188; Piggott, *Ruins* 118; Macpherson, *Poems of Ossian* 216: 'I am thoroughly convinced, that a few quaint lines of a Roman or Greek epigrammatist, if dug out of the ruins of Herculaneum, would meet with more cordial and universal applause, than all the most beautiful and natural rhapsodies of all the Celtic bards and Scandinavian Scalders that ever existed'.

[19] As Klancher puts it, '"literature" itself meant one thing in 1780—the whole array of educated genres from natural philosophy and history to poetry and drama—and something very different after 1820 (the restricted category of imaginative genres we know today)' (*Romanticism* 524). See also Siskin, *Work of Writing*, and Fulford, Kitson, and Lee, who recalibrate the start date of the romantic period to coincide with the *Endeavour* voyage in *Literature, Science, and Exploration*.

and insufficiently romantic. The second principle is the temporalization most influentially described by Foucault as an epistemic shift from the 'static' natural history of Linnaeus and Buffon to the comparative anatomy of Georges Cuvier. The historian of geology Martin Rudwick revisits this historical turn in *Bursting the Limits of Time* (2005), which argues strenuously for the impact of a newly empirical brand of 'erudite history' on the emergence of geohistorical science.[20] Rudwick returns again and again to this connection but develops it only in terms of the analogies available in Cuvier's and other geological texts. His work thus offers a strong incentive for more sustained efforts at recovering the connections between antiquarianism and natural history, as well as a salutary insistence on the international scope of these knowledge projects. Because Rudwick's 'erudite history' includes both archival and archaeological works, his concept also provides a bridge between the world of artefacts and the world of texts, too often segregated from each other.

III KNOWLEDGE WORK, OR, SCIENCES FROM THE MIDDLE

If 'romantic antiquarianism' demands some recovery work to be established as a contribution to knowledge, then 'knowledge work' might seem to require another kind of justification, an extrapolation into history of a present crisis. 'The academy can no longer claim supreme jurisdiction over knowledge', as Alan Liu warns us in *The Laws of Cool: Knowledge Work and the Culture of Information*. 'Scholars are themselves knowledge workers in a complete sense: they are intellectuals, but they are also middle managers responsible for an endless series of programs, committees, performance reports'—and for the education of students, Liu continues, who are themselves 'destined for service in the great, contemporary prosperity corps of knowledge work'.[21] As participants in the culture of information, university intellectuals come in for their share of the post-industrial alienation that Liu explores so thoroughly in his book; we too share somewhat unwillingly an imposed ethos of professionalism and our expertise, like that of beleaguered middle managers in IT corporations, is constantly threatened with obsolescence—but we share equally in the possibilities of resistance offered by 'cool'. The first Society of Antiquaries assured Queen Elizabeth, in their petition for a royal charter, that they would not 'interfere with the Universities', but from Elizabeth's time until Victoria's it would have been understood that the most dynamic and productive

[20] Foucault, *Order of Things*, cp. Rossi, *Dark Abyss of Time*; Rudwick, *Bursting the Limits of Time* 181–2 and *passim*.

[21] *Laws of Cool* 21.

knowledge work was carried on outside the universities in any case.[22] If there ever was a consensus that universities had 'supreme jurisdiction over knowledge', it was relatively short-lived. Venturing into socioeconomic territory typically mapped by analysts more closely tied to the corporate world, Liu departs from the usual terrain of literary scholarship. Though he grounds the concept convincingly in theories of social class as well as in management studies, the term 'knowledge work' itself was developed by consultants who were seeking to explain the impact of computer and internet technologies on traditional management structures. Scholars addressing the relationship between the university and the marketplace, by contrast, have more often focused on the corporate sponsorship of research.[23]

The contemporary scene of knowledge production offers some useful lessons for cultural history, including the dialectic between alienation and social mobility and the operations of knowledge production outside the university. I think a historical appropriation (not to say hostile takeover) of the term 'knowledge work' is warranted for several reasons. In Liu's usage (following a range of theorists), the terms 'work' and especially 'worker' carry the Marxian connotation of alienated labour, suggesting an ambiguous new space between bourgeoisie and proletariat. Similarly, eighteenth-century knowledge work was simultaneously understood as menial and scientific, though the ambiguity in this case is pre- rather than post-industrial. On the level of class, both types of knowledge work unsettle familiar hierarchies and distinctions. On the level of epistemology, knowledge work disrupts the 'totalizing' Foucauldian identification of knowledge with power: both the eighteenth-century research network and the IT corporation are at best loosely tethered to definable epistemic regimes. In his ethnography of the 'Tribe of Cool', Liu defines 'cool' as 'information designed to resist information' or a 'paradoxical gesture by which an ethos of the unknown struggles to arise in the midst of knowledge work'. Liu himself offers a justification of historical work when he writes that the role of art in this climate is to insist on history and disrupt the 'eternal "now" or self-evident presence of information'.[24]

[22] Antiquaries' petition to Elizabeth (1603) quoted in Gough, 'Introduction' iv.

[23] Liu, *Laws of Cool* 23–75, and especially the appendix 'Taxonomy of Knowledge Work' (391–3); Megill, *Thinking for a Living: The Coming of Age of Knowledge Work,* cp. Amar, *Managing Knowledge Workers.* Liu's articulation of 'knowledge workers' with Antonio Gramsci's 'intellectuals' is especially helpful (391). For examples of sociological critique interested in the conflicts between corporate sponsorship and intellectual freedom—far from a 'tired opposition', in my view (Liu, *Laws* 15)—see Kleinman, *Impure Cultures,* and Greenberg, *Science for Sale.*

[24] *Laws of Cool* 23, 179, 8. For a graphic comparison of the class positions of knowledge workers under a variety of socioeconomic theories, see Liu's Figure 1.1 (33). The disruptive force of history in this model seems indebted to the 'weak Messianic power' of history to recover moments of resistance to oppression from the historical record to serve present needs, as described by Walter Benjamin, who compares this operation to blasting (*sprengen*) in a mine (*Illuminations* 254, 261).

Analogously (if less dramatically), to recover the knowledge work of d'Hancarville, Schnebbelie, and their like is to disrupt the self-evident identification of certain areas of discovery with single individuals, such as Hamilton and ancient vases or Gough and medieval tombs—and to make these areas, once again, 'unknown'. Finally, I would argue that eighteenth-century knowledge work is becoming 'salient' as on object of inquiry, in the sense articulated by Lorraine Daston, because of the heightened visibility and diffusion of contemporary knowledge work on the Internet.[25]

The original Society of Antiquaries was dissolved by James I, who understood all too well the disruptive potential of antiquarian research. The Society was not revived until 1707 and finally received its royal charter in 1751, making it still the second-oldest royally chartered learned society in Britain. The scholarly publications associated with this and other societies often depended on the contributions of working-class subjects and fieldworkers, just as eighteenth-century natural history and ethnography depended on relationships between gentleman amateurs and provincial collectors or indigenous colonial guides. Although the Royal Society factors equally into a number of the works in question, my institutional history focuses primarily on the Society of Antiquaries of London, particularly in Chapter 6. A wider survey of the kinds of work involved in both natural history and antiquarianism will help to indicate the broad geographic and human range from which I have chosen the subset of knowledge projects and workers featured in the chapters to follow. My thematic focus on antiquity, along with some of the unique institutional features of the Society of Antiquaries, motivates my particular attention to this body. The Society's various collaborative preservation projects, in particular, provide a lens on both the historical turn and the changes in class structure that led eventually to the proliferation of more specialized disciplines—archaeology, art history, philology, and others—with their more rigidly specified objects and forms of publication and their new class-neutral rhetoric of disciplinarity. Compared to the disciplines ensconced in state-sponsored continental bodies and, increasingly, in universities, British antiquarianism was slow to develop such a rhetoric or to formalize its fundamentally collaborative—if hierarchical—methods. Following Calè, Craciun, and Martins, we might frame these 'unruly' working methods as 'predisciplinary encounters'.[26]

[25] Daston describes the genesis of scientific objects in terms of the 'cultural salience [that] made these objects visible', a point of view that has also influenced my argument for the proliferation of antiquities (*Biographies of Scientific Objects* 7).

[26] Calè, Craciun, and Martins, 'Romantic Disorder'. The Society of Antiquaries officially declared its pursuit of a 'better knowledge' of both 'antiquities and natural history' in the published *Queries* of 1754.

The recognition achieved by employees of the Society of Antiquaries, such as Schnebbelie, provides one measure of this fluidity. A 1792 review of Schnebbelie's independent book project noted that 'during the short course of his progress in science, it may with great truth be asserted, that no one laboured more intensely in his profession, or sought to gratify the enquirer with greater facility than Mr Schnebbelie'. The idea of the professional scientist is so thoroughly naturalized today that this language applied to Schnebbelie requires translation into a historical idiom. In this case and in others throughout this book, I prefer the equally anachronistic 'knowledge worker'. Despite the IT connotations, this twenty-first-century term facilitates the recovery of Schnebbelie's achievement, including his complex relationship with Gough. Above all it refers to the labour that is too often left out of histories of science.[27] Schnebbelie's reviewer implicitly recognizes both his mental and physical exertions, and both were necessary to the larger agenda advanced by Gough and his associates in the Society of Antiquaries. Schnebbelie and other paid knowledge workers supplied the empirical evidence that allowed history to become more than 'mere narrative', as Gough demanded. Schnebbelie's contributions helped to make antiquarian scholarship more inclusive, collaborative, empirical, and professional in practice, long before the study of the past was deliberately reorganized along these lines. Like Carter, though for a shorter time, he also gathered evidence that helped to support the growth of historic preservation.[28]

The Baconian empiricism that pervaded a whole range of knowledge projects from the mid-seventeenth century forward meant that there was always a demand for knowledge workers. In addition to architects and artists, antiquarian fieldwork also required labourers to excavate, erect scaffolding, and execute preservation plans. The Royal Society from its inception provided a role for merchants and travellers as fact-gathering foot soldiers in the campaign for knowledge, a practice that Mary Poovey highlights as a historical cause of the fact-value distinction in her *History of the Modern Fact*.[29] The Royal Society's most vocal early critic, a Quaker named Henry Stubbe, quickly pointed out how convenient it would be for the gentlemen fellows if they could monopolize the knowledge gathered in this way. Sydney Parkinson, the gifted young natural history painter who sailed with Captain Cook on the *Endeavour*, was also a Quaker, and natural history in general profited from the work

[27] 'The Antiquaries Museum' 282. For one recent treatment of the element of labour in science studies, see Harding, *Sciences from Below* Chapter 4. Harding also notes, however, an ideological tension between global knowledge work, in its twenty-first century sense, and other kinds of labour: 'the widespread praise for the increase in new knowledge workers... obscures its shadow of increases in exploited labour at the very sites where the new knowledge is being produced' (93).

[28] See further Chapter 6 below; for history as 'mere narrative', see Gough, 'Introduction' ii.

[29] *History of the Modern Fact* 93, 117–20.

of Quakers and members of other Dissenting sects. The *Endeavour* would have made a much less distinguished circumnavigation of New Zealand and (part of) Australia without the assistance of the Polynesian interpreter and guide, Tupaia, who aided Cook for close to two years until his death on Jakarta late in 1770.[30] The Duchess of Portland welcomed Banks, and his contributions to her specimen collection, on his return from this voyage, and her friend Mary Delany created a new art form to record these South Seas botanical specimens. For naturalists and tourists confined to England, the Dorset paleontologist Mary Anning was only the most prominent of a legion of rural guides, often women, who provided fossil specimens. Both naturalists and antiquaries relied on male and female subjects for their ethnographic work, and antiquaries too relied on local guides, who sometimes displayed an impressive range of knowledge.[31] Author-illustrators such as Parkinson, Schnebbelie, Carter, and Webster may be the particular type of knowledge worker who occupies most nearly the ambiguous position of the 'new new middle class' that Liu describes—though quasi-professional and upwardly mobile, these artists always remained the social inferiors of the scholar-patrons who depended on their skill.

C. Stephen Briggs presents a compelling example of this class dynamic that also draws attention to the crucial position of classical antiquity as a marker of social and methodological distinction. In his analysis of Richard Colt Hoare's *Ancient Wiltshire* (1812–1819)—one of Piggott's prime examples of properly archaeological research—Briggs focuses on the contributions of William Cunnington, 'a wool merchant and self-educated geologist' who used stratigraphic as well as archaeological evidence to establish a pre-Roman date for burial mounds on Salisbury Plain. In a letter to another member of the research team in 1808, Cunnington pondered the role of classical education: 'I must observe that I have ever had the highest respect for people who have had a liberal education like yourself, but I contend that the information to be gathered from the Roman & Greek historians will afford little information as data for illustrating Abury, Stonehenge, Marden &c &c, the works of an ancient people like the Celtic Britons.' Hoare subsequently 'turned the clock back on this intellectual achievement' with his conservative revisions of this chronology after the geologist's death, but he did acknowledge that Cunnington's fieldwork had made possible his famous declaration, 'We Speak From Facts, Not Theory.'[32]

[30] Sox, *Quaker Plant Hunters*; Williams, 'Tupaia'.

[31] In one typical case in 1772 a polite antiquary met with 'a very surprising person in the clerk of the parish, a man that, in the low occupation of a joiner, shews a wonderful genius for Antiquities' and turns out to have excavated hundreds of Roman coins and other objects (*ILH* IV.855). On local natural history guides, see Richard Hamblyn, 'Landscape and the Contours of Knowledge', Chapters 3–4; on Anning, see Torrens, 'Mary Anning'; and on Delany, see her *Autobiography and Correspondence*.

[32] Briggs, 'Prehistory' 229–31; Cunnington quoted in Briggs, 'Prehistory' 231. This view of Hoare differs substantially from Piggott's (*Ruins* 125–7).

This is a far cry from the Dutch anatomist Albinus, who referred to his illustrator, Jan Wandelaar, as 'a tool in my hands' in 1747. In addition to the knowledge workers themselves, subjects in the field often proved to be more than 'tools' in the inquirers' hands as well.[33]

These examples show that the work involved in antiquarianism and natural history had the power to unsettle class distinctions. My point is not that knowledge workers constitute a class, nor is it my intention to narrate a Whiggish history of scientific professionals displacing gentlemen amateurs. A significant body of historical scholarship takes up this question, and the model of progress is only one of several existing approaches.[34] My more modest aim is to assemble Parkinson, Fabris, d'Hancarville, Schnebbelie, and Webster as a meaningful group within this larger context. The patrons associated with them—Banks, Hamilton, Gough, and Englefield—form a more homogeneous group: all four were born into the landed gentry in Britain and were members of both the Antiquarian and Royal Societies, which in turn provide the common institutional factor in these knowledge projects. Moreover, they themselves blurred the line between patron and practitioner, unlike some of their counterparts on the continent.[35] The more varied national and educational backgrounds of their client intellectuals reflect the variety of the scenes and objects of inquiry: South Sea islands, Mount Vesuvius, Greek vases unearthed on the Bay of Naples, medieval English churches, and the strata of the Isle of Wight. The observational skills that they contributed, and the visual and verbal documents that made their careers, also reflect this diversity of objects. The particularity of these objects accounts for the 'knowledge' in 'knowledge work', and in a certain sense the objects themselves demanded these innovative framing concepts and techniques of documentation.

These previously (to some degree) unknown objects—peopled, as it turned out, by human subjects and human history—provide one of the sources of uncertainty that binds these actors into a group. Following Bruno Latour's account of Actor Network Theory, we may locate salient features of knowledge work around the five uncertainties that Latour uses to define the field. The third 'area of uncertainty', summed up in the vexing motto 'objects too have agency', has been most stimulating for my account. This mysterious agency is

[33] *Albinus on Anatomy* 18. For one intriguing case of a recalcitrant subject in the field, see Rawdon (Countess of Moira), 'Particulars' 93–4.

[34] See, for example, Rudwick, *Great Devonian Controversy* 63–8 —an account of the eventual recognition of William Smith that is especially apposite here. See also Hall, *All Scientists Now*, and Jardine, 'Whigs and Stories'.

[35] Winckelmann, for example, had patrons in Cardinal Albani and Herzogin Anna Amalia who did not themselves publish scholarship on antiquities. It is worth noting that three of the four (excepting Gough) also belonged to the Society of Dilettanti, signalling their simultaneous commitment to classical along with other kinds of antiquity. See further Kelly, *Society of Dilettanti*.

most apparent in objects that reveal themselves to be 'matters of concern' in addition to 'matters of fact', as Latour puts it. 'For instance', he explains, 'spermatozoids used to be obstinate little machos swimming forcefully toward the powerless ovule; they are now attracted, enrolled, and seduced by an egg the agency of which is becoming so subtle now that it can select the good sperm from the bad—or at least this is what is now *disputed* in developmental physiology'.[36] Likewise, fifth-century Attic vases dug up in Italy 'used to be' Etruscan curiosities and then became Greek masterworks. Latour's examples also include chimpanzees, formerly seen as altruistic, but now understood as 'fiercely competitive' by primatologists. In this book I am concerned with ancient artefacts that revealed themselves as art, including Etruscan vases that emerged as Greek, and relics of Gothic superstition (chantry chapels) that were recognized as architecture; and with emergent natural objects, such as 'ruined' strata that assumed an orderly fold structure, and volcanoes that came to instantiate an archetypal telluric agency. The force of these objects presents itself in the elements of surprise and accident that are characteristic of fieldwork, in failures of explanation, in aesthetic responses both scripted and involuntary, and in figures of otherness that as often as not are displaced onto the objects from the cultural encounters that provided the context for their discovery.

Latour specifies five 'occasions' that restore or enhance the 'visibility' of ostensibly 'non-social objects', beginning with an occasion that is thoroughly familiar from the history of science: 'the study of innovations and controversies has been one of the first privileged places where objects can be maintained longer as visible, distributed, accounted mediators before becoming invisible, asocial intermediaries'. Another one of these occasions is distance, including distance in time, which is especially crucial to the sciences of antiquity. Latour turns to palaeontology for an example that beautifully illustrates not only the uncertainty of objects but the continuity between past and present sciences of antiquity: 'Even the humblest and most ancient stone tools from the Olduvai Gorge in Tanzania have been turned by paleontologists into the very mediators that triggered the evolution of "modern man."' Similarly, d'Hancarville made ancient vases tell the story of the origins of art, a story that in some ways anticipates evolutionary anthropology. Of the other areas of uncertainty, the second—'no group, only group formation'—captures the improvised collectivity of knowledge work, and the fifth, which concerns the method of inquiry, is also relevant: 'the network does not designate a thing out there that would have roughly the shape of interconnected points, much like a telephone... "network." It is nothing more than *an indicator of the quality of a text* about the topics at hand', which 'elicits networks of actors' when it captures 'the

[36] *Reassembling the Social* 115–16.

ability of each actor to *make* other actors *do* unexpected things'.[37] In each of my chapters I set out to evoke these unruly networks of actors, including their newly salient objects. At the same time I suspect that 'agency' is too vexed a term to express what the objects do, and as a literary scholar I must be content sometimes—as Latour is not—to attribute phenomena to social causes (such as the 'context' of 'cultural encounter', a formulation that Latour would reject).

It is partly the emphasis on authorship that limits my engagement with the broader field of knowledge work that I have outlined, which certainly depended on a greater variety of 'actors' than those specified in the chapters that follow. Women naturalists and antiquaries remained outside the learned societies, and, unlike women writers in other areas, outside the literary marketplace, as did most non-European contributors to these knowledge projects.[38] Nonetheless, the term 'knowledge worker' aptly describes many women of the period, most famously perhaps Mary Anning, whose largely undocumented science has been reconstructed by generations of scholars and fiction writers—most thoroughly by Hugh Torrens and most recently by Tracy Chevalier, in the novel *Remarkable Creatures* (2010). (Fiction or 'scientifiction', Latour informs us, is another, final occasion for making scientific objects visible, and Susan Sontag's *The Volcano Lover* provides a relevant example here in its fictionalized treatment of Hamilton's work in Naples.) The term applies as well to the many non-Western people who entered the historical record primarily as guides to European explorers and colonists. Sacajawea is one example that overlaps both these categories. My cases are chosen on the basis that they left archives of correspondence with their patrons, on which I draw heavily in several of the chapters, and on the basis that these employers were also polymathic in their pursuits, broadly engaged with various aspects of both antiquity and (excepting Gough) natural history. Most important, the knowledge workers I have chosen to portray here all reinvented themselves to some degree as authors.

Despite the difficulties involved, scholars working on Banks and the Cook voyages have made substantive contributions to our understanding of the role of Pacific Islanders, such as Tupaia, as contributors to Enlightenment knowledge projects. My chapter on the *Endeavour* voyage touches briefly on this scholarship, and in my chapters on Hamilton's Vesuvius research I address the role of native Neapolitan guides and fact-gatherers, whose position in this moment of British hegemony is not unlike that of colonial subalterns. My account does not extend to the day labourers who excavated Herculaneum and

[37] *Reassembling the Social* 80, 81, 129.

[38] The literary marketplace was less exclusive than the learned societies, and among several noteworthy exceptions the German botanist and entomologist Maria Sibylla Merian (1647–1717) stands out for her successful, independent publication projects. The more typical marginalization of women is thematized in a variety of interesting ways by the essays in Gates and Shteir (eds), *Natural Eloquence*. On women antiquaries, see *Antiquaries* 69–79.

Pompeii, so memorably captured in Pietro Fabris's engraving of the Temple of Isis (Figure 2.2), though at least their names and wages were recorded at the time. Among other scholars engaged in thick description of knowledge projects, Holger Hoock has identified rich archival material for understanding archaeological practices in eighteenth-century India, and Beth Fowkes Tobin has explored the importance of slave kitchen gardens in Jamaica for British writings on colonial botany and agriculture. Their research illustrates the broad range of possibilities for knowledge as a concept. From a post-colonial perspective, Sandra Harding argues categorically that 'the development and modernization of the West were materially as well as ideologically *built on the exploitation*...of the societies which European expansion encountered...including its sciences and their philosophies'.[39] A full elaboration of the idea of the knowledge worker in the eighteenth-century context would require a much more exhaustive engagement with knowledge-producing subjects. This study offers something more like 'sciences from the middle', in addition to what I hope is a new paradigm for the proliferation of antiquities, and a revaluation of romantic antiquarianism in its relation to natural history.

IV TRANSFORMATIONS OF ANTIQUITY

Sciences of Antiquity revisits the historicity invested in the natural world in the course of the eighteenth century, as established by Foucault, Wolf Lepenies, and Paolo Rossi (among others) at a turning point in the history of ideas. The rise of an analogy between 'the history of the earth and the history of nations', in Rossi's phrase, appeared as a global factor in the emergence of modern geology, biology, and what Foucault calls the 'human sciences'. My account focuses squarely on antiquarianism as the framework that allowed the earth and its nations to be considered side by side in their historicity. Antiquary-naturalists such as Banks and Hamilton were more empirical in their approach than Rossi's chronologers, yet also less specialized than Foucault's 'modern' scientists. The Society of Antiquaries was a major force in eighteenth- and nineteenth-century intellectual life and left traces on projects that were central to its research mission, such as Gough's *Sepulchral Monuments*, as well as projects seemingly more remote from this mission, including Banks and Parkinson's Pacific natural history and Webster's Isle of Wight stratigraphy. Plate books such as *Description of the Isle of Wight*, *Sepulchral Monuments*, and *Campi Phlegraei* emulated the lavish publications that constituted one of the primary benefits of membership in the Society of Antiquaries, and, like

[39] Hoock, *Empires of the Imagination* 288–314; Tobin, *Colonizing Nature* 56–80; Harding, *Sciences from Below* 68.

those publications, they provided employment to contributors who found ways to adapt both the form and the content to other purposes. Copper engraving and state-of-the-art printing technology were among the material drivers that generated more affordable avenues for publication. These ranged from popularizing efforts pursued by Schnebbelie and others, who published larger works by monthly or quarterly numbers, and substituted etching for engraving, to specialized scientific writing as developed by Webster. Two predisciplinary factors thus linked antiquarianism and natural history in their production of more specialized disciplines: a similar division of labour, which became less visible in the later idiom of disciplinarity, and a literary marketplace that sustained an exceptionally wide range of contributors and publications in both fields.

My first chapter on Parkinson and the *Endeavour* voyage positions Parkinson's *Journal of a Voyage to the South Seas* at the head of a proliferation of Cook voyage narratives, which multiplied over the course of his three voyages and beyond to encompass official and unofficial narratives, scholarly and popular accounts, lavish quartos and cheap compendia. The seeds of this evolution are contained in the discrepancies between Parkinson's narrative and those of Banks and John Hawkesworth, in the intense conflicts between Banks and Parkinson's heir over the rights to his journal and to the specimens that informed it, and above all in Parkinson's pioneering contribution, not just as an illustrator but as an author unencumbered by classical and antiquarian commonplaces. In some of these respects he resembles Pietro Fabris, who illustrated and sold Hamilton's *Campi Phlegraei* and who occupies a central position in my second chapter. Another vital connection lies in the volcanic origin of Tahiti and other South Pacific islands, first recognized by Banks, and linked by him to the state of their civilization. So too Hamilton (whose friendship with Banks postdates the *Endeavour* voyage) insisted on the volcanic origin of the entire region of Naples and made inferences from it about the history of its population. This chapter also introduces the exceptionally complex network around Hamilton, including Antonio Piaggio, who supplied both words and pictures that Hamilton used in the later phase of his Vesuvius research. The order of the chapters is roughly chronological—although Hamilton's interest in Vesuvius began to take shape in 1766, two years before the *Endeavour* sailed, *Campi Phlegraei* was finally published in 1776, three years after Parkinson's *Journal*. More important, the *Endeavour* voyage cemented the association between exploration and antiquity that informs all the subsequent chapters.

The second section of the book—three chapters devoted primarily to d'Hancarville's *Collection of Etruscan, Greek, and Roman Antiquities* (1766–76)—turns from natural history proper to the deep past of art. In the preface to the third volume of a work that began as a catalogue of Hamilton's first vase collection, d'Hancarville observed that 'Antiquity is a vast country'. Although masquerading as an aristocratic grand tourist, 'Baron' d'Hancarville was increasingly influenced by exotic travel narratives in his last two volumes on classical antiquities, composed in 1773–75 when numerous narratives of Cook's and Bougainville's voyages were already in circulation. At the same time, simultaneous work on the geology of Vesuvius and the archaeology of Herculaneum, beginning in the mid-eighteenth century, reinforced the affinities between antiquarianism and natural history. Chapter 3 documents the production of d'Hancarville's book, which involved a team of eighteen collaborators, many of them artists familiar to d'Hancarville from his own experiment in publishing prints for the tourist trade that was fed by Vesuvius and by the region's exceptionally rich stock of architectural antiquities, including the recently rediscovered temples of Paestum. This chapter makes substantial use of d'Hancarville's letters to Hamilton, which incorporate dramatic autobiography with technical and fiscal details of his innovative use of the plate book form.

Chapter 4 engages with d'Hancarville's ambitious arguments in the book itself, elaborating a concept of 'ethnographic actualism' to describe his strategy of deriving explanations of the forms and representational content of the ancient vases from contemporary Neapolitan customs and manners. The term 'actualism', drawn from the history of geology, also signals a set of disciplinary and formal analogies between the *Antiquities* and *Campi Phlegraei*. Chapter 5 turns to the implications for art history of this ethnographic approach, coupled with d'Hancarville's expansion of the historical timescale and his geographic expansion of the project from southern Italy to Magna Graecia. Here I argue that d'Hancarville creates archaeological deep time by relocating the origins of art and ritual close to the moment of human origins. Like Parkinson on the *Endeavour*, d'Hancarville is a hired hand who deserves substantial credit for the achievements of a great collector and virtuoso, in this case Hamilton, the British ambassador to Naples from 1764–1800. Unlike Parkinson, d'Hancarville could boast of a good classical education, and of course the subject matter demanded this expertise. As suggested by d'Hancarville's sudden departure with the plates, which later appeared in a variety of published forms, classical learning played a more pecuniary role for him than for his employer.

The third section of this book concentrates on the English setting, and it is perhaps no coincidence that Schnebbelie and Webster, the subjects of the last two chapters, were more like Parkinson in this respect—indifferent classicists at best. Their patrons, Gough and Englefield, were both active in the earliest phase of the Gothic revival, 'pioneers of preservation', and they drew collaborators such as Carter, who developed a staunchly anticlassical line of defence for medieval antiquities. While Gough embraced classical scholarship, and Englefield (like Hamilton) collected Greek vases, their work with Carter and Schnebbelie may be understood as 'a self-conscious attempt to raise... British medieval antiquities to the level' of classical monuments, as Matthew Reeve has put it. Sam Smiles credits Gough with realizing that 'visual documentation of medieval architecture would not only provide a secure basis for scholarly investigation, as had happened with classical antiquities, but would also bring the medieval architectural legacy up to some sort of parity with classical remains'.[40] My sixth chapter focuses on the collaboration of Gough and Schnebbelie, who contributed the largest single pool of drawings for *Sepulchral Monuments of Great Britain* (1786–99). Unlike d'Hancarville, and remarkably for such a short career, Schnebbelie went on to develop independent publication projects that were not tied to aristocratic or institutional patronage. He did not contribute text to *Sepulchral Monuments* itself, but his images and his fieldwork made it possible for Gough to shift the direction of Gothic enthusiasm toward formal taxonomy and material culture. Although Carter placed this celebration of Gothic form in an anticlassical context, his collaboration with Gough and Schnebbelie is less antithetical than analogous to Hamilton's and d'Hancarville's vase scholarship. Both projects shared a strong ethnographic interest in premodern funerary customs. In both cases, too, 'curiosities' achieved the status of ancient art. This unexpected convergence is central to the transformation of antiquity, and warrants a thorough exploration of the rich archival material on Gough and Schnebbelie, which together with the print sources provides an exemplary record of collaborative knowledge work.

The final chapter returns to the conjunction of natural history and antiquities in a project that joined the Gothic with geology and launched the career of the first 'professional geologist'. By 1827, when he gave himself this title in a book prospectus, Thomas Webster had risen from relative obscurity to become the first salaried officer of the new Geological Society. Webster published his earliest geological observations by commission in Englefield's *Description of the Isle of Wight*, a work that held together historic architecture and picturesque landscape with scientific natural history even as it accommodated the more specific disciplinary idiom of the new science. Englefield hired Webster to illustrate this work because he was, like Schnebbelie, a topographical and architectural draftsman with a reputation for scrupulous accuracy. Unlike Schnebbelie, Webster had another relevant qualification, having created illustrations for England's first course of public lectures in geology, given by Humphry Davy in 1805. Webster's fieldwork on the Isle of Wight produced numerous drawings of medieval monuments and ruins to accompany Englefield's existing images. The geological part of this fieldwork exceeded Englefield's expectations. Webster drew such powerful conclusions about the structure of the chalk formation—he likened the region to a 'ruined arch or vault'—that Englefield credited him as co-author. Webster's training as an architectural draftsman enabled him to develop pioneering geohistorical insights from the antiquarian metaphor of 'ruin'. By attending to a new set of

[40] Reeve, 'Jacob Schnebbelie, Draughtsman' 81; Smiles, 'Data, Documentation and Display' 91.

ancient ruins, this chapter shows how strongly the traditional language and research questions of antiquarianism continued to shape geology even as it became a professional specialization.

Sciences of Antiquity recontextualizes all five publication projects as experiments that relocated disciplinary authority and allowed a set of historical sciences to define themselves against the arts and against each other. As cultural prestige was eroded from the single site of classical learning and redistributed in other fields, a growing corps of knowledge workers reshaped antiquity along the lines of a conflicted global modernity. Their contributions were increasingly required as visual documentation, scientific travel, and museum display all became central to the study of natural history and antiquities. As a consequence, these works began to ask new kinds of questions about antiquity: what counts as antiquity? How ancient must antiquities be, and how can their age be measured? How literally does the rubric of antiquity extend to natural objects? Must antiquities exhibit aesthetic values, or does antiquity also dignify commonplace artefacts, 'the household utensils of our ancestors' as Nietzsche called them? Relatedly, what are the geographic contours of antiquity, and how do the antiquities of advanced civilizations compare with 'primitive' productions of whatever age? The veneration of antiquity in the name of classicism led unexpectedly to archaeological and geological discoveries that forced an expansion of the whole rubric of antiquity. At the same time, the social energy of these discoveries fuelled the proliferation of antiquities in diverse disciplinary and commercial forms.

In a certain sense, the pursuit of non-classical antiquity—the gradual acknowledgment of historicity in post-classical works, in natural objects, in 'primitive cultures'—belongs to the history of romanticism. Once the non-classical and anticlassical tendencies in classicism became distinct as a discursive field, the name romanticism could be attached to them; hence the liminal status of many 'pre-romantic' works such as Lessing's *Laocoön* and much of Rousseau. Many of the naturalist-antiquaries examined here owed some of their ideas about the history of civilization to Rousseau, the first great promoter of non-classical antiquity. Rousseau's own lack of a formal classical education and his defiance of social privilege illuminate the social dimensions of antiquity in its expanding sense. Rousseau's uneasy dependence on aristocratic patronage did not deter him from an unorthodox and brilliant authorial career, as the 'father of romanticism', that took a different shape entirely from the careers of self-taught intellectuals who made contributions in natural history and antiquities. Their publications, however, provide a context in which authorship itself appears as a form of the knowledge work that cut across classicism and romanticism to transform the idea of antiquity along with the culture of patronage in which it evolved.

Part I

Natural History and Antiquity

1

Beyond Patronage: Knowledge Work, Professional Ambition, and the Competing Narratives of the *Endeavour* Voyage

1.1 THE NATURAL HISTORY CONTEXT

Chronologically, my story in this book begins with two self-taught intellectuals whose innovative knowledge work of the late 1760s is still widely remembered today: Sydney Parkinson and Pierre Hugues, 'Baron' d'Hancarville. Although Parkinson (1745–1771) was the younger of the two, he produced a substantial body of work before his early death. His words and pictures formed an important early chapter in the narrative of Pacific encounter—a body of narrative that made a tremendous immediate impact and influenced much of European thought in the ensuing decades, including the conjectural history of d'Hancarville, and of others, in the 1770s and 1780s. Parkinson remains especially well-known for the many accurate and beautiful drawings he made on Captain Cook's first voyage. He is also a representative figure for natural history in the last quarter of the century, and natural history launches the conceptual arc of this book. Natural history, in essence descriptive rather than narrative, is not explicitly a science of antiquity; to understand its implicit role in antiquarianism we must consider it first on its own terms. My first two chapters therefore concentrate primarily on natural history—of the South Pacific and of Vesuvius, respectively—even as they begin to argue that natural history and antiquarianism are intrinsically related as cognate sciences of antiquity. Parkinson's work, as colonial knowledge work, also helps to define the geographic arc of this book, which leads from British narratives about the South Pacific via British and European treatments of southern Italy back to the imperial centre in the final two chapters. Within this narrative framework, Parkinson's *Journal of a Voyage to the South Seas* exemplifies the discourse on primitive peoples that gradually made its way into eighteenth-century scholarship on classical antiquity and British history.[1]

At the distance of a quarter-millennium, the most obvious legacy behind the scientific name *Procellaria parkinsoni* (Parkinson's petrel) is that of Carolus Linnaeus, who established the modern system of binomial nomenclature in *The System of Nature* (1735). Linnaeus's work was already recognized as a landmark in natural history during his lifetime, but he had to contend with other towering figures, such as Buffon, for the limelight. Linnaeus made a strong impact in Britain, and directly inspired the participation of Joseph Banks in Cook's first voyage, on the *Endeavour* (1768–71). Because of the natural history contributions made by Banks and his well-chosen collaborators, especially Parkinson and a student of Linnaeus named Daniel Solander, this voyage became a scientific landmark in its own right. Parkinson's legacy is also recorded in this petrel's Linnaean name, though it is less obvious at this distance; by contrast, Buffon had several species named after him, even though he was critical of Linnaeus's system. Of the three, only Banks had a genus named after him, the *Banksia* genus of Australian shrubs and trees named in his honour by Linnaeus the younger in 1782, just eleven years after the *Endeavour* voyage. In 1788 Sir James Edward Smith could still refer without ambiguity to this voyage as 'that celebrated voyage round the world', even though several more had been made by that time. Smith made his remarks in an inaugural address to the newly founded Linnaean Society in London, celebrating the English contribution to Linnaean systematics by Banks and his associates, which helped to establish natural history as a colonial and global practice.[2]

The Linnaean tradition is one key factor behind the achievements of Sydney Parkinson, whose collaboration with Banks is the main subject of this chapter. Parkinson drew petrels and many other specimens, especially plants, on the voyage, though Parkinson's petrel was named in his honour much later, in 1862 (followed by *Ficus parkinsonii* in 1901). The other key factor at work in Parkinson's *Journal* is the literary tradition of natural history embodied for the eighteenth century by Buffon as well as by Gilbert White, whose work is probably more widely read by Anglophone readers today. This work also appeared in 1788, as *The Natural History and Antiquities of Selborne*, and

[1] Although my subsequent chapters are less directly concerned with colonialism, Parkinson's career may also be taken as a case study insofar as it resembles those of other specifically colonial knowledge workers, such as John Gabriel Stedman, George Forster, and Mungo Park. Tobin's *Colonizing Nature* provides a good analogue in her study of the artist William Hodges, hired for Cook's second voyage (117–43). Although she does not use the expression 'colonial knowledge work', her account of Cook as autodidact (144–67) also opens possibilities for thinking about the captain's role in these terms.

[2] Smith, *Discourse on the Rise and Progress of Natural History* 151. For an overview of natural history, see Heringman, 'Natural History in the Romantic Period.' On Banks and empire, see R. E. R. Banks (ed), *Sir Joseph Banks: A Global Perspective*, and Gascoigne, *Science in the Service of Empire*.

was at first largely overlooked as an obscure county history until it re-emerged in abridged form as *The Natural History of Selborne*: in this guise White's book, too, achieved the status of a landmark in natural history by the 1820s. White was not a part of the Linnaean movement, and drew heavily on the natural theology of an earlier generation, represented especially by John Ray. In spite of its intensely local character, White's book can be linked in various ways to global natural history. The forgotten antiquities embody another crucial link between White's project and the *Endeavour* journals. Both Banks and White were naturalist-antiquaries, and I hope to show that Sydney Parkinson, though he began the voyage primarily as a botanical draftsman, eventually fashioned an authorial persona for himself that appropriated and revised the 'naturalist-antiquary' model.[3]

Scholarship on Banks has seldom attended to his antiquarian publications and the antiquarian elements of ethnographic descriptions in both Banks's and Parkinson's journals have gone largely unnoticed. This forgetting is a by-product of modern scholarship, whereas White's predisciplinary study of antiquities is not so much forgotten as vanished, excised by his early editors under the influence of emerging disciplinary boundaries. Parkinson's *Journal*, published posthumously in 1773, was praised for its faithful, unvarnished description, as White's journal eventually was. Banks gave his own journal of the voyage to John Hawkesworth, who was paid to write the official *Account of the Voyages* (1773). Hawkesworth's account proved so controversial that Banks regretted his decision, and reviewers' praise of Parkinson may owe something to the fact that his *Journal* offered an alternative to Hawkesworth. When J. C. Beaglehole published Banks's journal manuscript in 1962 it proved to be a very substantial source on all three branches of natural history in the South Pacific—geology, botany, and zoology—and also revealed (retrospectively) how much Banks's influence on this voyage contributed to solidify the connection between these three branches and ethnographic description, which essentially constitutes a fourth branch of natural history in the early voyaging literature. Parkinson restricts himself mainly to the plants, peoples, and languages of the islands, but his prose on these subjects is more strictly and more consciously devoted to recording his observations, as my comparisons will show. At the same time, his conjectures on Oceanic civilizations relative to human antiquity clearly show Banks's influence. It would be sheer Whiggism to argue that Parkinson was a better observer than Banks because he set his own, more advanced professional standards up against Banks's gentlemanly ones. Comparative analysis of the two journals does, however, afford criteria

[3] I owe this term to Maria Toscano, 'Figure of the Naturalist-Antiquary' (see esp. 226 on Banks). On the publication history of and natural theology in *Selborne* (imprint date 1788), see Peterfreund, 'Great Frosts and ... Some Very Hot Summers'. On White and global natural history, see Menely, 'Traveling in Place'.

for distinguishing Parkinson's relatively independent approach to knowledge work from the traditional client-patron relationship.

This endorsement of the positive early response to Parkinson may appear problematic in light of recent cultural histories of scientific voyaging. Parkinson was, first of all, beholden to a larger colonial enterprise that has long been subject to ideology critique, sometimes in conjunction with Bruno Latour's paradigm of the 'centre of calculation' as applied to Banks's (and the Royal Society's) London-based global collecting enterprise. John Gascoigne has shown how Banks gradually created relationships between the Royal Society and the King, the Navy, and the Committee on Trade to promote 'a view of the Empire as an emporium of raw materials for the benefit of the mother-country' and specifically that of 'the great business of improvement' which concerned him as a landowner.[4] Parkinson himself cannot be so clearly implicated in the colonial politics of voyaging, especially as it evolved after his death. But recent scholarship has also questioned the accuracy of Parkinson's drawings in particular—the most famous part of his legacy—on the grounds that they sometimes betray a leaning toward either Enlightenment ordering principles or a nascent romantic aesthetics, particularly where the Pacific Islanders are concerned. Jonathan Lamb, for example, points out (drawing on anthropological studies) that the tattoo pattern on one of Parkinson's most famous images, *Head of Otegoongoon*, was probably transferred to the face from the subject's thigh for aesthetic effect. Nicholas Thomas similarly critiques the engravings of Oceanic 'Weapons and Ornaments' prepared from Parkinson's drawings: 'the things imaged are decontextualized and their uses become irrelevant'.[5]

Without dismissing these critiques, I want to shift attention first to the qualities of Parkinson's text. The absence of classical and antiquarian commonplaces sharply distinguishes Parkinson's ethnographic prose from that of the classically educated Banks and the self-consciously literary Hawkesworth. The engravings in Hawkesworth's *Account* are also based on Parkinson's drawings, but his images have a different set of effects in the context of his own prose in the *Journal*. Although they may fall short of a present-day scientific standard for accuracy, Parkinson's images gave wide circulation to a new set of natural and cultural materials, and his journal vividly illustrates the peculiar vitality of the specimens he collected.

[4] Miller in 'Joseph Banks, Empire, and "Centers of Calculation"' (cp. Fulford, Kitson, and Lee, *Literature, Science, and Exploration*, esp. 26–7, 45–8, 120); Gascoigne, *Science in the Service of Empire* 106; *Joseph Banks* 255.

[5] Lamb, *Preserving the Self* 106, drawing on the anthropological research of Alfred Gell; Thomas, *In Oceania* 101. Cp. also J. R. H. Spencer's skeptical 'Coastal Profiles and Landscapes,' esp. 270. Parkinson's drawings of plants and animals, on the other hand, continue to stand as contributions to the scientific record, as suggested by their treatment on the Natural History Museum (UK) website and in publications such as Zeidler, 'Drawings of Hyperiid Amphipods.'

Parkinson, Banks, Solander, and other members of the naturalist 'team' on board the *Endeavour* took great care to preserve natural history specimens, preserving small animals in spirits, drying and pressing leaves and flowers, and even restoring specimens damaged when the *Endeavour* collided with a reef in 1770. Although some specimens from Cook's and other eighteenth-century voyages of discovery remain in the British Museum and other institutions, many more have perished. The images produced by Parkinson and the artists who followed him have proved more durable. Parkinson's 2000-odd drawings and sketches recorded ethnography and landscape as well as natural history: of those eventually engraved on copper (more than 750), a handful have circulated widely ever since they were published in Hawkesworth's notorious *Account*. Perhaps the most famous of all is that singled out by Lamb, Parkinson's *Head of Otegoongoon, Son of a New Zealand Chief, the face curiously tataow'd*, reprinted in many modern works on South Pacific exploration.[6] Parkinson's ethnographic interest—initially a mere sideline to his paid work as botanical draftsman—created an added incentive for accurate drawing, since human subjects were (with a few rare exceptions) not only perishable but non-portable. Focusing more on ethnographic than botanical specimens, this chapter explores some of the relationships between the visual and verbal record and the specimens themselves. Broadly speaking, these relationships suggest that the record was partly motivated by a fear that these artefacts would perish, a fear that I will trace to the nascent architectural preservation movement associated with eighteenth-century antiquarianism. The specimens themselves circulated along with the images created to record them in the vibrant economy of European expertise and collection, as attested by bitter disputes over their possession and attribution.

Previous scholarship on voyages of discovery has seldom emphasized the antiquarian origins of associated practices including the concern with preservation, the visual recording of specimens, and monumental publication. In one famous case, the folio publication of a voyage narrative with natural history illustrations became the only monument of the commander, Comte de La Pérouse, who sent his manuscript materials overland to Paris from the Kamchatka Peninsula shortly before his ships were lost in 1788. Along with Louis-Antoine de Bougainville's employment of the botanist Philibert Commerçon in 1766, Parkinson's employment for the *Endeavour* voyage helped to set the precedent for La Pérouse as for all later voyages involving natural history. Banks and Parkinson did not have access to Bougainville's narrative, which was published in 1771, but it too partakes of the naturalist-antiquary

[6] Among more far-flung occurrences, the Virgin Atlantic Airways in-flight magazine *Hot Air* (Summer 2007) included one of Parkinson's images (uncredited) in an article on Tahiti. The magazine illustration, an image of Tahitian dancing girls, is adapted with very little change from the version in Hawkesworth, *Account* II, plate VII.

mode at times with its conjectures on the antiquity and primitive state of Pacific civilizations.[7] In Banks's case antiquarianism played an especially decisive role, as witnessed by his twelve publications in *Archaeologia* (versus only one in the *Philosophical Transactions*). In his decision to hire Parkinson, he was influenced by a coterie of landed gentry, including Thomas Pennant and Thomas Falconer, who retained personal draftsmen to record architecture and monuments both on their estates and on their travels. Another of these amateur archaeologists, Richard Gough, lamented in his richly illustrated *Sepulchral Monuments in Great Britain* that 'in a few years more we shall have no foundation left' for a study of Gothic architecture, and specified many architectural features of Gothic churches that were 'crumbling away without having been drawn' (*SM* I.i.3–4). Banks extended this anxiety to natural history specimens and his example helped to make it common practice for state-funded voyages of discovery to employ teams of shipboard artist-naturalists. Banks himself even snatched up the drawings of George Forster, done on board the *Resolution* at Admiralty expense, after a dispute developed between the Forsters and Cook after his second voyage.[8]

The common ground of scientific and antiquarian preservation becomes even clearer in the career of the British ambassador to Naples, Sir William Hamilton (the subject of my next chapter). Hamilton amassed Greek vases and volcanic rocks with equal assiduity around Vesuvius: although both sets of specimens were sent to London, many of the volcanic specimens survive only in illustrations.[9] These early collectors could not anticipate that the boom in scientific voyaging would cheapen their natural history specimens, which have been preserved with less care than classical artefacts during the intervening centuries. Surviving specimens, however, are now much easier to locate thanks to the online specimen catalogues developed in recent years by the Natural History Museum (London) and other institutions. This chapter includes a reassessment of the bitter dispute that developed over Sydney Parkinson's private collection of ethnographic and natural specimens from the *Endeavour* voyage after his death. In the preface to Sydney's posthumous *Journal*, his brother and editor Stanfield claims that Banks had embezzled Sydney's collection. History has sided with Banks, dismissing Stanfield as a mentally unstable

[7] La Pérouse's discussion of the colossal heads of Easter Island, which responds to Bougainville, Cook, and J. R. Forster, provides one symptomatic and illuminating example of the linkage between natural history and antiquities in this context (*Voyage de La Pérouse* II.85–8 and Plate XI). Rice is one of several scholars to point to Banks's employment of Parkinson as a European precedent, though this judgment seems to neglect Bougainville and other earlier examples (*Voyages of Discovery* 144).

[8] Rice, *Voyages of Discovery* 172. In addition to Rice, Beaglehole, Thomas, and Lamb, I am also indebted to the editorial matter in Lansdown, *Strangers in the South Seas*, for the history of voyaging as summarized in these pages.

[9] See Moore, 'Sir William Hamilton's Volcanology' 185–7.

illiterate. Advances in cataloguing may help to settle the issue of whether Banks appropriated Parkinson's specimens for his own vast collection, which he eventually gave to the museum. But even if these particular contested objects prove to have died and left no trace, the dispute testifies to intense and conflicting eighteenth-century desires to make them live. Banks wanted to preserve an unmatched physical inventory of the South Pacific, and his interest in the ethnographic materials suggests an antiquarian view of the local cultures, too, as dying. Stanfield Parkinson wanted the same specimens—circulated in visual form—to stand for the independent expertise he felt his brother had achieved in what he calls 'his profession of natural history'. This chapter puts Stanfield's claim to the test. Natural history specimens die into collections, but nonetheless they have their monuments: both physical monuments, such as the Natural History Museum, and the sepulchral professions dedicated to depicting, classifying, and explaining the creatures of distant worlds.

1.2 SYDNEY PARKINSON AS AUTHOR

Parkinson's arresting images, though produced in a short span of time, have themselves enjoyed a remarkable afterlife. The visual record that he created helps to justify Beaglehole's audacious claim that when Captain Wallis landed on Tahiti, two years before Cook, he 'had stumbled on a foundation stone of the Romantic Movement'. Alan Bewell has argued more systematically that the romantic idea of nature was derived from colonial voyages of discovery, pointing to the *Endeavour* voyage, with its rich legacy of observation, as the precedent.[10] The *Endeavour*'s team of naturalists and artists was assembled and paid for by Banks, who accompanied Cook out of sheer predisciplinary enthusiasm, but—as Bewell and others scholars have noted—almost all European voyages of discovery from that point forward included such teams of knowledge workers. After Banks fell out with the Admiralty, it quickly became common practice for the state to pay the salaries of these workers. Thus Parkinson contributed significantly to the emergence of a 'profession of natural history'. Most of the scholarship on Cook's voyages has considered Parkinson mainly in light of his client relationship to the great scientific patron and manager Banks, and his *Journal* is often mentioned only as a footnote to the controversy over Hawkesworth's official account of the voyage. Attention has rightly focused on Parkinson's images, which were engraved at

[10] Beaglehole (ed.), *Journals of Captain James Cook* I.xciv–xcv; Bewell, 'Romanticism and Colonial Natural History.' Parkinson's botanical images have been mounted on the web by the Natural History Museum (UK) at <www.nhm.ac.uk/jdsml/nature-online/endeavour-botanical/index.dsml>. Stanfield refers to Sydney's 'profession' in his preface to Parkinson, *Journal* vii.

Banks's direction for Hawkesworth's volume and in this form circulated so widely as to provide the European public with its dominant image of Tahitian and Maori natures and cultures. My argument, however, addresses Parkinson's professional contribution as a naturalist and author in order to recover an intellectual achievement too easily obscured by his class status. By his own tacit admission, confirmed by references in Cook's and Banks's journals, Parkinson was not a 'gentleman', but he enjoyed a higher status than the four 'servants' or 'trained collectors' who completed Banks's team.[11]

Parkinson's intellectual achievement is twofold. On the one hand, his *Journal* established an idiom for the scientific voyage narrative, a model of the professional plain style that eventually superseded Hawkesworth's genteel literary model as the standard for such accounts. His brother's editorial stance—probably filtered through the hired pen of William Kenrick, who was regarded by many as a notorious hack writer—also helped to shape this prototype of a new literary product for an expanding market. On the other hand—and this is the other neglected aspect of Parkinson's intellectual achievement—he was able to establish a model of professional identity soon more fully embodied by J. R. and George Forster, Ferdinand Bauer, and other shipboard artist-naturalists, because he consolidated the functions of the antiquarian draftsman and the colonial artist-naturalist that set the terms of his employment with Banks.[12] Parkinson's position in a network of Quaker and other Dissenting naturalists—including the Forsters as well as John Fothergill and William and John Bartram—also helped him to consolidate these roles. This section of the present chapter situates Parkinson within the tradition of domestic and colonial knowledge work to show how he transformed that tradition both as a writer and as a working naturalist. In the following sections I turn to Parkinson's collection and then to antiquarianism and antiquity as categories informing his conception of his intellectual work on the voyage.

[11] Unlike these four, but like Solander and the astronomer Charles Green, Parkinson had his own cabin near the captain's on deck. For the uses of 'gentleman', see Parkinson, *Journal* 21 and Cook's *Journal* (27 Jan 1771), quoted in Lysaght, 'Banks's Artists' 25. Lysaght also critiques the 'unfortunate' [but still prevalent] mischaracterization of Thomas Richmond, George Dorlton, Peter Briscoe, and James Roberts as 'servants' (11). This substantial essay provides a good introduction to the work and life of Parkinson, who grew up as the son of a Quaker brewer in Edinburgh.

[12] On Kenrick, see further below. The Forsters were responsible for natural history and illustration on Cook's second voyage (the *Resolution*); Bauer did the same work for Capt. Matthew Flinders on the *Investigator*. The sense of 'professional' here is naturally very different from that developed in Larson's influential *The Rise of Professionalism*, and I do not wish to argue that professionalism 'emerged' a century earlier than we thought. It is striking, however, that 'profession' and 'professional' were used in the period to describe not only Parkinson but most of the knowledge workers treated in this book. One of my secondary goals is to develop this earlier sense of the term.

When Parkinson's *Journal* was printed in 1772 it was at first suppressed and then appeared in 1773 under unfavourable circumstances, including the mental decline of Parkinson's brother and editor Stanfield, that had a negative effect on sales. Although it came too late to benefit the Parkinsons, a second edition appeared in 1784 under much more favourable circumstances and solidified the journal's long-term influence. A review in the *Gentleman's Magazine* provides an early vindication of Parkinson and illustrates some of these changed circumstances. The reviewer reminds us that in spite of a court injunction delaying its publication and of attempts to suppress it entirely, 'the work came forth, and was approved. With respect to authenticity, it remains unquestioned—whatever the journalist reports of the manners, customs, employments, pastimes, arts, genius, temper, and civilization of the inhabitants of the several islands at which he touched, he collected, not from the books and relations of others, but from his own attentive observation and judicious remarks'. This reviewer (probably John Duncombe) also celebrates the 'originality' of Parkinson's illustrations—two dozen of which were engraved for the original publication—by which he means that Parkinson shows Pacific Islanders who are not generic savages or stock classical figures but have 'all their distinctive features most strongly expressed'. The stakes become clear at the end of this review, which alludes to the scandal of Hawkesworth's romanticized second-hand account of the voyage, and to voyage narratives before and since, such as those of Bougainville and the Forsters, that also suffered from second-hand interpolations and/or literary pretensions. In closing, the reviewer anoints Parkinson's journal with 'a superiority over those of contemporary voyagers, who, being intent on gaining the characters of fine writers and elegant artists, have departed from the simplicity of Nature to give a scope to the decorations of art'.[13] Lamb and Thomas are right to point out the limitations inherent in Parkinson's approach, but his reputation for accuracy in the period is striking enough to merit further investigation.

When first published in 1773, Parkinson's *Journal* was almost completely overshadowed by Hawkesworth's government-sponsored *Account of the Voyages* of Cook and his predecessors, as the authors of the court injunction had intended.[14] By the end of the year, however, criticism of Hawkesworth had become so general and so severe that it was said to have hastened his death

[13] *Gentleman's Magazine* 55 (1785): 52, quoted in Joppien and Smith, *Art of Capt. Cook's Voyages* I.51–2. George Forster used his semi-official status to stake a claim against the unofficial histories of Cook's second voyage as 'marvellous histories,' too 'romantic' even for our ancestors (*Voyage Round the World* I.7)—which incidentally provides a glimpse of a whole industry surrounding the official publications, an industry largely ignored in my selective account.

[14] The *Monthly Review*, for example, makes only a brief mention of Parkinson's *Journal*, explaining that 'we have already so largely gratified the curiosity of the Public with regard to the voyage of the *Endeavour*, by the extracts which we have given from the journals of Captain Cook and Mr Banks, compiled and methodised by Dr Hawkesworth' (*MR* 50 [1774]: 411).

that same year at the age of 58. As Lamb has noted, Hawkesworth's moral relativism and his broader Enlightenment sympathies made him the target of a moral and religious opprobrium at least as general as, and more violent than, the objections to his ghostwriting and sham elegance restated here by the *Gentleman's Magazine.*[15] Parkinson's *Journal* has none of these liabilities, and its second edition, published within five years of Cook's death, benefitted from the celebration of his memory as well as (indirectly) from negative publicity associated both with Hawkesworth and with the publication of Cook's second voyage.[16] Although Stanfield Parkinson was ridiculed for complaining in print about the machinations of Banks and Hawkesworth, who (he asserted) abused and defrauded him of his brother's papers and collections, the late artist himself retained his unblemished character through all these disputes. There is a hint of class prejudice in the reviewer's suggestion that Parkinson, unlike Hawkesworth—who also rose from humble origins—knew his place and did not take advantage of his profession for undue social or financial gains. But it is hard to argue with the reviewer's empiricism, an editorial stance that led the magazine to promote the careers of other budding professionals of the artisan class, most notably John Carter: Parkinson really was in the South Seas and recorded his experiences in accurate plain prose. It is not clear if Parkinson himself had thoughts of publication, but even if he had, he might have avoided on religious grounds the ostentatious professions of plain truthfulness so common to voyage narratives. As Lamb suggests, these exaggerated professions register the habitual refusal of reading audiences to accept voyage narratives as fact, and Lamb goes on to claim that neither explorers nor readers were primarily interested in imperial fact-gathering in the first place.[17] Parkinson's refusal to make these professions, his immense artistic productivity, his virginal Franciscan persona, the utility and striking beauty of his journal, and his early death all contributed to establish his journal as a new standard, for literary products as well as scientific prose (as we might now distinguish them).

Although the editorial circumstances of Parkinson's *Journal* are richly revealing in terms of class conflict and cultural politics, his prose stands on

[15] Lamb, 'Circumstances Surrounding the Death of John Hawkesworth'.

[16] The second edition was actually a reissue, with added front matter, of 400 unsold copies of the first edition. Beaglehole oversimplifies matters when he states that 'Fothergill charitably bought up the remainder of the edition' of Parkinson's *Journal* after his death (*Journals of Capt. Cook* I.ccliv). John Nichols characterizes the Forsters—who were responsible for much of the conflict around the publication of Cook's second voyage—as ungrateful foreigners, a character in pointed contrast to the one created here for Parkinson (*LA* III.90–93n.).

[17] 'To insist that the point of such voyages was exclusively the production of truth for imperial purposes is to ignore two things: first, the confusing and sometimes inexpressible experience of extreme conditions endured by the voyagers themselves; and second, the reasons for the currency of their stories in the metropolitan book market' (*Preserving the Self* 7; see further 80–1, 100, 106–8).

its own merits as a deliberately professional production that transcends its origins in the context of a client-patron relationship. The preface, probably ghostwritten for Stanfield Parkinson by William Kenrick, emphasizes the 'accuracy he aimed at, not only in the particular walk of his profession of natural history, but also in describing the persons, languages, customs, and manners of the natives of the several islands and continents they visited'.[18] In light of the Parkinsons' Quakerism, the primary connotation of 'profession' here may be religious, and the Quaker traditions of truth-telling and of indifference to social rank provide important clues to Sydney Parkinson's narrative approach. At the same time, Stanfield's implicit distinction between natural history and ethnography points to the high degree of specialization Sydney had achieved as a painter of plant and animal specimens prior to the voyage. For this reason Stanfield appears surprised to find much ethnographic content in the journal at all. Parkinson threw himself so avidly into ethnographic drawing and description largely because Alexander Buchan, the painter employed by Banks to draw figures, dresses, and landscapes, died very early in the voyage. Parkinson's fame today rests equally on his 1300 illustrations of flora and fauna and on the several hundred drawings of native peoples, architecture, and artefacts that he created to fill the vacancy left by Buchan's death. Almost all the images initially circulating in print came from the second group.

According to a shipmate, Parkinson 'frequently sat up all night, drawing for himself or writing his journal' (*Journal* xi). One of the most remarkable intellectual products of these lucubrations is a theory of Polynesian migration with no equivalent in the journals of Banks and Cook or in any published account of the voyage. Parkinson's new professional responsibilities following the death of Buchan clearly posed a larger intellectual challenge, and he made multiple uses of the time spent drawing native subjects and practices. His widely praised vocabularies of Tahitian, Maori, Malay, and other languages, for which he created his own system of transliteration, occupy close to one-quarter of the *Journal*. He must have spent many hours in conversation, especially in Tahiti, gathering material for these vocabularies, for his account of funerary customs and his ethnobotanical catalogue, and collecting material culture, especially objects of everyday use. He even 'underwent the operation' of tattooing, along with a few of the seamen, to improve his visual and verbal grasp of the practice—all while other members of the ship's company, as Stanfield piously notes, 'were indulging themselves in . . . sensual gratifications' (vii). Parkinson's attention to languages, however, allowed him to reach conclusions on the migration of South Pacific peoples that evidently escaped Banks and Cook and therefore Hawkesworth as well. Concluding his Maori

[18] *Journal* vii. Parkinson's *Journal* is cited parenthetically in the text hereafter. On Kenrick's involvement, see Beaglehole (ed.), Endeavour *Journal of Sir Joseph Banks* I.61.

vocabulary, Parkinson notes its close kinship with the Tahitian language, as the others had done, but goes on to argue that 'one place was originally peopled from the other, though they are at near two thousand miles distance, and nothing but the ocean intervenes, at least to our knowledge; and such a long navigation, we should hardly believe, could be practicable in their small canoes, the only vessels that they appear to have ever possessed' (125). For Parkinson the similarity in customs and manners clinches this theory, while one particular difference allows him to specify the direction of migration as west to east: the Maori, he reports, are 'totally unacquainted with the use of bows and arrows . . . whereas the people of [Tahiti] use them with great dexterity, having doubtless discovered the use of them by some accident after their separation; and it cannot be supposed that the New-Zealanders would have lost so beneficial an acquisition, if they had ever been acquainted with it' (125).

Archaeologists now believe that, conversely, New Zealand was colonized from Polynesia, but Parkinson's reasoning, based on his own empirical research, holds true in other respects, and it differs symptomatically from the accounts of Banks and Hawkesworth.[19] What I am calling Parkinson's professional style emerges as he declares his intellectual independence from the antiquarian lore and the grand style of conjectural history that cloud the gentlemanly accounts of this issue. Banks states that he 'firmly believes'—without giving a reason—that both populations came from a single place of origin to the west, and Hawkesworth embellishes this supposition with a problematic piece of 'folklore': They have both a tradition that their ancestors, at a very remote period of time, came from another country; and, according to the tradition of both, that the name of that country was 'HEAWIJE'. He then elaborates Bank's conjecture:

> But supposing these islands, and those in the South Seas, to have been peopled originally from the same country, it will perhaps for ever remain a doubt what country that is: we were, however, unanimously of opinion, that the people did not come from America, which lies to the eastward; and except there should appear to be a continent to the southward, in a moderate latitude, it will follow that they came from the westward.[20]

Parkinson, by contrast, posits a migration by canoe, parallel to the *Endeavour*'s track, a hypothesis borne out by modern archaeology (apart from the direction of travel).

[19] Though he does not mention Parkinson, Brian Durrans provides an overview of the larger issue, in both anthropological and historical terms, in 'Ancient Pacific Voyaging'.

[20] For ease of comparison, I cite here and elsewhere the carefully prepared Internet texts of Banks's *Endeavour* Journal and Hawkesworth's *Account*, available on line as part of the National Library of Australia's *South Seas: Voyaging and Cross-Cultural Encounters in the Pacific 1760–1800*. For these three passages, see <http://southseas.nla.gov.au./journals/banks_remarks/250.html>, <http://southseas.nla.gov.au/journal/hv23/473.html>, and <http://southseas.nla.gov.au/journals/hv23/477.html>.

Hawkesworth depended on Banks's observations and gentlemanly style of learning for the authority of his own account, and systematic comparison shows that Parkinson too sometimes emulated Banks's methods, perhaps for some of the same social reasons. Historians of science have shown that the consolidation of a gentlemanly style precedes and enables the professionalization of emerging disciplines, a finding that helps to contextualize both narratives.[21] The case of Parkinson—like that of other knowledge workers—further proves that the education and social status of a gentleman were not prerequisites for legitimate contributions to knowledge, even before the advent of scientific professionalism. The use of Banks as a model provides evidence for a correlation between social and scientific status, but also shows that it was partly a matter of style. Hawkesworth achieved social mobility by means of his literary prose style; Parkinson's prose style is his own, but he achieves a kind of scholarly mobility (rather than socioeconomic) when he emulates Banks's conceptual style or adopts his intellectual priorities. An added difference between Parkinson and Hawkesworth is that the former carried out his research in the field; hence the stylistic and structural parallels between his journal and Banks's unpublished one reveal not only moments of emulation but also points of divergence, as in the case of the migration thesis.

More important, the comparison also reveals Banks's dependence on Parkinson and other collectors for his material. Banks indirectly acknowledges Parkinson in this instance (as in others) when he writes that 'the greatest part of the southern Language was not taken down by myself and I am inclind to believe that the person who did it for me made use of more letters in spelling the words than were absolutely necessary'. Banks's successful management of knowledge workers may be seen in its own right as an 'insistence on professionalism,' as Averil Lysaght has put it.[22] Systematic comparison of Banks's *Endeavour Journal* with Parkinson's suggests a certain amount of unacknowledged borrowing, though there is no strong evidence to support Stanfield's suspicion that Banks was responsible for the mysterious and total disappearance of the fair copy of Parkinson's journal manuscript. At the same time, Parkinson's intense ethnographic interest developed at Banks's direction following Buchan's death, and many of his particular subjects, too, such as costume, dance, and theatrical performances, were dictated by particular drawing assignments from Banks. Parkinson, however, very rarely mentions his labour as a draftsman, especially in his main area of botany, and instead offers interpretations of culture that sometimes follow and sometimes depart

[21] See, for example, Rudwick, *The Great Devonian Controversy: The Shaping of Scientific Knowledge among Gentlemanly Specialists.*

[22] Bank's discussion of Parkinson's Pacific vocabularies is quoted, again, from the text on *South Seas* (249). Lysaght insists on Banks's 'professionalism' in botanical matters as taking precedence over his 'gentlemanly'/Grand Tourist inclinations as pointed out by Bernard Smith ('Banks's Artists' 30–1).

from those of Banks. When he does turn to botany in his long catalogue of Tahitian plants, his descriptions are ethnobotanical, dictated in part by the needs of the crew, in part by Banks's larger economic interests, but in part also by his own infectious curiosity about plants used for food and medicine. The catalogue is particularly noteworthy for its long digressions on the flavour of breadfruit and coconut and for its mention of several plants only described to Parkinson by the natives and never seen by the ship's 'botanical gentlemen', as he calls them.[23]

Parkinson's antiquarian interests provide another rich locus of contrast and continuity with the gentlemen's accounts. Banks himself was an active Fellow of the Society of Antiquaries and followed the suggestion of a fellow antiquary, Thomas Falconer, when he hired a draftsman for his planned excursion to Lapland in 1768. William T. Stearn observes that Banks's father and grandfather, too, had employed antiquarian draftsmen, so his antiquarianism may be seen as securing his place among the upper landed gentry, to which he was a relative newcomer.[24] Banks's colleague Thomas Pennant employed a draftsman for his work in both natural history and antiquities, Moses Griffith, whom he had discovered in a village on his estate and called his 'treasure'.[25] Parkinson, however, came to Banks with a specialized education and a budding London reputation, and was less a retainer than an intellectually ambitious autodidact with his own programme of reading in aesthetics and poetry as well as antiquities—as attested by the catalogue of Parkinson's personal shipboard library recorded in his sketchbook.[26] One of his earliest projects for Banks was to copy the remarkable drawings of animal specimens from Ceylon done by Pieter de Beveren, a half-Sinhalese artist employed by the Dutch governor Johan Gideon Loten. Both the antiquarian tradition of draftsmanship and the high quality of de Beveren's models, with their obvious exchange value as documents that circulated among patrons of natural history, must have informed Parkinson's profession as he understood and practised it.

Rüdiger Joppien and Bernard Smith argue that Parkinson's shipboard reading of Hogarth's *Analysis of Beauty* visibly influenced his ethnographic depictions, and likewise his reading of antiquarian and historical texts on board seems to have influenced his representation of Tahitians as noble savages who are, 'in

[23] Gentlemen are those who 'amuse themselves,' or so Parkinson's frequent use of that verb to describe the activities of Banks and Solander might suggest (e.g. 21).

[24] Stearn, 'Account of the Botanical Explorations.' Stearn cites communication with Falconer concerning Banks's planned excursion to Lapland in 1768, but since Banks gave up this plan as soon as he learned of the opportunity to sail with Cook, it seems fair to credit Falconer for Banks's engagement of Parkinson for the *Endeavour* voyage (Parkinson probably would have been his choice for the Lapland voyage too). Banks's paternal grandfather was the family's first major landowner.

[25] *Literary Life of the Late Thomas Pennant* 10.

[26] Reproduced in Joppien and Smith, *Art of Capt. Cook's Voyages* I.51–2.

constitution, what the ancient Britons were before their civilization. Unhappily for us, the athletic constitution of our ancestors is not to be found amongst us, being enervated by excesses of various kinds' (*Journal* 23). Hawkesworth, following Banks, takes a classicizing route instead of Parkinson's romantic one: 'It is scarcely possible for those who are acquainted with the athletic sports of very remote antiquity, not to remark a rude resemblance of them in this wrestling-match among the natives of a little island in the midst of the Pacific Ocean: and even our female readers may recollect the account given of them by Fenelon in his Telemachus, where, though the events are fictitious, the manners of the age are faithfully transcribed.'[27]

Some scholars have emphasized Parkinson's romanticism and/or Eurocentrism, and despite his intellectual independence, Parkinson's prejudices are no less visible than those of Banks and Cook.[28] At the same time, scholarship has also confirmed what contemporaries already noticed about Parkinson's depiction of South Pacific peoples. Joppien and Smith have argued that Parkinson's drawings are relatively free of both classical and ethnographic visual stereotypes, a quality noticed by reviewers as authenticity or 'originality'; they and others have also shown how some of Parkinson's drawings were greatly altered on their way to becoming engravings. Referring to the latter group, one reviewer in 1773 noticed that 'some' of the engravings in Hawkesworth 'are executed with such a degree of elegance, as rather evinces the abilities of the artists who produced them, than represents the subjects with fidelity'. Art historians have pointed specifically to plates by the fashionable engraver Francesco Bartolozzi as not based directly on Parkinson originals, but rather on modified drawings prepared by G. B. Cipriani.[29]

The very prominent and detailed figures in Figure 1.1, from Parkinson's *Journal*, differ markedly from the idealized figures in Figure 1.2, one of Bartolozzi's contributions to Hawkesworth. Bartolozzi's figures are much smaller, and carefully framed in the picturesque manner, but once adjusted for scale it is clear that the anatomy and physiognomy of these figures are a pastiche of classical and picturesque elements. Parkinson's focus on the Tierra del Fuegans' anatomical difference from Europeans may be distorting in its

[27] Joppien and Smith, *Art of Capt. Cook's Voyages* I.35–41; Hawkesworth, *Account of the Voyages* II.120. This print edition is cited parenthetically in the text hereafter as *Account*.

[28] See especially Lamb, *Preserving the Self*, and Thomas, *In Oceania*, as well as Lysaght, 'Banks's Artists' 18, 30–1, 66.

[29] *Critical Review* 37 (1774): 10; this is the end of the third and last installment of this lengthy review. The *Gentleman's Magazine* review cited earlier also anticipates this art historical distinction between more and less faithful engravings after Parkinson, pointedly contrasting Parkinson's 'originality' with that of 'elegant artists' who 'have departed from the simplicity of Nature'. According to Joppien and Smith, the *Endeavour* artists' lack of specific training in figure drawing was a formal disadvantage, but 'their freedom from the classical visual stereotypes of figure drawing was a challenge, especially for Parkinson', that produced more innovative and in some ways more immediate representations (*Art of Capt. Cook's Voyages* I.21).

Fig. 1.1. *A Man, Woman & Child, Natives of Terra del Fuego, in the Dress of that Country.* Engraving by T. Chambers from a drawing by Sydney Parkinson. Parkinson, *Journal of a Voyage to the South Seas* (1773), Plate I. Image courtesy of Grinnell College Libraries Department of Special Collections.

own right, but this engraving's claim to the status of visual documentation (rather than classical elegance or picturesque beauty) is immediately clear. Among other salient differences, the furs worn by the figures in Figure 1.1 are obviously more suitable to the climate at 55° S latitude than the draperies worn by the semi-nude figures in Figure 1.2. Some differences between the two images must also be attributed to the significantly lower engraving budget available to Stanfield Parkinson; Thomas Chambers, for example, was a successful engraver but not a celebrity like Bartolozzi. Another image from Parkinson's *Journal* underscores this contrast: Stanfield somehow obtained Buchan's drawing of the Tierra del Fuegans' dwellings (Plate II), which are more conical and rugged in form than the artfully rounded 'primitive hut'—possibly a visual allusion to Marc-Antoine Laugier's description of that original human dwelling—in Bartolozzi's engraving. It is notable that the word 'hut' is used in the caption (Figure 1.2), as opposed to 'hovel' in Hawkesworth's text. Joppien and Smith also reproduce an original sketch by Parkinson, *A Young Woman in a Dancing Dress*, side-by-side with a well-known Cipriani/Bartolozzi image entitled *A View of the Inside of a House in the Island of Ulietea, with the Representation of a Dance to the Music of the Country.* The bare-breasted young dancer in the latter image follows Parkinson reasonably closely in details of costume, but the pose and physiognomy of this figure are altered in a classicizing manner, while the interior, with its classically ranged columns and group of spectators (in togas), seems full of freely invented additions.[30]

Buchan's death and Banks's interest in the subject made it possible for Parkinson to develop his competence, both visual and verbal, in the ethnographic part of natural history. To explain what is new and distinctive about Parkinson's *Journal*, though, it is even more important to understand the independent choices and earlier training that contributed to this understanding of what Stanfield aptly called his 'profession of natural history': his extensive training as a draughtsman, probably at William de la Cour's drawing school in Edinburgh, as Lysaght has argued; his early exposure to the economic and cultural contiguity of antiquarian and botanical draftsmanship, and hence of human and nonhuman subjects; and his own ambitious course of reading, along with less tangible aspects of temperament and religious belief.[31] All these factors account for the distinctive and memorable quality of Parkinson's verbal and visual sketches of a wide variety of topics, including the perennially hot ones of tattooing and cannibalism, as well as picturesque landscapes incorporating ethnographic elements. My assessment of Parkinson as an author continues in the later sections of this chapter on antiquarian motifs.

Parkinson's authorship is one important aspect of his development as a 'naturalist-antiquary', to borrow Maria Toscano's term again. In Toscano's account, the naturalist-antiquary's collection defines his vocation at least as strongly as his writing, and I pause here to examine Parkinson as a collector. The intense dispute between Stanfield Parkinson and Joseph Banks over Sydney's papers and large specimen collection was essentially a debate over whether Banks was his patron, and therefore entitled to all his work, or a legal employer, and thus entitled only to the contracted botanical drawings and not to his professional resources. By the time of Cook's second voyage, the shipboard artist-naturalist team was a pair of well-paid civil servants who also managed, despite careful contractual stipulations, to earn money from the sale and publication of independent visual and verbal accounts. The Forsters' success and subsequent voyaging history confirm the importance of both precedents set on the *Endeavour*: Banks's patronage as well as Parkinson's independent vision of the naturalist's profession.[32]

[30] See, respectively, Buchan, *An Indian Town at Terra Del Fuego*, reproduced in Joppien and Smith, *Art of Capt. Cook's Voyages* I, Fig. 4 (cp. the frontispiece of Laugier, *Essai sur l'architecture*); Parkinson, *A Young Woman in a Dancing Dress* (Joppien and Smith, Fig. 84a); and Hawkesworth, *Account of the Voyages*, vol. 2, plate VII.

[31] For Lysaght's hypothesis concerning Parkinson's training, see 'Banks's Artists' 13–15.

[32] The evolution of the natural history professional may also be traced fiscally from the starting point of Parkinson's annual salary of £80, which Banks felt gave him the right to appropriate all Parkinson's drawings after his death in 1771. When the *Resolution* returned in

Fig. 1.2. *A View of the Indians of Terra del Fuego in Their Hut.* Engraving by F. Bartolozzi from a drawing by G. B. Cipriani (based on originals by Buchan and Parkinson). John Hawkesworth, *Account of the Voyages . . . in the Southern Hemisphere* (1773), Vol. II., Plate I. Courtesy of the Division of Special Collections, Archives, and Rare Books, Ellis Library, University of Missouri.

1.3 PARKINSON AS COLLECTOR, OR, THE AFTERLIFE OF ETHNOGRAPHIC SPECIMENS

Less than a year after Banks returned to England on the *Endeavour*, Stanfield Parkinson charged him with stealing ethnographic and zoological specimens collected for personal use by his brother Sydney, who died of dysentery on the homeward voyage. Some half-a-dozen historians have dismissed the charges brought by Stanfield, who did receive £340 compensation, and even the

1775 Banks bought the drawings of George Forster directly from the artist (Rice, *Voyages of Discovery* 172). Banks had previously offered Johann Zoffany £1,000 to serve as painter on board this voyage, but ultimately the Admiralty appointed J. R. and George Forster as naturalist and artist, respectively, with a total compensation of £4,000—equivalent to the whole cost of the *Endeavour* voyage (Beaglehole, ed., Endeavour *Journal* I.73, 80). Parkinson's *Journal* may also be seen as the precedent for subsequent quasi-professional voyage accounts by both Forsters (1777, 1778), William Wales (1778), and many others.

significance of the disputed objects.[33] The peculiar afterlife of these objects, however, encourages renewed investigation both of this conflict and of more positive aspects of Sydney's collaboration with Banks, who strongly influenced the artist's increasingly independent pursuit of natural history and antiquities. Even inorganic objects can live in strange and uncomfortable ways, as illustrated by Lorraine Daston, Bruno Latour, and other contributors to the volume *Biographies of Scientific Objects*. Daston argues that scientific objects come into being when cultural factors make them salient: 'they become more richly real as they become entangled in webs of cultural significance', she writes, but then eventually 'slip back into the wan reality of quotidian objects, which exist but do not thicken and quicken with inquiry'. If Daston's objects live by dying into discourse, then Latour's more radical concept of nonhuman 'agency' seems to aspire to an autonomy prior to discourse, collapsing the boundary between the real and the constructed.[34]

Parkinson's disputed collection merits a special chapter in the rich collective biography of objects collected on the *Endeavour*. Cook voyage artefacts continue to generate new research and new questions. The disputed objects, though they cannot be positively identified now, participate in this rich collective afterlife because Stanfield recorded them in some detail. Moreover, the collaborative nature of research on the *Endeavour* and the joint pursuit of natural history and antiquarianism both require a special emphasis on the temporality of these objects. Wordsworth's lament that 'we murder to dissect' and Foucault's theory of 'the fragmentation of the surface upon which all natural beings had taken their ordered places' must both give way here to a new understanding more closely tied to the naturalist-antiquary's way of working.[35] Naturalists on the *Endeavour* seem to have operated in both phases of the Foucauldian epistemic shift at once, collecting plants and animals under the aegis of Linnaean natural history and artefacts under the sign of a proto-ethnographic antiquarianism. This dual agenda, aptly described as 'typological collection' by Jeremy Coote, somewhat loosens the potential hold of any death-dealing regime of classification. By collecting pairs or sets of 'type specimens' in ethnography as in natural history, shipboard naturalists marginalized the artificers themselves but assimilated their artefacts into the emerging chronology of deep time.[36] Pacific artefacts became 'ancient' by virtue of being 'natural.' At the same time the burgeoning market for these

[33] See Beaglehole (ed.), Endeavour *Journal* I.59–61; Wilfrid Blunt, 'Sydney Parkinson and His Fellow Artists' 40; Denis J. Carr, *Sydney Parkinson* xi–xiii; Maurice Holmes, *Captain James Cook* 53; Lamb, *Preserving the Self* 97; Lysaght, 'Banks's Artists' 27–8; and Edward Smith, *Life of Sir Joseph Banks* 29.

[34] Daston, *Biographies* 13; Latour, 'On the Partial Existence' 253.

[35] Foucault, *Order of Things* 275.

[36] Coote, Gathercole, and Meister survey earlier scholarship by Gathercole and Kaeppler on the shared methodology of ethnographic and natural history collecting ('Curiosities Sent to

'curiosities'—one factor behind the collecting compulsively pursued by Cook's crew on the second voyage, and seemingly a major factor behind Stanfield Parkinson's complaint against Banks—animated many Pacific objects (Daston-style) as they travelled through collections and museums and eventually into the scholarship on colonialism and the history of science.[37]

Parkinson's collection appears somewhat ghostly in light of the diverging disciplinary histories of natural history and ethnography. In empirical terms, the curatorial record for artefacts associated with the *Endeavour* voyage, and with Banks in particular, is more imperfect than the record for subsequent voyages. In disciplinary terms, Parkinson's botanical drawings may have had the most robust afterlife among all *Endeavour* objects, with a continuous history of reproduction and publication. Banks spent between £7,000 and £12,000 and employed at least eighteen artists over eleven years just to copy and engrave 743 of these drawings and thereby preserve his Pacific *Florilegium*. Banks's plant and animal specimens, with the drawings, entered systematically into the British Museum's study collection and were transferred to the separate Natural History Museum in Kensington around 1880.[38] He distributed the ethnographic material—by contrast—more widely and much earlier, after less careful study. In a 1784 letter, first noticed by Adrienne Kaeppler, J. C. Fabricius reports that Banks gave a large ethnographic collection from Cook's third voyage to the British Museum, where it was neglected for years by the keeper of the Department of Antiquities. Kaeppler also found that original documentation survives for only a fraction of the extant Cook voyage artefacts—particularly those from the *Endeavour*—suggesting a more pervasive neglect. Coote acknowledges that 'there is still little that can be traced' to the *Endeavour* voyage, but has recently identified a small collection from this voyage that Banks donated to Christ Church College, Oxford before 1773, and another well-documented first-voyage ethnographic collection was

Oxford' 183). Coote describes the practice as 'typological collection' in his 'Interim Report' (115). On deep ethnographic time, see Heringman 'Deep Time in the South Pacific'.

[37] Thomas analyses Cook's dismay over the 'passion' or mania for collecting that possessed his crew on the second voyage (*Cook* 183). There is no direct evidence concerning the market value of Pacific artefacts before the return of this voyage in 1775 (see, among others, Kaeppler, '*Artificial Curiosities*' 44). Collecting among the ship's crew played a less significant role on the *Endeavour*, but 'curiosities' from that voyage must have sparked sufficient interest to help provoke this 'passion' for collecting on later voyages.

[38] The online catalogue (<www.nhm.ac.uk>) returns, among other examples, a surgeonfish (*Naseus olivaceus*) and a grouper (*Anthias sacer*) matching specimens that Parkinson painted at Madeira in 1769, still present in the museum's study collection 240-odd years later. On the production and cost of Banks's *Florilegium* (left unfinished until the 1980s), see Adams, *Flowering of the Pacific*, esp. 140, 147–9, 176. On the acquisition history of both specimens and illustrations see the introduction to <www.nhm.ac.uk/nature-online/art-nature-imaging/collections/endeavour-botanical/>.

passed on to Trinity College, Cambridge by the Earl of Sandwich in 1771 after Cook handed it over to the Admiralty.[39]

One reason for revisiting the ethnographic project of the *Endeavour* is thus paradoxical: the massive effort of Banks and his staff to preserve biological specimens took its impetus from the antiquarian conviction that objects are perishable and must be recorded. The ironic result of these preservation efforts was that the biological specimens became integral to a modern monument, the Museum of Natural History, while the artefacts achieved at best a murky afterlife in the Department of Antiquities. Coote's research on the well-documented Forster collection at Oxford has shown that neglect was not universally the fate of early museum collections of Pacific artefacts, and his examination of the Christ Church collection suggests that Banks too may have taken them more seriously than was previously recognized.[40] The Forsters donated the Oxford collection in 1776, but later tried to sell artefacts, and in 1782 the Electors of Hanover paid over £120 for a collection of Pacific artefacts that Johann Friedrich Blumenbach requested for the university museum in Göttingen. This sequence of events, culminating in the purchase of a collection from a specialist dealer, George Humphrey, tends to confirm Stanfield Parkinson's premonition that the market value of Pacific artefacts would increase—if not as dramatically as he anticipated.[41] The market value of the artefacts, and particularly of Sydney's shells, may have played a minor role, but there was more at stake in Banks's decision to amalgamate Parkinson's collection with his own. Banks appropriated several other collections—perhaps including that of Tupaia, Cook's Polynesian guide, as Paul Tapsell intriguingly suggests—and his scientific motives should not be minimized, but he did not provide the careful documentation that the Forsters and

[39] Fabricius quoted in Kaeppler, 'Tracing the History' 174; Coote, 'Interim Report' 111. Fabricius goes on to say that Banks, on hearing this news, allowed him to reclaim the collection and divide it with his friends. On the Sandwich collection, see Kaeppler, *'Artificial Curiosities'* 39–40.

[40] Coote points out that the Forsters used essentially the same Linneaean logic of 'typological collection' to assemble the roughly 220 objects that were carefully accessioned and immediately noticed at Oxford. See further Coote et al., 'Curiosities Sent to Oxford' (esp. 183); Coote, 'Interim Report,' where he implies that Banks's prior gift to Oxford might have inspired the Forsters' gift of this collection (119); and Coote, 'Joseph Banks's Forty Brass Patus,' which suggests that Banks developed a sophisticated political understanding of artefacts as trade goods. In Kaeppler's contrasting view, Banks's ethnographic collecting had more to do with 'prestige' than with 'science' ('Tracing the History' 167).

[41] Urban, 'Acquisition History of the Göttingen Collection' 65–6. Urban summarizes the market for Pacific artefacts based on this transaction (66) and goes on to discuss the Forsters' role in this marketplace (66–9). He also shows that J. R. Forster had been collecting Pacific artefacts since 1771 (57), but the earliest direct evidence for a market is that Humphrey visited the *Resolution* as soon as it docked in 1775 and 'bought whatever he could' to sell to collectors (Kaeppler, 'Göttingen Collection' 91).

Humphrey, respectively, provided to the institutional recipients of their collections.[42]

Through all the conflicts over ownership, uncertain documentation, and periods of curatorial neglect, the objects themselves have played an essential role in the recovery of historical knowledge about islanders as well as voyagers. Attributing any kind of non-human agency to these ethnographic 'specimens' would seem to involve a highly problematic suppression of the human agency of their original makers and owners. Thomas's critique of Parkinson's 'decontextualized' 'Weapons and Ornaments' helps to explain why this is so. I am mobilizing them here to underscore their refusal to die into collections: they remain an uncomfortable presence in the historiography of the *Endeavour* voyage, and thanks in part to Parkinson's *Journal*—one of the most widely cited primary sources in pre-contact Pacific ethnography—they are helping to supply a record of their makers as well. The peculiar afterlife of these disputed textiles, weapons, utensils, and other artefacts illustrates the residual pressure of antiquarianism on a historical narrative that seeks to adopt Banks exclusively for the history of science. Even today's more critical post-colonial historiography retains the anachronism of Banks as a scientist whose importance lies with the Royal Society and natural knowledge rather than the antiquarian field in which he published his research. This partial view implies that Parkinson's artefacts would have been too insignificant to the scientist–statesman to be worth stealing, but also provides an alibi in the assumption that independently collected artefacts rightfully belong in one final representative collection.[43] The continuing importance of Parkinson's *Journal*, and of the fieldwork behind it, militates against these suppositions and supports a reconsideration of the antiquarian context.

There are good reasons, then, for reviving a story dismissed by historians as a 'sordid . . . dispute', a 'tedious business', and an 'unedifying squabble'.[44] Stanfield Parkinson, the executor and editor of the deceased artist-naturalist, opens his case with a forceful contrast between Sydney's method of collecting and that of the 'oppressor' Banks, as Stanfield calls him (*Journal* xxii). The critique of colonialism may not permit such fine distinctions, but it is undeniable that Sydney's journal records a more extensive set of Tahitian and Maori customs, and especially languages, than any other period source, and his prose bears witness to a refined ethnographic sensitivity with roots in his family's

[42] Tapsell, 'Footprints in the Sand' 102–3. Tapsell also speculates: 'If Tupaea had reached England, would his drawings, observations, and accounts have been accepted with equal validity alongside Cook and Banks? And would the *Endeavour*-associated curios have been more accurately credited when placed in museums?' (107).

[43] This contradiction, among others, is already in place in Fothergill's apologia for Banks ('Explanatory Remarks' 6–7).

[44] Blunt, 'Sydney Parkinson and His Fellow Artists' 40; Beaglehole, Endeavour *Journal* I.59; Lamb, *Preserving the Self* 97.

profession of Quakerism. According to Stanfield, Sydney Parkinson conciliated the native peoples, 'engaging their attention by the powers of his pencil, disarming them of their native ferocity, and rendering them even serviceable to the great end of the voyage, in chearfully furnishing him with the choicest productions of the soil and climate, which neither force nor stratagem might otherwise have procured' (*Journal* vii). 'By such honest arts and mild demeanour', Stanfield continues, Sydney accumulated not only his widely praised vocabularies, but also 'a choice and rare collection of curiosities, consisting of garments, domestic utensils, rural implements, instruments of war, uncommon shells, and other natural curiosities, of considerable value' (vii–viii). The account of this collection also lends some credence to Stanfield's sense of Banks as the oppressor: he defends his publication of Sydney's *Journal of a Voyage to the South Seas*—in the editorial preface I am quoting—as his only recourse against Banks's appropriation of this collection, which he regarded as the better part of his brother's estate. All but one of the scholarly accounts have been so eager to vindicate Banks's 'undoubted generosity', as Beaglehole puts it, that many circumstances favouring Stanfield's version of events have been ignored.[45] Whether we see the publication as an attempt to recapture a legacy unjustly withheld, or merely as a scheme to make money, it is clear that without Stanfield's intervention Sydney Parkinson's journal would almost surely have been lost, since Banks and the Admiralty saw it as competing with the official account of the voyage.

Readers tend to become impatient with Stanfield's account of his travails not only because it is unflattering to Banks, but also because of its excessive, perhaps obsessive, level of detail. As Stanfield tells it, Banks notified him of his brother's death soon after the *Endeavour* returned and promised him Sydney's effects in accordance to his will. His long delay in fulfilling this promise made Stanfield suspicious, and by his third visit there was a heated exchange, specimens and manuscripts were flung about, and Stanfield ended up with a mysteriously reduced collection of damaged and substitute goods that did not even match the inventory Banks himself provided. Stanfield went to members of the ship's company for information and found a number of allies, who rehearsed from memory what was missing from the collection as they now saw it, including 'a quantity of seeds of curious plants, many birds and animals preserved in spirits, many lances, bludgeons, and other weapons used in war, likewise household utensils and other instruments, purchased of the natives' (xi). One of these men also told Stanfield—so he claims—that Sydney had

[45] Beaglehole, *Journals of Captain James Cook* I.ccliv. Two points in Stanfield's favour include the erasure of Sydney's self-portrait from what became plate 2 of vol. II in Hawkesworth, first noticed by Joppien and Smith (*Art of Capt. Cook's Voyages* I.26–8); and a letter from Banks (12 January 1773), also cited by Joppien and Smith, that gives directions for Sydney's name to be—as Stanfield put it—'meanly and invidiously suppressed' from the Hawkesworth plates that were based on his designs.

made many drawings of landscapes and figures on his own time to give to his friends, and that he kept a journal regarded by many as the best on board. Stanfield threatened a lawsuit and in November 1772 John Fothergill, a Quaker physician and botanist well-known to Banks, stepped in as mediator. According to later accounts by Fothergill and others, Stanfield had begun to show signs of the paranoia that heralded his rapid decline into madness in 1772; he died in an asylum, apparently in 1773 within months of the journal's publication. As Joppien and Smith have noted, it is small wonder that Parkinson became suspicious when there was such a strong appearance of collusion in the actions of Banks and Fothergill.[46]

As the dispute proceeds, Stanfield places increasing emphasis on the artefacts, attempting to mark boundaries between the human and the natural and between collaborative (or salaried) and personal collecting—areas that tended to merge in the naturalist-antiquary's practice. In his 'Explanatory Remarks' Fothergill defends Banks's right to all Sydney's objects as justified by the 'promotion of knowledge.' In light of this view, which Banks shared, Fothergill's argument to Stanfield seems disingenuous: to return the collection so that Banks could purchase individual items needed to fill gaps in his collection (xiii).[47] Instead of looking over the collection once more, Banks immediately proposed (through Fothergill) a settlement whereby Sydney's heirs would receive £500, including the last two years' salary (£160) and what amounted to a death benefit. As a condition, Stanfield would have to resign his right to all of Sydney's effects and even his status as executor. Stanfield's inventory of the items he returned dwells on what he regarded as personal effects: forty-five items of textiles or clothing and 'a great number' of fish-hooks and other 'instruments used by the people on the southern islands,' along with 'a very large parcel of curious shells' (xiii–xiv). Stanfield may have overvalued the shells, and he may have undervalued Banks's settlement. As to the latter, he had no choice but to accept it. As to the former, Stanfield evidently included the shells with the artefacts because he felt they were not part of the *Endeavour*'s natural history collections. In his critique of the settlement, Stanfield argued persuasively for his right not only to collections but also to manuscripts and drawings relating to anything other than botany and zoology, the specified area of Sydney's salaried employment. But he did not carry his point concerning either the collections or the surprisingly modern idea of his brother's professional status (vii).

[46] *Art of Capt. Cook's Voyages* I.53.

[47] 'Explanatory Remarks' 6–7. Joppien and Smith present visual evidence that Banks had already integrated some of Parkinson's ethnographic specimens with his own on the return voyage from Batavia, after Parkinson's death (*Art of Capt. Cook's Voyages* I.56–9, 161). Banks, then, felt entitled to these objects. But a surviving letter from Parkinson to his cousin Jane Gomeldon, also quoted by Joppien and Smith (I.51), proves that he was collecting for her, and therefore probably for himself and others as well.

Stanfield did, however, secure a loan of what appears to have been a draft of Sydney's journal—the fair copy of which Banks claimed was lost, and which has never been found. Stanfield's claim that he had no immediate plans for publication has been dismissed too readily. He points out that the 'foul papers' he obtained were inadequate and included no drawings whatsoever (xvin.) He claims that he pursued publication only after his return of the manuscript failed as an incentive for Banks to return Sydney's collection, and that he then used manuscripts and drawings supplied by members of the ship's company who responded to his advertisement (xvii). As noted above, Fothergill bought up four-fifths of the copies and added a lengthy prefatory essay which Lysaght, among other historians, regards as 'completely exonerating Banks and... Fothergill'. These 400 copies were republished with Fothergill's preface only after his death in 1784, but the *Gentleman's Magazine* review (1785) carefully reconstructed the history of the book in such a way as to cast doubt on Fothergill's asseverations 'that nothing could be more generous than Mr Banks's dealing with Mr Parkinson; nor more humane and friendly, than Dr Fothergill's interposition in his favour'. It is surprising, then, that Beaglehole and Lysaght take Fothergill as their primary source in retelling the story of the journal's publication. Both accounts agree, in the absence of any real evidence, that Stanfield Parkinson was simply deluded about the money value of his brother's collection. But as Joppien and Smith point out, in the one sympathetic account to date, Banks's generosity 'must remain a matter of opinion', even if 'most historians agree in exonerating' him.[48]

The contested objects speak for themselves most directly in two plates conceived and executed after Parkinson's death, at several removes from the composition of his journal. The engravings render them as static specimens, or even monuments, but the text overflows with evidence of their temporality. This plate (Figure 1.3) and a companion image (*Journal* pl. XXVI) were engraved from drawings made by Samuel Hieronymous Grimm of objects that may or may not have belonged to Parkinson, and the accompanying text is marked as an editorial insertion. Stanfield commissioned these engravings as illustrations for the *Journal*, but they also stand in for the collection that Banks refused to return. Some of these 'instruments and utensils' may have been provided by the same crew members who supplied the loose manuscripts and drawings. The two plates in question are two of only four in the published *Journal* that do not bear Parkinson's signature as delineator, so Grimm must

[48] Lysaght, 'Banks's Artists' 28; Fothergill (quoting an unnamed friend), 'Explanatory Remarks' 22; Beaglehole, Endeavour *Journal* I.61 and n. 3; Joppien and Smith, *Art of Capt. Cook's Voyages* I.54. Fothergill and his editor, John Coakley Lettsom, offer specific rebuttals to many of Stanfield's points. Fothergill's estimated value of £50 for the entire disputed collection ('Explanatory Remarks' 7–8) seems absurdly low; there is very little positive evidence before 1782, but one period source does place the price of Pacific shells as high as £5 apiece (G. Forster quoted in Urban, 'Acquisition History' 67).

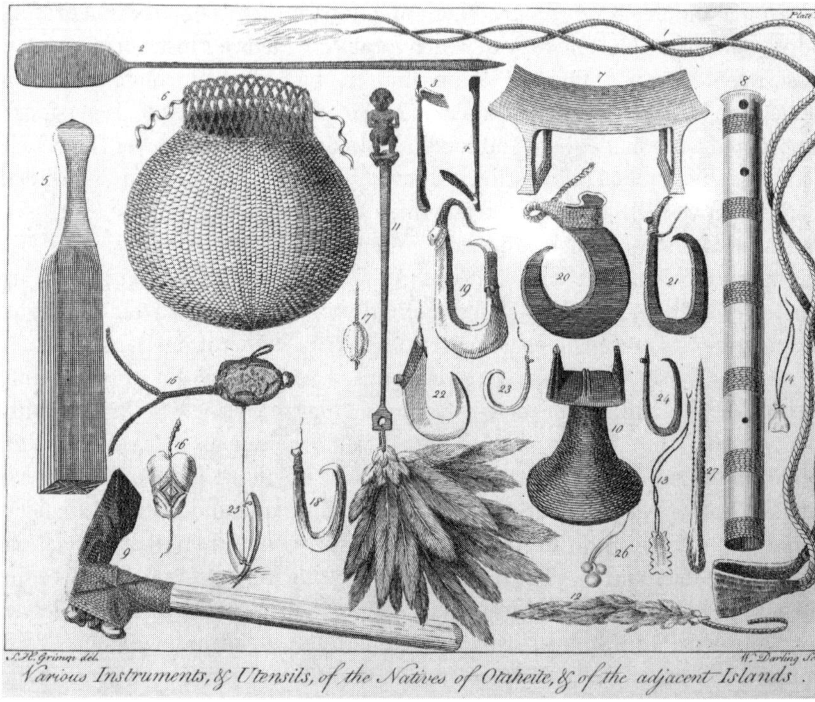

Fig. 1.3. *Various Instruments, and Utensils, of the Natives of Otaheite, and of the Adjacent Islands.* Engraving by W. Darling from a drawing by S. H. Grimm. Parkinson, *Journal of a Voyage to the South Seas*, Plate XIII. Image courtesy of Grinnell College Libraries Department of Special Collections.

have worked directly from the objects rather than from sketches by Parkinson (as Thomas's critique assumes).[49] The lack of certainty over their ownership forcibly calls attention to their history as trade goods and to the removal of these objects from their original human context.

The engraving, then, not only decontextualizes but also recontextualizes these artefacts as artificially outside time and subject to formal principles associated specifically with antiquarianism and antiquarian engraving. Grimm's arrangement of these objects is intended to fill the space, much like the heterogeneous objects in a composition he created for Gough's *Sepulchral Monuments*. Grimm began his career as an antiquarian draftsman in 1768 and

[49] The other two (out of 27 total) are Pl. II, J. Newton's engraving of Buchan's sketch of the 'huts' at Tierra del Fuego (mentioned above); and B. Longmate's engraving of Cook's map of New Zealand (Pl. XXV). The compositions may have been influenced by Parkinson's sketches, even if based primarily on the artefacts themselves; likewise, the verbal keys to these images—which remain significant to Pacific anthropology—probably came from Parkinson's notes, but were clearly given their final form by another hand.

forty views by him appeared in Francis Grose's *Antiquities of England and Wales* in 1773, the same year that Parkinson's *Journal* was published. In the 1780s Grimm made numerous drawings for Gough. Among those engraved for *Sepulchral Monuments,* one resembles Grimm's Parkinson plates in its compositional density and its strange juxtapositions—in this case, architectural details, coffins with human remains, an ancient chalice and other artefacts, all seemingly arranged according to a spatial logic on one page (*SM* II.lxviii). The logic of curiosity insisted on by Thomas is not the only logic operating in the Grimm/Parkinson image.[50] The predisciplinary logic of antiquarian natural history informs the engraving and even its spatial logic is not without meaning. There are three symmetrical pairs of fish-hooks, for example, with the points facing each other in the two central pairs. The four-foot long sling, drawn roughly to scale, is wrapped around the composition as a border, and, more importantly, arranged artfully to conform to the 'line of beauty' as described in Hogarth's *Analysis of Beauty*.[51] Economy not just of space, but of labour and materials, was a more pressing factor for Stanfield than it was for Gough or for Hawkesworth, and thus reinforced the formal logic of specimen illustration.

In the antiquarian context this style of composition may connote plenitude rather than the 'vacuity of meaning' that it suggests to Thomas. The density and incongruity of the objects—for example, the incongruity of placing 'domestic and warlike instruments' together—also speaks of superabundance, underscoring the loss of Sydney's collection, on which Stanfield, at least, set a high value. His decision to add these two images also problematizes Thomas's sharp distinction between scientific curiosities and commodities, instead linking them together as they were linked for collectors of classical antiquities (or for connoisseurs of the 'ancient systems' of Pacific craftsmanship, as J. R. Forster put it). In the case of these two plates, at least, the 'scientific enclave' is not created to assert a form of social distinction between the gentleman-naturalist's collecting and that of the crew, but rather to link curiosity with market value.[52] The added materials enrich the journal's

[50] Thomas, *Entangled Objects* 130 (quoting *Journal* 98); Thomas, *In Oceania* 97–119. See also Daston and Park, *Wonders and the Order of Nature*, esp. 135–72 and 303–28.

[51] Hogarth, *Analysis of Beauty* 48–59. Grimm paid some regard to scale, since most of the individual objects are depicted on a scale ranging from 1:5 to 1:6. One notable exception is the headrest or 'stool used as a pillow' (*Journal* 76), listed as 12.5 inches long and therefore on a scale of roughly 1:7. The general conformity of scale, though hardly exact, suggests that in this respect, too, the assemblage here is not totally haphazard.

[52] Thomas, *In Oceania* 101, 116; Forster, *Observations* 196. Thomas uses the verbal key to *Journal* pl. XXVI (100) among other sources to help support his later claim that 'the abstraction of artefacts . . . authorized the natural philosopher's travel and collecting by making the particular claim that a curiosity was a specimen, something in a scientific enclave, rather than an object of fashion or mere commodity' (*In Oceania* 116). In the engravings, he argues, a 'vacuity of meanings emerges . . . from the extreme decontextualization of the pieces depicted' (101). In the

narrative, which occasionally refers to some of these cultural objects and related practices by way of references to the plates added in square brackets and occasional editorial footnotes. At the same time they hint at the development of exchange on a much larger scale, in which even objects deemed rare according to recent scientific standards became more common as Europeans returned again and again to the South Pacific. The two companion plates together also construct a conventional distinction between peaceful Tahitians and warlike Maori, since only a very few small items here (Figure 1.3) might be considered weapons (the spear point just to the left of the flute, and probably the sling), whereas the New Zealand plate includes four large and two small 'warlike instruments' along with the 'domestic' ones (128).

Even in this decontextualized form, the objects depicted here continue to enable modern anthropologists to recover the cultural life suppressed by the 'rigidly neutral' tenor of the engravings (Thomas) and their accompanying lists. Gundolf Krüger, for example, singles out ten of the twenty-seven objects in this image as analogues for objects in the Göttingen museum collection and repeatedly cites Parkinson's explanations for their use. Krüger also distinguishes various means by which such goods were exchanged, including theft (or supposed theft) by islanders and its counterpart, the retaliatory confiscation of goods by Europeans. These artefacts may therefore have a longer and more complex history of contested ownership than Stanfield realized. Krüger counts the flywhisk here as one of only four known examples of anthropomorphic carving on a flywhisk handle from Tahiti, the other three being actual specimens from Göttingen and Vienna that resemble this one.[53] He refers to Sydney's own comments on this subject (*Journal* 16) and also cites the *Journal* as a primary authority for the distinction between a headrest (or 'stool, used as a pillow') and a stool used for sitting—the latter a rare luxury item. The other eight objects cited in Krüger are all fish-hooks, a type of object he describes as 'a special gift of great value', as Banks also noted.[54] Krüger's description seems to justify Stanfield's strong sense that his brother had merited the affections of the islanders as well as his particular interest in the fish-hooks. Krüger's citations also confirm, at least indirectly, eighteenth-century judgements concerning the fidelity and accuracy of the *Journal*. George Humphrey's 1782 catalogue of the Göttingen collection cites Parkinson's *Journal* over eighty times, more than any other source.[55]

case of Parkinson's *Journal*, however, there is a very simple reason why the text fails to offer 'narrative positioning' and 'adjudication' of the depicted artefacts: the images were created independently of Parkinson after his death.

[53] Krüger, 'Tahiti and the Society Islands' 147–8; 166–7.

[54] Krüger, 'Tahiti and the Society Islands' 165; *James Cook: Gifts and Treasures* 76, 292.

[55] Urban, 'Acquisition History' 62. Humphrey's catalogue is reproduced in its entirety in Little and Ruthenberg (eds), *Life in the Pacific of the 1700s* II.110–42.

Stanfield Parkinson contributed new layers of aesthetic and economic connotation when he added this posthumous visual record of his brother's 'curiosities,' silencing them in one respect but allowing them to speak in others. Sydney did not write specifically about these objects. They were added as an afterthought, as Stanfield tells us, and as Grimm's signature (in lieu of Parkinson's) confirms. As afterthoughts they reflect Stanfield's preoccupation with valuable curiosities, and as such they have caught the attention of scholars such as Thomas and Krüger, even though, as Thomas argues, this type of design may have been intended to frustrate specifically ethnographic interpretation.[56] In one poignant passage in his preface, Stanfield Parkinson mourns the loss of a 'great number' of shell and ivory fish-hooks that is made clear when Banks returns 'quite empty' a wainscot box that had been full of them (*Journal* xv, xvi). Fish-hooks feature prominently in both the companion plates depicting artefacts (XIII and XXVI). It is unclear at what point in the sequence of his unruly exchanges with Banks that Stanfield commissioned these images, but the absence of any fish-hooks is notable in the corresponding plates in Hawkesworth's official *Account* of the voyage (Figure 1.4).[57] Banks had similar fish-hooks drawn by John Frederick Miller, but these were never published. The food-pounder and cloth-beater from the plate in Parkinson's *Journal* (1.3) are also present in the sparser composition of the Hawkesworth plate, though these may be different specimens. The most telling absences, however, are those of the headrest and the basket, which appear only in the *Journal* and in one earlier source: this was a drawing made within weeks of Parkinson's death by a seaman called on to replace the dead artist in the last months of the voyage.[58] Banks may have suppressed the publication of these artefacts from Parkinson's collection that remained in his hands after Stanfield Parkinson called attention to the affair.

[56] In *In Oceania* (100), Thomas quotes the text that accompanied the engravings under discussion here (*Journal* pl. XIII, XXVI), but reproduces other, later engravings that are similar to it (98–9) in the process of defining a type of ethnographic engraving. In the earlier book *Entangled Objects*, Thomas not only quotes but explicitly engages Parkinson in his argument concerning curiosity (130), an aspect of which is then developed in the later work.

[57] Did Parkinson secretly retain these fish-hooks even as he complained of being defrauded of them? Or did he go to members of the ship's crew to acquire replacement specimens that would allow him to reconstruct his brother's collection enough to illustrate the volume? Related questions may be asked of several plates in the *Journal*.

[58] Both this drawing attributed to the seaman, C. H. Praval, and Miller's unpublished drawing of the fishing hooks are reproduced in Joppien and Smith, *Art of Capt. Cook's Voyages* I.59, Fig. 99, and Fig. 170. 'Many of these objects' depicted by Praval, 'such as the stool, pestle, cloth beater and basket, appear in a similar form in S. H. Grimm's drawing for the engraving by W. Darling in Parkinson' (*Art of Capt. Cook's Voyages* I.161); see Figure 1.3. Similar items appear in unpublished drawings made for Banks by John Frederick Miller, but 'the Tahitian basket [among others] seems only to have been depicted here'. The disappearance of this basket could be seen to support Fothergill's claim that Stanfield did in fact retain some artefacts ('Explanatory Remarks' 14).

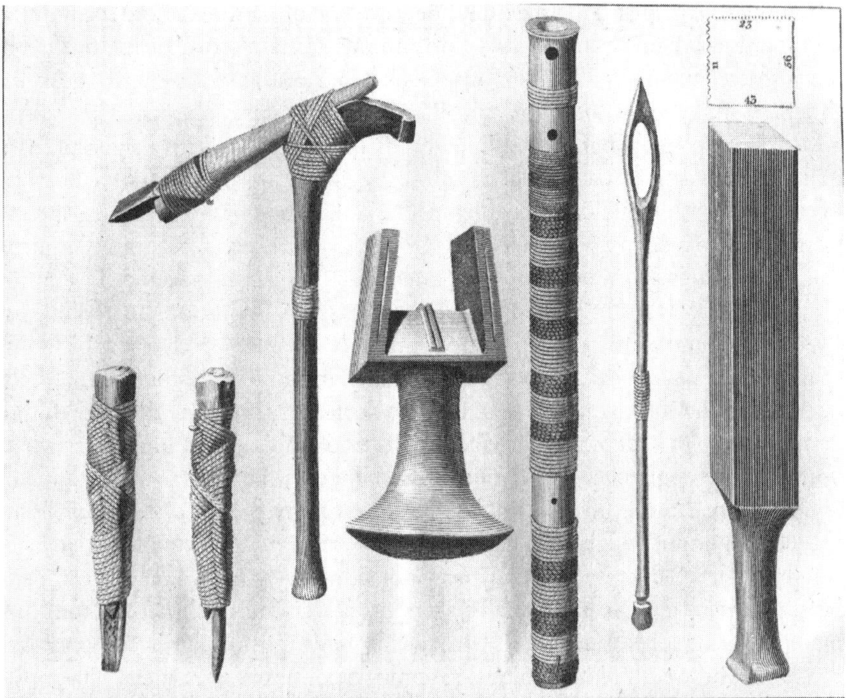

Fig. 1.4. *Tools and Instruments from the Society Islands.* Engraving by John Record from a drawing by J. F. Miller. Hawkesworth, *Account of the Voyages . . . in the Southern Hemisphere* (1773), Vol. II, Plate IX. Courtesy of the Division of Special Collections, Archives, and Rare Books, Ellis Library, University of Missouri.

The engravings help us to conjure these objects that refuse to die. Their history, and the larger history of the *Endeavour* collections, shows that typological collecting and collaborative research meant different things to Banks and the Parkinsons. These differences have long-term consequences, better understood when decoupled from the history of science. Kaeppler has pointed out that the *taniko* border on one of the Maori cloaks in the British Museum's Cook Collection resembles the one appearing in Benjamin West's famous 1773 portrait of Banks, and more recently Coote has matched a specific cloak in the Christchurch collection with the portrait.[59] Banks was aware of the added value attaching to garments given as ceremonial gifts; could one of these cloaks have belonged to the collection that Stanfield was forced to return to him, with its fifteen New Zealand garments? Or would a dog-skin cloak have been a gift from the Maori to Tupaia, as Tapsell argues? The same questions may be asked of other objects in the portrait, such as the

[59] Kaeppler, 'Tracing the History' 171; Coote, *Curiosities* 10.

canoe paddle. Exact matches are difficult, and often poor documentation means that such matches cannot be ruled out, even if it also makes them hard to establish.[60] But even the available fragments of evidence serve to confirm that as often as these objects have been wished away they have reasserted their peculiar life. Denigrated in their own time as trivial and savage curiosities and later on as the fetishistic objects of a petty squabble, Parkinson's South Seas artefacts were anything but dead. Rather than 'slip[ping] back into the wan reality of quotidian objects', in Daston's words, they continue to thicken with inquiry.

1.4 ANTIQUARIANISM ON THE *ENDEAVOUR*

So far I have made a point of emphasizing Sydney Parkinson's intellectual and artistic independence from his patron, Joseph (later Sir Joseph) Banks. A closer examination of their working relationship will help to balance the account: Banks's antiquarian outlook determined many of Parkinson's particular drawing assignments and must have reinforced the connection between antiquity and ethnography in Parkinson's narrative. Banks is famous today both for his botanical work and as the public, political face of science under George III, but in his lifetime he was also well-known as an antiquary, one of the most prominent among the substantial number who were Fellows of both the Royal and Antiquarian Societies. The disciplinary and epistemological affinities of antiquarianism and natural history appear vividly both in Banks's and in Parkinson's voyage narratives. As noted above, the importance of retaining a draftsman was brought home to Banks by other antiquaries and naturalist-antiquaries such as Falconer and Pennant. It is also worth repeating here that Parkinson had to do double duty for most of the *Endeavour* voyage after the death of Alexander Buchan, the painter hired to record landscapes and figures. Yet Parkinson's own intellectual inclinations appear to have aligned themselves neatly with Banks's interest in antiquarian pursuits. The larger antiquarian context, with its elite and popular, as well as classical and non-classical forms, returns us to the question of Parkinson's style: his writing at times betrays influences shared in common with Banks, and at times points to sources of his own, including material on native

[60] Tapsell, 'Footprints in the Sand' 103. The mezzotint of West's 1773 portrait of Banks, showing the cloak, the paddle, weapons, and a Tahitian 'mourning dress' is reproduced in Beaglehole (ed.), Endeavour *Journal* I.68. Hauser-Schäublin notes that clothing should be understood as specifically a ritual gift, as opposed to trade goods (*James Cook: Gifts and Treasures* 20–5). For a match between a Parkinson image and a *Tewhatewha* (Maori battle-axe) from Banks's collection, see <www.nla.gov.au/nla/pic-an6396511>.

British antiquities.[61] Overall he resorts less often to the practice of comparing living Pacific Islanders with any set of ancient people. If the lack of such literary habits makes his record more authentic, as many readers have felt, it is less authentic in the sense that it avoids any mention of his own primary occupation of drawing specimens—presumably because that work was done by others in the typical naturalist-antiquary narrative.

Antiquarianism as a science and practice has been largely discredited since the advent of more specialized successor disciplines and the professionalization of history. As I argued in my introduction, however, antiquarianism is not so easy to dismiss as a sociopolitical phenomenon. Antiquarianism bridged elite and popular practices and made a complex contribution to the colonial project. Its role on the *Endeavour* is a case in point. While some antiquarian projects led to the expropriation of artefacts, most famously those of classical antiquity, the ethos of preservation also heightened ethnographic awareness and opposed the erasure of other cultures. In the case of the *Endeavour* voyage, antiquarian pursuits appear to have participated in a complex reciprocal rendering of native colonial peoples and native Britons.

Parkinson's *Journal* shares the ethnographic impulses of antiquarian researchers who were increasingly occupied with documenting popular customs and manners in Britain and Europe. One such work, Pietro Fabris's *Collection of Various Costumes and Trades of the Kingdom of Naples* (1773), is examined in my next chapter. Like the native Pacific Islanders encountered during the *Endeavour* voyage, rural Britons, especially along the Celtic fringe, appeared to antiquaries as living fossils who displayed traditional, nominally ancient customs and manners. Parkinson's delineations of the languages, rituals, dress, culinary habits, and architecture of the islanders parallel antiquarian research into these practices among ancient Britons and their descendants. According to a notebook entry analyzed by Joppien and Smith, Parkinson's shipboard library included James Macpherson's pseudo-Gaelic *Fragments of Ancient Poetry* (attributed to Ossian) and William Robertson's *History of Scotland*, as well as a history of ancient peoples in the East, Charles Rollin's *Ancient History*. Of the twenty-seven plates included in the *Journal*, six depict 'native dress,' five depict architecture, six depict customs (such as the *haka*, fighting stances, or tattooing), and over fifty 'implements' or 'utensils' are illustrated—all images of perennial interest to the antiquary. Comparable images of ancient and medieval British dress, architecture, and artefacts feature prominently in the pages of the journal *Archaeologia* and in scores of antiquarian books. Much of Parkinson's text is devoted to cataloguing the customs, arts and crafts, food plants, and perhaps especially the languages, of the peoples encountered during the voyage, an approach that

[61] For an account of Parkinson's reading in history and other fields, see Joppien and Smith, *Art of Captain Cook's Voyages* I.51–2.

would have satisfied the authors of the *Queries Proposed to Gentlemen in the Several Parts of Great Britain* published by the Society of Antiquaries in 1754. I have discussed already some of the evidence suggesting that Parkinson must have spent all his free time, especially on Tahiti, gathering these data. It was this ethnographic aspect of antiquarianism that led the prehistorian Stuart Piggott, among others, to claim it as the precursor of modern archaeology.[62]

The differences among Hawkesworth's, Banks's, and Parkinson's accounts of the *Endeavour* voyage may be attributed in part to their differing conceptions of antiquarianism. Whereas Banks, and consequently Hawkesworth (whose account generally follows Banks's unpublished journal), appeal to a traditional scholarly preoccupation with classical antiquity, Parkinson appeals to an emergent, and arguably more democratic, concern with native British antiquities.[63] Banks's articles in *Archaeologia* are notable for their particular specialization in the display and description of Roman antiquities found in England. In some ways Banks's tastes were more closely aligned with the explicitly classicizing and aristocratic Society of Dilettanti than with the most active researchers in the Society of Antiquaries. 'Antient Britons' posed particular problems for the collector of classical antiquities.[64] Their 'rude' artefacts, 'barbaric' practices (including human sacrifice), and tribal conflicts appeared in stark contrast to the *beau ideal* of classicism. As suggested by Banks's own presentations to the Society of Antiquaries, an intense desire to align the ancient Britons with the more civilized and civilizing practices of classical antiquity coloured some of the antiquaries' findings, and not surprisingly, these kinds of investigations often favoured aesthetic objects over utensils, dress, and manners. Banks's own essays emphasize the fine and expensive workmanship displayed in the relics of Roman Britain. Parkinson's focus in his journal diverges, then, from Banks's own interests and appears indebted to an alternative antiquarian focus on British history that attempts to engage with 'antient Britons' on their own terms, using objects as a means by which antiquaries can reconstruct everyday life.

[62] In Piggott's view, the returning emphasis on the study of the material culture of the past led to a decisive break within antiquarianism between the empirical work that paved the way for modern archaeology and the work on heraldry, documents, and texts more related to philology and textual disciplines. See his *Ruins in a Landscape* 117–29.

[63] The popular antiquarianism of John Brand, Joseph Ritson, Joseph Strutt, and William Hone among others—concerned with the customs and manners of the common people—helps to reinforce the distinction between the elite tradition of classical study and the study of British antiquity. For an overview see Butler, 'Antiquarianism (Popular)'.

[64] Banks's differences with the medievalist antiquaries became most acute in 1799 when he backed the controversial election of James Wyatt to fellowship in the Society of Antiquaries and then tried to influence the election of new officers, which won him the epithet 'Magog Banks'. See Sweet, *Antiquaries* 103–5 and Chapter 6 below. On the 'ancient Britons,' see Smiles, *Image of Antiquity*.

In practice, both approaches reciprocally influenced each other and both Parkinson and Banks attend closely to artefacts, customs, and manners. As a result, the idea of native peoples as savages begins to be displaced by a more detached examination of culture. This turn is more obvious in Parkinson, who (for example) largely avoids the inaccurate descriptor 'Indians' so heavily used by Banks. The affinities to popular antiquarianism are also more readily apparent in Parkinson, even his landscape studies, which are aptly described by Joppien and Smith as creating a 'hybrid picturesque' of Parkinson's own devising.[65] These images show affinities with the antiquarianism of many English local and county histories by juxtaposing the ruins of nature and the ruins of time, or situating built works within a landscape. His 'View of an Arched Rock', showing a Maori *hippa* atop a colossal natural arch within the coastal cliffs, is one striking example (*Journal* plate XXIV). A typical popularizing work that Parkinson might have read, John Owen's *History of the Ancient Britons* (1743), demonstrates the ethnographic tendencies that made antiquarianism such an important and multivalent disciplinary framework for the *Endeavour* voyage. Owen derives the name 'Briton' from the Celtic word *Brith*, meaning 'painted', and signifying the practice of painting or tattooing prevalent among early inhabitants. Parkinson admits that he himself 'underwent the operation' of tattooing on Tahiti to improve his understanding of the practice (25). Owen attributes the comparative health and longevity of the ancient Britons to their 'robust and manly diversions', much as Parkinson conjures up 'the athletic constitutions of our ancestors' when observing life on Tahiti.[66] Likewise, Owen and Parkinson both include descriptions of funeral rites, tools, weapons, songs, and costumes of ancient Britons and Pacific Islanders, respectively.

Parkinson's slightly younger contemporary William Blake, exposed to many of the same influences, created memorable images of ancient Britons in his poetry and painting, including some that were directly influenced by John Speed's portrayal of blue-skinned 'Britaines' in *The History of Great Britain* (1611). On the one hand, Parkinson's descriptions belong to the same body of early romantic representation; on the other hand, analogies to the ancient Britons crop up occasionally even in the earliest literature on North American encounter, though classical references are much more common. It is far from being a simple case of authenticity, then, but the relative novelty of the British, non-classical paradigm at least mimics the novelty and freshness of Pacific encounter. Peter Ruthenberg makes a similar point about the sketches, as opposed to the finished drawings, of John Webber, the artist on Cook's third voyage: 'the work-overs back home in London

[65] *Art of Captain Cook's Voyages* I.26.

[66] Owen, *Compleat and Impartial History of the Ancient Britons* 6; Parkinson, *Journal* 23. I owe the Owen reference and parallels to Crystal Lake.

created a fantasy world, and amalgam of "classical arcadia" and "noble savage paradise", that does not accurately reflect Pacific reality', whereas the sketches done *in situ* reflect Webber's 'overwhelmed reaction' before these 'filters were applied'.[67]

I have already noted the striking contrast between the 'ancient Britons' invoked in Parkinson's account of Tahitian athleticism and the ancient Greeks and Romans appearing in Banks's and Hawkesworth's accounts. In the late Enlightenment context, the absence or presence of classical references is neither a sufficient criterion for differentiating antiquarian accounts nor is it a wholly reliable marker of social distinction. In the wake of a 250-year tradition of European exploration narratives, any reference to antiquity is already overdetermined.[68] It is fair to say, I think, that Parkinson is less prone than most to the habit of collapsing cultural with temporal distance. In the case of Polynesian migration, he is willing to consider Pacific peoples as having an ancient history of their own. In the case of Tahitian athleticism, where the comparison to ancient people is present, it serves Parkinson's moralizing distinction between virile islanders and 'enervated' moderns (*Journal* 23). But such moralizing digressions in Parkinson are rarer and briefer than they are in Hawkesworth, Banks, and elsewhere. In this respect his prose resembles that of Sir William Hamilton, who explicitly set limits to such digressions in his letters to the Royal Society in order to focus on the factual matter at hand.[69] The comparison to ancient Greek sport also serves a different purpose in Hawkesworth and is entirely absent from both Cook's and Banks's journals for this day (5 May 1769). As noted above, Hawkesworth offers up his comparison between Pacific and Greek sports as one that 'even our female readers' will grasp. Though condescending, this remark can be taken as evidence of a broad consensus concerning the physical fitness of the ancients, and Parkinson's Britons exemplify another version of this thesis available to non-classical readers. At the same time, however, Hawkesworth turns a relatively impartial remark of Cook's into the decidedly pejorative one that the contestants 'grappled without the least dexterity or skill' (*Account* II.119).

Parkinson at times refrains consciously from mythologizing Pacific peoples. Banks makes use of the athletic contest, which, he notes, drew a crowd of at

[67] On Blake, see Worrall, 'Blake's *Jerusalem*', and Brylowe, 'Of Gothic Architects'. On Webber, see Ruthenberg, 'John Webber'. In Webber's case, one of these 'filters' was his academic training as a portrait painter, which Parkinson did not have.

[68] Even the appeal to ancient Britons transferred occasionally from antiquarian scholarship to voyage narratives, most famously in Theodor de Bry's illustrations to Thomas Hariot's *Briefe and True Report of the New Found Land of Virginia* (1590). (Strictly speaking, de Bry compares the Americans to Picts.) It is not clear to what extent Parkinson participated in the daily reading of voyage narratives that occupied the hours after breakfast in Banks's and Solander's days aboard the *Endeavour*.

[69] See, for example, 'Account of an Eruption' 59.

least 500, to suggest that his and Cook's visit to this Tahitian 'chief,' Tootaha, was a diplomatic event sanctioned by a public ceremony. Hawkesworth uses the episode to promote his own idea of a rude but ethical society, describing Tootaha as an 'ancient patriarch' and emphasizing the good sportsmanship of the wrestlers. Parkinson does not describe the wrestling match himself, but his reflections on athleticism, and on the social state of the Tahitians in general, occupy the entries in his journal (4–5 May) that were written while Cook and Banks were making this visit. Banks's own summary remarks on the 'Manners and Customs of the South Sea Islands' include another classical reference (Virgil's *Georgics* II.458) in their treatment of breadfruit and the accompanying fiction that the islanders did not have to labour for their bread. Parkinson, setting these general questions in the context of his Tahitian-Briton analogy, seems more inclined to challenge some of the utopian assumptions associated with the myth of the noble savage. He is almost surely alluding to Rousseau (first *Discourse*) when he observes that the seeming absence of any concern over property and theft 'might induce a celebrated writer to conclude, though erroneously, that mankind . . . are necessarily rendered worse . . . by the cultivation of the arts and sciences' (*Journal* 23); the problem is not 'civilization' but modern European 'intemperance and debaucheries' in contrast to the ancient/native avoidance of 'superfluities'.

Parkinson's restraint, and his stance of witnessing, also characterize his handling of more controversial topics such as sexuality, tattooing, and cannibalism. Parkinson is comparatively reserved on these topics, particularly sexuality, and this restraint is one key feature of the plain style of his narrative, which resembles that of more strictly Baconian contemporaries, such as Hamilton, and anticipates that of Charles Darwin. Sexuality may be a special case for Parkinson, especially if, as his brother averred, he avoided the 'sensual gratifications' pursued by most of the company.[70] It is certainly a special case relative to Tahiti, especially evident in the name New Cytherea, attached to the island by Bougainville and influential on Hawkesworth if not yet on the *Endeavour*'s company. Parkinson's one general comment on sexuality in Tahiti occurs shortly after the allusion to Rousseau's 'erroneous' view concerning the arts and sciences, in what turns out to be his most sustained general description of Tahitian culture (22–26).[71] The description as a whole is

[70] A recent biographical study by David Sox cites one slight piece of evidence that 'even Parkinson fell to the charms of the Tahitian women', as Daniel Solander evidently recalled it (*Quaker Plant Hunters* 94). The phrase 'Mrs Patine' [as a euphemism for Parkinson's supposed Tahitian lover] seems to be a fabrication on this author's part, however. Parkinson's Tahitian vocabulary includes no words for the genitals or for intimate acts apart from *Whainaow*, 'to beget' (61).

[71] As Joppien and Smith point out (*Art of Captain Cook's Voyages* I.25), Kenrick's involvement in the publication of the *Journal* raises the possibility that certain passages are interpolated by him; Kenrick knew Rousseau well, and was writing (unlike Parkinson himself) after the

economical and relatively detached, incorporating some commonplaces, but it is notable too for the interspersed personal anecdotes that might be seen as confirming Parkinson as a reliable witness. His comment on the superabundance of food, for example, includes a characteristic first-person footnote detailing the surprising strength of the slender coconut stalk (23n.), and he reports that he was unable to imitate certain native facial expressions (25).

This meditation on 'moral theory' and customs is decidedly less utopian, and less relativist, than most others on the subject of Polynesian chastity. He acknowledges that 'the married women have not a very delicate sense of modesty' (itself a point debated in the subsequent literature), but is inclined to judge the Europeans who seek out sexual contact: 'Most of our ship's company procured temporary wives amongst the natives,... an indulgence which even many reputed virtuous Europeans allow themselves, in uncivilized parts of the world, with impunity;... which is to suppose, that the obligation to chastity is local' (25–26). Nonetheless, he follows Banks's sexual negotiations quite dispassionately in his daily entries (32), using the same tone to report these adventures, Banks's and Cook's 'survey of the island' (26 June–1 July 1769), and other diplomatic/scientific excursions. Banks himself, early on declaring the island an 'Arcadia' populated by figures from Greek myth, employs his classical sensibility conventionally enough as an alibi for sexual adventure. As Neil Rennie has argued concerning the infamous 'Point Venus scene', a performance of public sex described by Cook and Hawkesworth, the 'discovery of Tahiti' was simultaneously 'a rediscovery... of the lost paradises of biblical and classical myth'.[72] Parkinson and Banks both comment on another well-known episode involving a woman who 'exposed herself quite naked', seemingly as a matter of 'ceremony' (27). Neither of them mention the 'Point Venus scene', though Parkinson alone reports seeing 'a person who had the appearance of an hermaphrodite' on that same day (14 May).

Parkinson's brief account of sexual mores follows from his digression on Tahitian 'moral theory', which leads to a rapid survey of agriculture, technology, and customs. The art of tattooing occupies a middle ground in this survey, between remarks on warfare and remarks on music and dance, from

publication of Bougainville's voyage. This is a salutary caution, but even Joppien and Smith assume that they are citing Parkinson's own views when they quote the *Journal*, and in this first sustained narrative on Tahitian culture (*Journal* 22–6) it seems unlikely that much is pure interpolation.

[72] Rennie, *Far-Fetched Facts* 93. Parkinson's descriptions of Banks are neither censorious nor as deferential as one might expect in the account of a patron, though this passage in particular suggests that Parkinson did not anticipate Banks reading his journal. On Banks's classicism in this context, see Joppien and Smith, *Art of Captain Cook's Voyages* I.21. Rennie's thoughtful chapter on the 'Point Venus scene' (83–108) is one of many scholarly works on sexuality in early Pacific encounter. Rennie (253n79) also refers helpfully to William Wales's sceptical defence of Tahitian women's chastity; a more locally informed version of this defense is mounted in *The Journal of James Morrison*, rpt. Lamb et al. (eds), *Exploration and Exchange* 134–5.

which the discussion of sexuality soon follows. Parkinson has become probably the most influential early European authority on tattooing. His and Banks's accounts together contribute significantly to the initial understanding of cultural differences between Tahitians and Maori, as well as finer distinctions—registered in differing visual patterns—among ethnic and regional groups. On this subject the two accounts show more similarity than difference, and happen to be relatively free of either antiquarian or moralizing digressions. Hawkesworth, on the other hand, expresses far stronger moral disapprobation than either, condemning tattooing as a 'horrid deformity' (III.449).

Parkinson's authority is partly visual, owing to the wide dissemination of his images (particularly the portrait of Otegoongoon), and partly personal, since he was one of the very first educated Europeans to 'undergo the operation'. It also comes from his aesthetic, practical, and comparative engagement with this subject—*tataowing*, according to Parkinson's more exact transliteration—in the *Journal*. Initially Parkinson describes the technique and the instruments used (25); helpful cross-references are inserted by the editor that direct readers to images of these instruments in two of the plates (see Figure 1.3). He also includes the candlenut, from which their tattooing ink is derived, in his ethnobotanical catalogue (44). Banks develops a more sustained narrative of tattooing as an initiation rite and offers a detailed description of the forms with some aesthetic judgment, but he too is more strictly empirical in this first engagement than in the later more digressive account (by both) of Maori tattooing. Parkinson recorded the typical solid black buttock pattern in his sketchbook, but the only image of Polynesian tattooing in the *Journal* is an engraving showing facial tattooing, which Banks reports observing only in a single case. To judge from an analogous case, in which Banks expressly took Parkinson on an excursion to Raiatea to sketch scenes of a *heivo*, or dramatic performance (74), their shared ethnographic interest in tattooing was initially fostered by Banks and became increasingly central for Parkinson during their circumnavigation of New Zealand.

Parkinson and Banks both admire Maori tattooing, but paradoxically what they describe as its greater refinement, relative to Tahiti, is subsumed under a more general distinction observable in so many Pacific voyaging narratives between the beautiful (Tahiti) and the picturesque (New Zealand). They never remained anchored long off New Zealand and thus had little opportunity to observe agriculture or social customs that might have mitigated the perceived savagery of the Maori. Though geometrically more sophisticated and hence more 'curious' than tattooing on Tahiti, the art for both Parkinson and Banks becomes associated with the roughness and ruggedness of both the mountainous New Zealand landscape and the seemingly uncivilized behaviour of the Maori. As Bernard Smith first noted, Parkinson uses 'curious' and its derivatives more than once in describing Maori arts in particular, and the concept becomes a focal point for Thomas as well in his engagement with the *Journal*.

In his second attempt to describe Maori tattooing, Parkinson notes that 'it is done very curiously in spiral and other figures; and, in many places, indented into their skins, which looks like carving' and appears to be associated with high status (97). He immediately opposes it to the 'nasty custom' of face painting more typical for 'servants and women', and notes in passing that the women 'do all the drudgery', even though they have 'very good features, and not the savage countenance one might expect' (98). Parkinson's 'curious(ly)' carries some of the force that it carries in eighteenth-century volumes of the *Philosophical Transactions* (where it occurs very widely); in addition to this 'scientific detachment', as we might call it today, there is the gentlemanly sense, also analysed by Thomas, that Parkinson must have seen as linking his own discourse to aristocratic connoisseurship.[73]

This admiration resurfaces all the more strongly in a visual idiom unique to Parkinson, whose aesthetic judgment is more substantive and more marked than Banks's in this instance:

> The men have a particular taste for carving: their boats, paddles, boards to put on their houses, tops of walking sticks, and even their boats valens, are carved in a variety of flourishes, turnings and windings, that are unbroken; but their favourite figure seems to be a volute, or spiral, which they vary many ways, single, double, and triple, and with as much truth as if done from mathematical draughts: yet the only instruments we have seen are a chizzel, and an axe made of stone. Their fancy, indeed, is very wild and extravagant, and I have seen no imitations of nature in any of their performances. (98)

Here again the description returns to more 'savage' elements as if to compensate for the admiration expressed for the refinement of these arts. The contradiction is manifest in the suggestion that these people lack a capacity (mimetic drawing) that obviously requires less skill than one they have ('mathematical' drawing). In addition to this ambivalence, which Parkinson shares to some extent (along with the idea of 'wild' imagination) with Banks, these passages also demonstrate the kind of reflection that had become habitual for Parkinson since he became responsible for recording ethnographic subjects at the death of Buchan six months earlier. The word 'curiously' in this passage may reflect an effort at detachment, but it also carries a sense of the struggle to understand aesthetic form in the context of ethnographic prejudices that seem to militate against it, as well as a conflict that arose between Parkinson's increasing attentiveness toward culture and the increasing physical and moral exhaustion of the voyage.

Banks also offers a full account, which makes comparison on this point especially rewarding. His account too displays the contradiction between perceptions of aesthetic refinement and of social ruggedness, or 'elegance'

[73] Smith quoted in Thomas, *Entangled objects* 130; Thomas, *In Oceania* 101.

and 'ugliness' in Banks's terms. But it differs in its deliberate manifestation of antiquarian and sexual desire. The first appears in the context of a metaphor that links the arts of war and of ornament with reference to a metalworking practice ('chasing' or embossing) known among ancient as well as European peoples:

> This may be done to make them look frightfull in war . . . Yet ugly as this certainly looks it is impossible to avoid admiring the immence Elegance and Justness of the figures in which it is form'd, which in the face is always different spirals, upon the body generaly different figures resembling something [*sic*] the foliages of old Chasing upon gold or silver; all these finishd with a masterly taste and execution . . . nor do I remember to have seen any two alike, for their wild imaginations scorn to copy as appears in almost all their works.[74]

The term 'foliage' here may suggest a specific reference to medieval European works—evoking the warlike passions and vigorous imagination associated with the age of chivalry—but the reference is vague and deliberately suggestive. (Hawkesworth makes its range even broader by adding a reference to 'filagree work'.) Banks's metalworking metaphor also renders the Maori as subjects and objects simultaneously.

Banks and Parkinson both, in other words, admire Maori workmanship while seeking to objectify them in other ways. Both turn to their contrasting practice of 'painting themselves with Red Ocre' for this purpose, but Banks's comment here displays an aristocratic rakishness that is foreign to Parkinson's style and renders the women in particular as sexual objects. Joppien and Smith's contrast between the two, though perhaps overdrawn, helps to underscore the distinction: Banks was born into the landed gentry, while Parkinson was 'born to serve, not to command'.[75] Banks writes that the practice was 'condemnd by us, for if any of us had unthinkingly ravishd a kiss from one of these fair Savages our transgressions were wrote in most legible Characters on our noses'. Hawkesworth carefully preserved this and many remarks in a similar spirit in the *Account*, though often enough he bore the blame himself when readers disapproved of this rakishness. Parkinson, as we have seen, concentrates instead on the women's subjection to 'drudgery'. Banks, too, admires the contrasting artifice of tattooing. But facial tattooing in particular poses an obstacle to the classicizing gesture that Banks indulged on Tahiti, where he compared the islanders to the sculptures of Phidias.[76]

[74] Banks, Endeavour *Journal* <http://southseas.nla.gov.au/journals/banks_remarks/209.html>, cp. Hawkesworth, *Account* III.453. Banks and Parkinson (127) both conflate tattooing with *ta moko*, the specifically Maori form that involves impressing (or 'indenting') rather than puncturing the skin.

[75] Joppien and Smith, *Art of Captain Cook's Voyages* I.24.

[76] <http://southseas.nla.gov.au/journals/banks_remarks/210.html>; Beaglehole (ed.), Endeavour *Journal* II.331. Lamb contrasts tattoo patterns with facial contours in *Preserving the Self* 106.

For a variety of reasons, Parkinson too responds less fully to the humanity of the Maori, and it is important to remember that his most influential document of tattooing is the portrait of a prisoner of war. Though increasingly attentive to cultural particularities, and (like Banks) perhaps somewhat less influenced at this stage by antiquarian and other commonplaces, Parkinson also becomes less critical of violence against indigenous peoples. Near the beginning of his journal, Parkinson exclaims, 'What a pity, that such brutality should be exercised by civilized people upon unarmed ignorant Indians!' (15). But at the scene of the Poverty Bay Massacre (10 October 1769), Parkinson merely reports that Cook, Banks, and Solander 'killed and wounded several' Maori (88), whereas the other journalists expressed considerable remorse and Cook 'heavily revised' his entry to extenuate his culpability.[77] The prisoners taken on this occasion gave Parkinson his first opportunity to observe tattooing and to compare the two languages (88–89); his vocabulary for New Zealand is much shorter (126–28), suggesting that there were few such opportunities, and the relatively rapid circumnavigation of the North and South Islands also meant that he encountered more linguistic variation. The same kind of violent encounter (25 November) leads to detailed observations, by both Banks and Parkinson, on regional and local variations in *ta moko* patterns (108). The *Head of Otegoongoon, Son of a New Zealand Chief, the face curiously tataow'd* (Pl. XXI) is the trophy of a more dangerous encounter four days later. Banks, who was botanizing on shore, 'had like to have been apprehended by one of the natives'. When the marines fired on them, Otegoongoon (or Otegoowgoow) was wounded and taken on board (109). Parkinson observes somewhat laconically that 'these barbarians . . . would have been a formidable enemy, . . . but they had no kind of order or military discipline'.

Nonetheless, they are 'accustomed to the carnage of war from their infancy', as Parkinson reflects in the last sentence of his New Zealand narrative, composed 'about ten leagues N.W. of Charlotte's Sound, or Cannibal Bay' (124). It had been 'near seven weeks' since they had departed on their circumnavigation of the South Island, taking leave of a people who helped them navigate and offered (as trade goods) human heads that had been prepared as trophies of military victories. The entry for 6 February 1770 records that 'we left the bay, which we called Cannibal Bay . . . the captain called it [Queen] Charlotte's Sound' (117). Although more reserved than the other journalists, as usual, Parkinson reacts with 'surprise' and indignation at

[77] Bohls and Duncan (eds), *Travel Writing 1700–1830* 519n. In his entry for the previous day Parkinson also notes that the natives seemed unconcerned about their dead (*Journal* 88). On a later occasion, however, he is more sympathetic when a man is shot for suddenly sailing away without completing his part of a bargain: 'What a severe punishment of a crime committed, perhaps, ignorantly!' (104). The increasing frequency of these incidents (cp. 102) at least partly explains his increasingly laconic treatment of them.

these 'adequate proofs' of Maori cannibalism (115–16). The first 'dressed and picked' human bones were observed by a small party who landed the pinnace near the dwelling of a family in 'a little cove'—the word itself seems to anticipate a much more traumatic proof of cannibalism encountered in nearby Grass Cove during Cook's second voyage—who explained that these were the bones of their recently slain enemies (115). Parkinson goes on to say that this group more than once brought human bones on board to sell, and 'evinced that they thought human flesh delicious food'. The same group also offered the prepared heads or skulls, but the price was too high. Parkinson carefully describes the method by which these skulls were prepared, but concludes the incident of these 'horrid midnight repasts' with a moralizing digression on cannibalism as 'one of the worst of moral vices' (116).

As George Forster observed in his narrative of Cook's second voyage, it is surprising that the explorers on board the *Resolution* still doubted the existence of Maori cannibalism, since Cook himself had witnessed the same evidence that Parkinson saw in 1770. Nonetheless, J. R. Forster and William Wales engaged in debates about the practice in their narratives of the second voyage and sought out mitigating circumstances even after they learned of the Grass Cove Massacre in 1774.[78] Hawkesworth's account of the *Endeavour* voyage, as often, provided Wales and the Forsters with a point of departure. Hawkesworth's account gives rhetorical fancy dress to the moral indignation and horror that we see in Banks and Parkinson. The emphasis on revenge as a motive in Hawkesworth comes directly from Banks, who takes pains to show that cannibalism is not in the course of nature by citing examples of many other mammalian species. Hawkesworth adds, however, that hunger or 'famine' is an even more plausible motive for the practice, and also suggests (characteristically) that cannibalism may be only 'relatively . . . pernicious' (*Account* III.448). Both Forster (senior) and Wales take up the motives of hunger and revenge in their discussions, and Wales in particular settles on a motive that originally comes from Parkinson rather than Hawkesworth: the idea that human flesh is 'delicious food'. They both reject Hawkesworth's conjecture that hunger was the original motive for cannibalism, but Forster develops the motive of revenge and situates it within a stadial history of human social development. Forster distinguishes 'better regulated societies' as those in which 'anthropophagy has been in disuse' for a long time, inferring that 'this barbarism' is more or less universally 'one of the steps' in the development of all societies. As Thomas puts it, 'the Maori merely possess,

[78] George Forster, *A Voyage Round the World*, in Bohls and Duncan (eds), *Travel Writing 1770–1830* 459. For an introduction to these debates, and a brief summary of the Grass Cove Massacre of 18 December 1773, in which nine of Cook's men (on the second voyage) were killed and eaten, see Lamb et al. (eds), *Exploration and Exchange* 93.

in more extreme form, the warrior virtues of early European populations such as the Germans'.[79]

The general framework of antiquarianism enabled the study of customs and manners on Cook's voyages. Where Parkinson's ethnographic observations differ from those of Banks or Hawkesworth, they reflect a more popular, less classical mode of antiquarianism. Parkinson's comparative restraint, especially on hot topics such as sexuality and cannibalism, also differentiates his prose from that of more self-consciously literary and philosophical voyagers. Banks and Parkinson did, on some level, share the perception of natives as ancients, and that perception motivates, in part, their careful attention to the subset of customs and manners we now call material culture.

1.5 PARKINSON, JOSEPH BANKS, AND MATERIAL CULTURE

Forster's reference to Germanic tribes signals the importance of the voyager's choice of ancient cultures in a specific comparison, a choice that varies by account as well as by voyage and by subject matter. Unlike Banks and Parkinson, Wales and Forster (and some later voyagers) make the link between Maori cannibalism and ancient military culture, and in this case that link may be a product of reflection on the first-contact narratives. Both Wales and Forster were scholars, too, unlike Banks and Parkinson, and their differences may reflect a larger Gothic versus Classical antithesis. Wales rejects Forster's Gothic paradigm of cannibalism but offers a relatively sophisticated classicizing framework for Tahitian performance art (which Banks himself had described loosely as Homeric). Wales displays the same ambivalence that we have seen in Parkinson concerning the erotic parts, but is in earnest when he compares one actor favourably to Garrick and concludes with an analogy referring to the legendary originator of Greek tragedy: 'the Drama is advanced in these Islands, very far beyond the Age of Thespis'.[80] This was the same type of performance, or *heiva*, that Banks had made a point of taking Parkinson to sketch on Raiatea four years earlier. The European understanding of dramatic performance may have given rise to some of the hostilities off New Zealand, where the *haka* was at times misconstrued as friendly entertainment, rather

[79] My overview of the cannibalism debate depends on *Exploration and Exchange* for the quoted excerpts from Wales (110), Forster senior (96), and Thomas's headnote (93). Bernard Smith and others have noted the visual and verbal parallels made on Cook's third voyage, again with respect to militarism and military culture, between modern Hawaiians and ancient Greeks. See Smith, *European Vision* 120, 127–30, and cp. Krüger, *'Sportlicher Wettkampf' auf Hawaii.*

[80] Wales, *Observations* (1778), in Lamb et al. (eds), *Exploration and Exchange* 108.

than a potentially hostile declaration. Parkinson, observes, for example, that one group, who had earlier 'behaved with great civility', later 'began to heivo and dance in their manner, which was very uncouth; nothing could be more droll than to see old men with grey beards assuming every antic posture imaginable, rolling their eyes about, lolling out their tongues, and, in short, working themselves up to a sort of phrenzy' (*Journal* 96). *Heivo*, seemingly a cognate of the Tahitian *heiva*, may have been Tupia's translation. Parkinson's attitude in these passages is hard to reconcile with the related images (Plates XVII, XVIII)—the latter of which includes a brandished human head—where he uses neoclassical elements of composition to achieve an effect far more warlike than ridiculous.[81]

Parkinson's observations share the ethnographic spirit of Owen's and other popularizing works on native antiquity, including Shakespeare's plays, but elite antiquarian works aimed at a more scholarly audience also began to place a value on more authentic records of vernacular traditions and material culture.[82] Ethnographic work on board the *Endeavour* owes something to the 'first' or eighteenth-century Gothic revival in both its elite and popular versions. Richard Gough, who became Director of the Society of Antiquaries as the *Endeavour* was on its home stretch in 1771, focused much of his own research effort on gathering 'faithful representations', both visual and verbal, of the funerary sculpture in England's Gothic cathedrals. These material remains, he argued in *Sepulchral Monuments of Great Britain,* are the most reliable evidence available on 'our manners, habits, arts, national taste, and style of Architecture' in bygone ages. On the one hand, Gough derives some British traditions from Greco-Roman antiquity, quoting copiously in Greek and Latin; on the other hand, he uses a Cook voyage narrative to corroborate Homer's account of Greek customs, displaying the reciprocal influence of different versions of antiquity.[83] John Carter, who began his antiquarian career as an architectural draftsman for Gough, developed his own account of Gothic cathedrals into a much more popular nationalism in the forum of his influential *Gentleman's Magazine* column, 'Pursuits of Architectural

[81] Though much indebted to the excellent discussion of the second of these plates in Joppien and Smith, *Art of Captain Cook's Voyages* I.37–41, I disagree with their conclusion that the overall effect is comic. The disjunction between text and image reflects the choices made by Stanfield and the publisher; these images were never intended by Parkinson as illustrations for his text. For the brandished head, see Joppien and Smith 1. 199, it is not included in Parkinson, Plate XVIII.

[82] In a fascinating essay, Michael Neill suggests that Parkinson's depictions of Maori were 'partly molded by familiarity with Shakespeare,' a copy of whose works was brought on board 'probably at the insistence of Parkinson himself' ('From the Editor' v). Parkinson's shipboard library also included the poetry of Chaucer and Spenser.

[83] See, respectively, Gough, *Sepulchral Monuments*, vol. 1 (preface) 5–6; and vol. 1 (introduction), p. i. In the second instance Gough uses J. R. Forster's account of funeral rites in his *Resolution* narrative to corroborate Homer's account of the funeral of Patroclus in the *Iliad.* This is one of several Cook voyage references in the work. See further Chapter 6 below.

Innovation'. Carter escalated the Gothic versus classical dispute into a war. Parkinson was not explicitly anticlassical, but like Carter he lacked both classical education and Royal Academy training, along with the prejudices that accompanied them. Both Gough and Carter insisted on preservation as the ultimate goal of studying material culture, and the same motive can be traced in different ways in Parkinson's and Banks's ethnographic work in the South Pacific.

Banks adopts the idiom of architectural antiquities in his discussion of a monument on Raiatea (or 'Ulietea') that he visited on 29 June 1769. Architecture and funerary customs belong to the subset of topics, along with botany, that seem to be of special interest to Parkinson and especially Banks, as compared to the journalists who followed them (unlike tattooing and cannibalism, for instance, which attracted more detailed attention by their successors). Banks was especially active in his investigation of 'morai' or *marae*, gathering places first noticed by the British on Tahiti, who understood them primarily as 'burying-grounds ... [and] places of worship' (*Account* II.166). As Banks and Cook explored the south-western portion of the main island they found two sites of architectural interest, the first a smaller burial site containing a pyramid as well as the first piece of stone carving they had seen in ten weeks on the island. If this was impressive, the second was astonishing: 'we no sooner arrivd there than we were struck with the sight of a most enormous pile, certainly the masterpeice of Indian architecture in this Island so all the inhabitants allowd. Its size and workmanship almost exceeds beleif, I shall set it down exactly'. This monument, terraced and shaped like a large pitched roof (267 feet long by 71 wide and 44 high), belonged to Oborea and was taken by Banks as an index of her status. It seems likely that the structure predated her hegemony, however, since Banks cites evidence that could indicate its antiquity, a noticeable curvature in the middle of the stone steps on one side (suggesting subsidence or wear over a long period of time). For Banks the workmanship is 'ancient' primarily in a technological sense: 'it is almost beyond belief that Indians could raise so large a structure without the assistance of Iron tools to shape their stones or mortar to join them, which last appears almost essential as the most of them are round; it is done tho, and almost as firmly as a European workman would have done it'. The use here of 'pile' and especially 'architecture', which occurs very rarely in the *Endeavour* narratives, evokes the period's burgeoning discourse on both stone circles and Gothic architecture.[84]

Banks's engagement with this monument suggests a concern with preservation as well as with material culture. The insistence on recording dimensions ('I shall set it down exactly') is a part of this effort. Hawkesworth's version of

84 <http://southseas.nla.gov.au/journals/banks/17690629.html>.

the passage (II.166) adds one detail from Cook that escaped Banks in his enthusiasm: the ornaments on top include large carvings in wood and stone, the latter of which is broken (evoking the then-current deterioration of medieval built works). Without making a direct argument concerning their antiquity and preservation, Banks notes that the *marae* are threatened as well by military devastation, as suggested by a coastal battlefield strewn with human bones very near the site. Finally, Banks records that he had drawings made of another *marae* just eight days prior to the *Endeavour*'s departure: 'After dinner we went ashore to the Marai no Dootahah of which I was desirous to have a drawing made and had not yet done it'.[85] Parkinson, characteristically avoiding any mention of his own work as a draftsman, does not include the incident, but a drawing is preserved in his sketchbook, reproduced by Joppien and Smith. He refers to Bank's 'enormous pile' only briefly, at second hand, as a 'pyramidal' monument (35).

Parkinson was, however, the first to use the word 'morai', and the first to describe such a site among the *Endeavour*'s journalists. He then developed his observations in both visual and verbal form when Banks asked him to draw a *marae* on Raiatea on 21 July. Parkinson does not record this assignment either, but it was engraved for his *Journal* and amplifies the ethnographic detail that, as often, distinguishes his description from Banks's. We find the first occurrence of 'morai' in Parkinson's entry for 6 May, three weeks after the *Endeavour*'s arrival, when he recorded that 'in walking through the woods we saw the corpse of a man laid upon a sort of bier, which had an awning over it made of mats, supported by four sticks; a square piece of ground around it was railed in with bamboos, and the body was covered with cloth. These burial places are called Morai' (*Journal* 26). Subsequent accounts by Banks and Parkinson elaborate on these basic elements, which are also present in the larger, more elaborate funerary sites that they encounter later: the enclosed ground, the trees surrounding the site (*Etoa*), the altar and the platform for the body (both *ewhatta* in Banks). Parkinson makes a point of distinguishing the bier (*whata*) from the altar (*whatee*) when he describes the Raiatean 'priest' (*heiva*) and his attendants, 'who assist him in placing the hog and fish for the Ethooa; as also in strewing the body of the defunct with leaves and flowers of bamboo' (70n.). The engraving (Plate X) emphasizes the bird life (a heron) and trees associated with the *marae*. Several platforms are clearly visible at the center of the composition, but the perspective is too distant for their contents to appear clearly. (Skulls and human remains are included in later depictions of these sites.) The next image, however, shows in great detail the head-dress and other garb of the 'priest', providing the first detailed study in an area of particular interest for early Pacific ethnography, mourning dress (Plate XI). Parkinson

[85] <http://southseas.nla.gov.au/journals/banks/17690712.html>.

surveys the workmanship of this dress along with the 'solemnities' and certain other features, such as the plant life, in more meticulous detail than Banks.

Banks, however, participated in a funeral on Tahiti, and gives a vivid subjective account of some of the same 'solemnities' there. Unlike Parkinson, Banks puts some of his observations on funerary customs into a systematic concluding description, retaining some of his field notes and altering or omitting others. His journal account of the *marae* on Raiatea is more cursory than Parkinson's but diverges curiously on one subject that he describes in detail, 'Ewharre no Eatua or god houses which were made to be carried on poles', which Parkinson takes to be cages for the 'reception' of the heron and another sacred bird. Banks writes: 'One of these I examind by putting my hand into it: within was a parsel about 5 feet long and one thick wrappd up in matts, these I tore with my fingers till I came to a covering of mat made of platted Cocoa nut fibres which it was impossible to get through so I was obligd to desist, especialy as what I had already done gave much offence to our new freinds'. Banks does not mention this incident when he returns to this site, 'Tapodeboatea', in the more formal 'Remarks' that follow his journal entries. Parkinson's more detached observations emphasize the stages of the funeral process and the construction of the sites, particularly the 'sort of stone pyramid' that becomes the permanent resting place of the bones after decomposition. In his 'Remarks' Banks argues explicitly that some of the sites are 'of great antiquity', citing the use of 'immensely large' stones assembled in 'rough' fashion without mortar.[86]

Both Banks and Parkinson present themselves as naturalist-antiquaries in these episodes, recording 'primitive' customs with an eye to antiquity and preservation. Banks's antiquarian agenda is evident in his stress on the importance of drawing the *marae*. The collaborative nature of this work, which like Parkinson's natural history illustration depended on Banks's directions and discussion, made its mark in both journals. It might even be reasonable—given Parkinson's interest in British antiquity—to frame this aspect of the voyage as a quest for ancient Britons, since they did not leave much in the way of material culture themselves (apart from seemingly impenetrable monuments such as Stonehenge). Banks's narrative, as usual, owes more to conjectural history and classical lore, though in the case of funeral rites he also engages with non-classical antiquity and with feudalism in particular. Concluding his account of funerary customs, Banks reflects on the class distinctions expressed and maintained by these customs, which amount to a surprisingly 'regular form of Government': 'the Subordination which takes place among them very much resembles the early state of the

[86] <http://southseas.nla.gov.au/journals/banks_remarks/184.html>. D'Hancarville uses the same criteria to confirm the legendary antiquity of the walls of Tyrins (*AEGR* I.108). Hawkesworth refers to the first piece of stone carving (discovered 29 June), not unrelatedly, as 'very rude workmanship' (*Account* II.166).

feudal laws by which our Ancestors were so long governed'. Although apparently lacking in written codes and 'Cloakd in mysteries', the religion practised on these sacred sites concentrates on 'invisible deities', Banks notes approvingly.[87] Hawkesworth elaborates on this theme, declaring the Tahitians 'not . . . guilty of idolatry' and even crediting them with a rational belief that 'disgraces the Christian' (*Account* II.242, 241). He introduces two additional authorities to support his noble savages here, Apollonius Rhodius and the Hebrew Bible.

Robertson's *History of Scotland*, which Parkinson had on board the *Endeavour*, voices the prevailing sentiment that 'Nations, as well as men, arrive at maturity by degrees, and the events, which happened during their infancy or early youth, cannot be recollected, and deserve not to be remembered'. Johann Joachim Winckelmann's neoclassicism, by contrast, not only recollects but privileges the infancy of one particular nation, the Greeks, by understanding it as pure Nature. Winckelmann's and related versions of neoclassicism are detailed in my chapters on d'Hancarville (Chapters 4–5), though certainly the idea of a 'Homeric' Greece as 'Nature' had some influence on Cook voyage narratives as well. Robertson, however, was reacting against the Rousseauvian thesis of the noble savage, as Adam Smith and Adam Ferguson, among other Scotsmen, had done.[88] Yet Ferguson, especially, followed Rousseau in drawing extensively on travel narrative, especially from North America, for his account of the four historical stages of civil society. Banks and Parkinson, in their different ways, were also deeply influenced by the Rousseauvian desire, modified by Scottish empiricism, to study the ancient past empirically through the 'primitive' present. (Kenrick's probable intervention in Parkinson's prose complicates matters further, since Kenrick achieved real distinction as the translator of Rousseau's *Julie*.) But where others, such as Robertson, saw in the customs and objects of primitive cultures either the ruins of a time best forgotten or the symptoms of degeneration, Banks and Parkinson saw material cultures that demanded critical analysis, even as their authenticity remained a problematic ideological construction.

When Peter Pindar (John Wolcot) satirized Banks's science as 'eating knowledge', he hit somewhat unwittingly on another interest shared by Banks and Parkinson, one that united their botanical pursuits with the study of customs and manners. Banks's antiquarian publications concentrated primarily on Roman Britain, but in one case he published a transcription of an early seventeenth-century manuscript that included representative menus for a nobleman's house. Banks annotated this part of the manuscript extensively, pointing out that 'the List of Birds and Fowls here served up at Table' includes

[87] <http://southseas.nla.gov.au/journals/banks_remarks/185.html>,/175, html,/183.html.

[88] Robertson, *History of Scotland* I; Ferguson, *Essay on the History of Civil Society* 80. See further Meek, *Social Science and the Ignoble Savage* 37–67.

many that 'are considered at this time as being rank carrion'.[89] Several times in the course of the *Endeavour* voyage the crew encountered peoples whose food appeared to them to be 'rank carrion'. One of their earliest attempts to barter with Maori off New Zealand, for example, yielded only fish that 'stank intolerably' (*Journal* 94), and in Cannibal Bay human bones were offered as 'delicious food' (115). Parkinson reports that on Tahiti, Cook, Banks, and Solander all found roasted dog 'the sweetest meat they had ever tasted', though he himself objects to its 'disagreeable smell' and notes that 'the rest of our people could not be prevailed upon to ate any of it' (20).[90] 'Eating knowledge' served the botanical project well from the earliest stages of the voyage: 'We found also many curious plants in the sallading that was sent to us; and desired the people that brought it to procure us, if possible, all the different sorts that grew upon the island [in the harbour of Rio de Janeiro]' (4). From Brazilian 'sallading' to Tahitian breadfruit, Banks was always on the lookout for applications in economic botany. In this respect as well, Banks's agenda shaped Parkinson's legacy. Parkinson's breadfruit, for example, was the only botanical drawing to be engraved for Hawkesworth's *Account* (Pl. XI), while the extensive botanical appendices in Parkinson's *Journal* also reflect Banks's botanical emphasis.

At the same time, Parkinson's individualized treatment of food and food plants allows him to break free of some of the constraints of both antiquarianism and natural history. His comparatively strong effort to master the language on Tahiti gave him privileged access to a variety of plants and 'eating knowledge'. The longest entries in his ethnobotanical catalogue are devoted to food plants, such as coconut (47–49) and breadfruit (45–46). His descriptions of flavours are particularly vivid and constitute some of the most compelling examples of situated knowledge in any of the *Endeavour* journals. As Stanfield pointed out in his preface to the *Journal*, Parkinson's 'facility' with language, and the 'confidence' bestowed on him as a result, 'paved the way also to his success' as a naturalist and collector (viii). Parkinson's Tahitian vocabulary includes eight words relating to breadfruit or coconut, including a verb 'to drink cocoa-nut liquor' (60), which he calls elsewhere 'one of the most refreshing and agreeable liquors in nature' (48). He records the character and flavour of different parts of the nut, in differing states of ripeness, minutely, and with a reverence that evokes his debt to the tradition of natural theology (cp. 11). The *Endeavour* embarked from Tahiti with not only breadfruit, but plantains, yams, taro, and another arum species (*Eape*) 'to serve us instead of bread' (78), and in New Zealand Parkinson adds fern root to this

[89] Banks, 'Copy of an Original Manuscript' 384. Thanks to Crystal Lake for this reference. On Wolcot's satires, see Heringman, 'Peter Pindar, Joseph Banks, and the Case against Natural History'.

[90] Cook, by contrast, declares that all but a 'few' of the people found Tahitian dog nearly as delicious as English lamb <http://southseas.nla.gov.au/journals/cook_remarks/013.html>.

category (105). Though a substitution is implied here, the emphasis on variety and on full representation of Pacific biodiversity works against the European norm of 'bread'. On the Endeavour River in Australia, Parkinson emphasizes the superabundance of food plants within the larger set collected by Banks, Solander, and himself (144), using Tahitian names for plants that he recognizes.

Parkinson's own knowledge work is also recorded in these biographies of scientific objects. As a botanical draftsman he focused on sensory experience of the plants; although the labour is submerged in the descriptions, it is there. Off New Zealand, Parkinson's attention to food and eating led him to posit the 'delicious' taste of human flesh as the cause of cannibalism, a thesis adopted and elaborated by William Wales on the second voyage. His *Journal* demonstrates that the antiquarian agenda of the voyage, at least as he developed it, also led to a focus on food plants as a key to customs and manners, the ethnographic branch of antiquarian research that he adopted as his own. One of the best-known anecdotes of Parkinson comes from a Cook journal entry (19 April 1769), in which Cook reports that they were 'incessantly tormented by the flies, which, among other mischief, made it almost impossible for Mr Parkinson, Mr Banks's natural history painter, to work; for they not only covered his subject so as that no part of its surface could be seen, but even eat the colour off the paper as fast as he could lay it on'.[91] Parkinson mentions the flies, but leaves his painting out of it (16). There was ample reason for him to be self-effacing: in spite of his place on deck, he did not include himself with the ship's 'botanical gentlemen' (21, 87) and imagined himself collecting artefacts to gratify family members (such as Jane Gomeldon) rather than the Department of Antiquities. But in the course of writing his *Journal*, and surely influenced in part by other voyage narratives carried on board, Parkinson fashioned himself in prose as an observer and occasional scholar—neither a gentlemanly naturalist-antiquary nor simply a draftsman.

Parkinson declared his independence from the economy of knowledge work by re-inventing himself as an author and a collector. At the same time he was a good collaborator for Banks, at that time in his twenties and himself engaged in parallel processes of self-fashioning. Despite the practical importance of their collaboration, both parties minimize the role of the other in their journals (though admittedly neither anticipated publication). Banks's journal includes a few passing references to Parkinson at work drawing and one more substantive reference, which upbraids Parkinson's curiosity as excessive and naive.[92] Parkinson mentions Banks more frequently, but does not thematize their

[91] <http://southseas.nla.gov.au/journals/hv23/097.html>.

[92] <http://southseas.nla.gov.au/journals/banks/17700920.html>. Of eight total references, four are passing mentions of Parkinson at work and two are passing mentions of his final illness and death in January 1771.

working relationship and subordinates Banks's and even Cook's activities to his narrative of places. Their collaboration, however, left its mark on both journals, which produce alternate versions of the naturalist-antiquary persona. The two approaches are complementary. In his study of architecture and funerary customs, Banks enlists Parkinson in the service of his concern with antiquity and preservation, while Parkinson takes the opportunity to build his knowledge of language and manners. In his emphasis on food plants, too, we can see Parkinson altering the contours of natural history and antiquarianism, which constrained but also enabled his work.

The chapters that follow turn their attention to more lavish plate books of natural history and/or antiquities. One such work, Banks's unfinished *Florilegium,* might provide a better example of what Thomas calls the 'scientific enclave' in which natural specimens could be stripped of their commodity status.[93] The *Florilegium* might also make a more apt comparison with these other volumes (instead of Parkinson's *Journal*), especially with Hamilton and d'Hancarville's *Collection of Etruscan, Greek, and Roman Antiquities* (the subject of Chapters 3–5). Brian Adams's account of the production of the *Florilegium* illustrates many of the features that these two works share in common. Like the *Collection*, the *Florilegium* employed numerous artists (mainly employed in finishing or engraving Parkinson's drawings)—though unlike Hamilton, Banks directly supervised this work, which was carried on in his house. The roughly 520 engravings for the *Collection* were produced over ten years, while Banks claimed that after eleven years' work (1773–84) he had nearly 770 plates for the *Florilegium*. Banks also invested far more of his own money—as much as £12,000, according to Adams—than did Hamilton, though the text remained unfinished when Solander died in 1789, and the images were not published in full until 1989.[94] Parkinson's *Journal* was produced much more cheaply and quickly, but even this simple volume included its set of key players, particularly Stanfield Parkinson, William Kenrick, and a number of artists, the most prominent of whom was probably Samuel Hieronymous Grimm. The *Journal* may serve to introduce the kinds of knowledge work involved in the production of late-eighteenth-century scholarship on natural history and antiquities, which proliferate along with the specimens and the number of volumes themselves as time goes on. It also offers rich literary testimony of the intellectual versatility and vivid observational powers of Sydney Parkinson, who has tended to be remembered more often for his commissioned botanical images than for his authorship. The goal of this chapter has been to right that balance.

[93] Thomas, *In Oceania* 116. The *Florilegium* itself, because never completed or offered for sale, also manages to evade commodity status.

[94] Adams, *Flowering of the Pacific* 140, 147–9, 176.

2

Campi Phlegraei and the Neapolitan Pursuit of 'Most Remote Antiquity'

2.1. ANTIQUITIES AND NATURAL HISTORY IN NAPLES

Shortly after embarking on his duties as British ambassador to Naples in 1764, Sir William Hamilton found a congenial companion for his exploration of the kingdom's classical territories in Pierre François Hugues, who styled himself Baron d'Hancarville. Their explorations proved fruitful, and within a year d'Hancarville found himself supervising a lavish publication intended to document the hundreds of Greek vases Hamilton had amassed from excavations in the region and from the purchase of existing collections. One hundred pages into the substantial introduction that d'Hancarville wrote for the first volume of this work, dated 1766, he recalled one of his earliest archaeological expeditions with Hamilton to the newly rediscovered ancient city of Paestum. A nominally Etruscan inscription accidentally discovered within the city's wall produced the impetus for d'Hancarville's wildly ambitious conjectural prehistory of art. Recalling this moment of discovery, he exclaims:

> Is it probable that the Grecians who bragged of having invented all the Arts, wou'd have . . . employed the Architecture of the Tuscans, as they did even to the very Citadel of Athens and in the Temple of Minerva . . . , if the Tuscans to rob them of the merit of being the inventors of the Art, had done nothing more than strip the Dorick order of the ornaments belonging to it? . . . Ought not the very great simplicity of the Tuscan Order, to lead one to think that it preceded all the others?[1]

Hamilton asked a similar rhetorical question concerning the prehistory of volcanoes in an essay written three years later: 'Surely there are at present many existing volcanoes in the known world; and the memory of many others has been handed down to us by history. May there not therefore have been

[1] *AEGR* I.96, 94. D'Hancarville indirectly challenges here the philhellene assumption that the Doric order is the most ancient. On this discovery (or supposed discovery) see Chapter 5.

many others of such ancient date as to be out of the reach of history?'[2] D'Hancarville's and Hamilton's shared pursuit of 'most remote antiquity' marks the deep, predisciplinary affinity of archaeology and earth history.[3] By using objects, either volcanic or ceramic, to push back the limits established by traditional chronology, these two ostensibly neoclassical projects also precipitate the fragmentation of antiquity into historical, prehistoric, and pre-human domains.

Modern scholars have frowned upon Hamilton's voracious collecting of artefacts, volcanic sublimity, and, later, sexual scandal. Kevin Nevers declares that 'the sublime' in Hamilton's accounts of Vesuvius 'at once endorses and enforces a colonialist agenda of human hegemony'. According to David Nolta, Hamilton resolutely stamped his 'collector's mark on the actual landscape settings depicted' in his *Campi Phlegraei*, 'and that collection includes the artist and his work as well'. Nolta also argues that Emma Hamilton's famous Attitudes 'mark both the performance and the performer as Hamilton's own'. Susan Sontag's *The Volcano Lover*, though a fictionalized psychological study, carries the same vein of ideology critique in its mock-deferential epithet for Hamilton, referred to throughout simply as 'the Cavaliere'. From an art historical vantage, Michael J. Vickers has claimed that Hamilton's first vase collection was simply 'a marketing job'.[4]

There is surely some truth in all these charges, but they miss the complexity and variety of Hamilton's projects, which depended far too much on a loosely orchestrated network of unruly agents, human and otherwise, to generate predictable outcomes. First there is Vesuvius, which erupted 'obligingly frequently' in the late eighteenth century, as Roy Porter memorably put it. The vases, too, with their provocative paintings, may be seen as 'actants' within the network in the sense that Bruno Latour ascribes to these terms.[5] Then there are the human collaborators—the peasants and *lazzaroni* on whom Hamilton depended for so much evidence and legwork; Ferdinand IV, King of Naples, who found a loophole for Hamilton in the law barring the export of antiquities; literally dozens of artists whom he patronized in his thirty-seven years at Naples, from the brilliant painter of volcanoes and local costume, Pietro Fabris, to the numerous draftsmen and engravers co-coordinated by d'Hancarville, and later by Wilhelm Tischbein; the learned but often ill-beneficed

[2] Hamilton, *Remarks upon the Nature of the Soil of Naples* 39. Cited parenthetically in the text hereafter. The two questions both appeal to an origin predating recorded history ('Tuscan' and pre-classical antiquity in d'Hancarville's case).

[3] The expression occurs at least twice in Hamilton (*Remarks* 13 and *Campi Phlegraei* 10); d'Hancarville uses 'remotest antiquity' (*AEGR* I.98).

[4] Nevers, 'Immovable Objects' 19; Nolta, 'Body of the Collector' 111–12; Sontag, *The Volcano Lover* 11 and *passim*; and Vickers, 'Value and Simplicity' 6.

[5] Porter, *Making of Geology* 123; Latour, *Reassembling the Social* 63–86. In this chapter Latour makes his case that 'objects too have agency'.

Catholic clergy who eagerly collected data for Hamilton, particularly Antonio Piaggio; and the obstreperous d'Hancarville himself, who fled the authorities in Naples with thousands of pounds worth of engraved copper plates, only to have the plates—fortunately for Hamilton and posterity—impounded by new creditors at his refuge in Florence.[6] This initial survey of Hamilton's network begins with d'Hancarville and then focuses on the collaboration with Fabris, concluding with an overview of the actors at work in the later part of Hamilton's career. Hamilton (1731–1803) was neither the first nor the last connoisseur of classical antiquity who found d'Hancarville irresistible in spite of his bad reputation.

Nonetheless, their shared pursuit of 'most remote antiquity' is surprising in light of the manifold differences between d'Hancarville's monumental folios and Hamilton's deliberately unpretentious letters to the Royal Society. D'Hancarville (1719–1805) spent ten years and thousands of pounds (much of it Hamilton's) publishing the four fifty-centimetre folio volumes of the *Collection of Etruscan, Greek, and Roman Antiquities* (*AEGR*). He paid elegant lip service to Hamilton's intention of 'hasten[ing] the progress of the arts' by providing a 'collection of exquisite Models' for artists—an intention borne out as early as 1769 by the founding of Josiah Wedgwood's Etruria Works—but instead of describing these models he inserted his own theoretical ambition of 'disclosing their true and first principles' in hundreds of pages of recondite conjectural history (*AEGR* I.vi). D'Hancarville insists that the vases themselves, though meticulously illustrated, only serve 'the pleasure of the sight' and need to be contextualized by his discovery of the aesthetic 'sistem the Ancients followed' in order to put the artist in the position of having 'the Originals themselves in his possession' (vi). Hamilton, anticipating the Baconian rhetoric of nineteenth-century geology, invokes 'system' as a term of derision when distinguishing his scrupulous reporting of the facts concerning Vesuvius from then-current theories of the earth. Hamilton's careful observations have garnered only praise from geologists and historians of geology, though cultural critics have been more sceptical. D'Hancarville's extravagant speculation has earned him the animosity of generations of art historians, though more recently he has also earned some half-reluctant admiration, particularly for his mastery of book production.[7]

The rogue antiquary and the diplomat-naturalist did share an interest in actualism. Hamilton's arche-geology is notable for its vocal insistence that the present is the key to the past, almost twenty years before James Hutton

[6] This story has been retold many times, though large gaps remain in our knowledge of the incident and more generally of Hamilton's and d'Hancarville's working relationship. The most thorough account is probably Jenkins, 'Contemporary Minds' 49–50.

[7] For scholarship in English, see Haskell, 'The Baron' as well as Jenkins and Sloan (*V & V* 40–51, 96–100, 146–55). The only book-length study is Pascal Griener, *Le Antichità Etrusche*. For Hamilton's rejection of 'sistems' and theories of the earth, see *CP* I.5.

declared that he saw 'no vestige of a beginning—no prospect of an end', and some two generations before Charles Lyell gained widespread acceptance for this uniformitarian doctrine in Britain.[8] The pursuit of 'most remote antiquity' led back again and again to the present (hence 'actualism'), in work not only on Vesuvius but also on ethnography and antiquities. In the case of d'Hancarville's boastful Grecians, the classical past (with its present prestige) is the key to a surprisingly familiar pre-classical past, in which 'Inventors having met with the same difficultys' as in Greece, their 'Industry... was obliged to employ the very same means'—that is, we can be sure that art originated in the same way in civilizations predating recorded history (*AEGR* I.52). For both Hamilton and d'Hancarville, the history recorded in place names and in classical literature reinforces the continuity between prehistory and the present. Hamilton's empirical bent leads him to look to contemporary Neapolitan customs and manners for further evidence of this continuity in the context of his natural history; within d'Hancarville's domain of antiquarianism such references were already conventional. This ethnographic actualism, the focus of Chapter 4, stems from their immersion in local culture and the art scene in particular. Italian antiquaries, painters of picturesque *vedute* for the tourist trade, the authors of a state-sponsored survey of regional costume—all claimed an especially strong continuity between the ancient and modern populations of the Bay of Naples (*V & V* 167–68, 241).

This chapter sets the proliferation of antiquities—including the study of contemporary popular culture—in the context of the networks and collaborations that nourished it. As is often noted, Vesuvius put Naples on the map in the eighteenth century: its frequent eruptions made it a tourist attraction and its past activity preserved whole cities, the rediscovery of which permitted a more immersive experience of Roman antiquity than Rome itself. Yet the fertile exchange between natural history and antiquarianism around Naples owes more to the forms and means of knowledge production than to the locality itself. The sublime ensemble of a highly active volcano and three ancient buried cities made this exchange particularly visible. But it was driven by the same tensions—between aristocratic collector/patrons and the increasingly independent 'client' intellectuals who laboured to produce their knowledge—that were common to other contemporary research projects. The many common products of these tensions, or rather of the networks that made them productive, included gorgeous folio volumes with engraved illustrations (Hamilton alone sponsored three); increasing specialization, and with it a variety of other more modest forms of publication pursued by individuals; and a fragmentation of the unified view of history associated with the Enlightenment and with both neoclassical and romantic aesthetics. By framing my

[8] Hutton, 'Theory of the Earth' 304.

account of vases and volcanoes between chapters on South Pacific exploration (Chapter 1), and on the preservation of Gothic cathedrals (Chapter 6), I aim to foreground these structures of historical epistemology.

While elaborating this larger argument, my account also attends to the peculiarities of Naples and neoclassicism: my specific aim here is to recover the unconscious anticlassicism at work in Hamilton's projects, especially *AEGR* and *Campi Phlegraei*.[9] The very privilege attached to antiquity by the neoclassical assumptions of Hamilton and many of his collaborators, including d'Hancarville, Tischbein, and others, motivated a rigorous analysis that severed historical, prehistoric, and pre-human domains and unsettled neoclassical consensus. Hamilton's pre-geological notion of deep time, built on the shared experience of Vesuvius in action, contributed to this fragmentation. Vesuvian time marks the study of vases as well, as I will try to show by restoring the dialogue between the first vase collection (*AEGR*, 1766–76) and Hamilton's collaborative studies of volcanism (published in journal article and then in folio form between 1766 and 1780, with two further articles in 1786 and 1795). This chapter also touches on the collaboration with Tischbein that produced the second vase collection, *A Collection of Engravings* (1791–1803), but my focus is on the earlier period.

Guided by the rubric of natural history, my account begins with a particular emphasis on *Campi Phlegraei* and on the collaborators who contributed to Hamilton's work on volcanoes. Turning more directly to classical antiquity, my next three chapters focus mainly on d'Hancarville and *AEGR*. Though subdivided in this way, all four Neapolitan chapters together (Chapters 2–5) aim at an overview that will illuminate gradual changes in the production and organization of knowledge about the past against the unified background of Hamilton's remarkable career in Naples. The complex entanglement of antiquarianism and natural history in Naples is more exceptional than representative; by placing this extended treatment against individual chapters on three other knowledge projects (Chapters 1, 6, 7), however, I hope to show specific connections as well as create a broader European and neoclassical context for the other projects. Previous scholarship has not adequately addressed either the affinities between antiquarianism and natural history or the independent motives of Hamilton's collaborators. Both these factors contributed to the fragmentation of classical antiquity as an ideology and an object of knowledge.

[9] Hamilton and especially d'Hancarville have traditionally been seen as early or 'pre' romantics by scholars. A typical instance is Burton Feldman in his introduction to the facsimile reprint of d'Hancarville's 1785 *Recherches sur l'origine, l'esprit et les progrès des arts de la Grèce*. This chronological scheme obscures the extent to which classic and romantic are part of the same dialectic, in which anti-classicism is not just a romantic project but also an aspect of classicism itself.

2.2. SIR WILLIAM HAMILTON'S COLLABORATORS

Creating a dialogue between *Campi Phlegraei* and the vase publications aids in recovering the antiquarian dimension of natural history as well as the natural history content of antiquarianism. Of the three major publication projects, *Campi Phlegraei* most obviously combines antiquities and natural history. Fabris's illustrations to this work make the relationship between culture and nature vividly apparent, and even the place names make the natural history of these fields of fire inevitably cultural. Fabris incorporates quarrying, viticulture, and other traditional rural practices into his volcanic landscapes, and in one instance even depicts the ongoing excavation of Pompeii. Partly because traditional practices or 'popular antiquities' form a part of the broad ethnographic domain of antiquarianism, Hamilton's text also refers to these practices for evidence of the social consequences of volcanism over time. Readers of *AEGR*, however, have not typically noticed the important part played by natural history in d'Hancarville's text, comparable in interest and importance to the antiquarianism present in Hamilton and Fabris's work on Vesuvius. Four of the five letters to the Royal Society later incorporated into *Campi Phlegraei* were written between 1766 and 1769 while d'Hancarville was working on the first two volumes of *AEGR*. The last and longest of these letters, dated 1770, develops more fully a number of themes that also appear in d'Hancarville's first two volumes.[10]

This final letter evinces an unusual emphasis on both the modern and the ancient populations as holding the key to volcanic and geological phenomena. The ongoing excavations at Pompeii and Herculaneum provided important clues to the nature and extent of volcanic rock, primarily *tufa* [tuff] and lava.[11] 'About two years ago', Hamilton reports, 'I saw the head of an antique statue dug out of this matter within the theatre of Herculaneum; the impression of its face remains to this day in the tufa', confirming that the material blanketing the city in CE 79 was initially a liquid mud composed of ash (*Remarks* 8). Hamilton also mentions impressions of human bodies of the kind captured so memorably by Felicia Hemans's poem, 'The Image in Lava' (1827). The ancient and the modern meet again at Cumae, 'allowed to have been the most ancient city of Italy' (32). By showing that the city rests on the same volcanic bedrock as everything else on the Bay of Naples, Hamilton can establish the certainty of eruptions beyond the reach of history, as he set out

[10] This letter, dated 16 October 1770, first appeared in *Philosophical Transactions of the Royal Society of London* (vol. 61); then, in book form, as *Remarks*, the version cited here (n. 2); then as letter V in *Observations on Mount Vesuvius, Mount Etna, and Other Volcanos*; and finally, also as letter V, in *Campi Phlegraei*.

[11] Hamilton introduced the word *tufa* into English. Now 'tuff' is used instead to signify the same compacted pyroclastic rock, while *tufa* is used for rocks formed by chemical deposition. In this chapter I will retain Hamilton's usage.

to do. In 1777 he would publish an article on Pompeii in the Society of Antiquaries' journal, *Archaeologia*, that underlines this point by drawing attention to the volcanic paving stones of the city, quarried from basalt produced by an unrecorded earlier eruption. Not content with archaeological evidence, Hamilton cites current local belief according to which Nero's Bath, a volcanic hot spring near ancient Cumae, 'is reckoned a great specifick in that distemper which is supposed to have made its appearance at Naples, before it spread its contagion over other parts of Europe' (32–33). D'Hancarville achieves a similar time-lapse effect as he shuttles back and forth via classical authors between theory-starved modern art and Etruscan autopoesis. Hamilton, in good Baconian fashion, did not hesitate to consult—and to cite—'the peasants in the neighbourhood of my villa' (40) on several volcanic particulars. D'Hancarville would have found confirmation of the eternal recurrence of animism in Hamilton's occasional lapses into humoral physiology: lava, for example, turns out to be 'foul vitrified matter' (7) in the same sense that 'foul humours' (11) are expelled from the body by eruption.

D'Hancarville, given his affinity for the *philosophes* and conjectural history, is more apt to take his contemporary animism from farther afield: sacred standing stones, the antecedents of sculpture, were 'consulted' by the ancients 'just as the negroes of Guinea still consult their Manitous and the Lapons their Tambours' (*AEGR* IV.26–27). Hamilton's notorious discovery of 'remains of the worship of Priapus lately existing at Isernia' provides another case in point—this time the active Catholic rite of St Cosmo presumed to derive from an ancient Roman fertility cult.[12]

Geological actualism plays a role in d'Hancarville's text as well. If mountains could be dated (as he gathered from Hamilton), then surely vases could be dated as well, and not just with reference to the written record. At his most empirical, d'Hancarville attempts to distinguish various local clays in an effort to place the origin of particular vases (I.154). In his chapter on Etruscan history, d'Hancarville asserts that prior to the arrival of the Greeks, sites in the Campania Felice already had 'Phoenician names, which according to the spirit of the Antient languages of the East indicated Clearly the Property of the Places which bore those names' (64). In a very lengthy excursus note to the passage, he details many of these place names, including Vesuvius, which, he claims, originally meant the place of flames (*ubi flamma*)—an etymology at least borne out by the ample evidence of prehistoric eruptions. In this note

[12] Richard Payne Knight begins his 1786 *Account of the Remains of the Worship of Priapus* with Hamilton's previously unpublished letter to Sir Joseph Banks on this subject, but the bulk of the volume is filled by his own 200-page essay, a comparative study of myth and religion for which d'Hancarville, then in London, served as 'mentor' (Haskell, 'The Baron' 41). Manitous are Algonquin and have nothing to do either with Guinea (as d'Hancarville says here) or with the Caribbean (as stated by his probable source, Octavien de Guasco; see further Chapter 4 below).

d'Hancarville delivers his own field observations of the 1766 eruption and theorizes the generation of scores of *monticules* by the parent mountain. When Johann Joachim Winckelmann visited Naples the next year it was d'Hancarville, not Hamilton, who led the tour of Vesuvius documented in several ecstatic letters by Winckelmann.[13] D'Hancarville thus shared with the 'modern Pliny' some of the authority from their intimacy with the mountain, and his writing drew heavily on the *Natural History* of Hamilton's ancient namesake, Pliny the elder, though not so much for volcanology as for its account of Greek art.

D'Hancarville's is in many ways a natural history of art. If art is endemic to all primitive cultures, it is because nature confronts them with the same necessities, and for this reason columns retained the form of tree trunks, bronze candelabra retained the form of young branches stripped of their leaves, and so on, as souvenirs intended to '[recall] the history of the Art, the origin of things, and the equality nature has placed amongst Men' (I.76). The candelabrum and the log/column appear with other examples in two of d'Hancarville's remarkably inventive title vignettes (Figures 5.2–3 in Chapter 5). The lack of a signature on these plates and the detailed verbal key to the images (177) may suggest his especially active involvement in these two designs. The art-historical function of formal features remains a central thesis throughout all four volumes of *AEGR*. Similarly, in preliterate civilizations, 'a Well, an Oak, a Field even, . . . were the warrants of the History which Fathers taught their Descendants' (116). In this passage, d'Hancarville also draws on Buffon's account of the origin of the continents for the sequence in which these primitive cultures populated the globe (114). Even the plates of the vases function somewhat like the specimens of volcanic matter and large transparencies that Hamilton sent to London to illustrate his natural history: the hand-coloured engravings, with colours and proportions minutely calculated, will enable his reader, d'Hancarville assures us, 'to judge of what we say . . . just as well as if he had the monuments themselves in his hands' (152).

Padre Antonio Piaggio, employed by Hamilton to keep a daily log at the foot of Vesuvius from 1779 to 1794, was also an antiquary, with ambitions to rival d'Hancarville's and with qualifications, or at least field evidence, superior to those of the Baron. Long before he began his anchoritic natural history, Piaggio had assisted the royal antiquary with the excavations at Herculaneum and Pompeii, and he is best remembered today for inventing a machine that unrolls ancient papyri without damaging them. The government, however, 'scarcely [paid] him a very miserable pension' and refused to sponsor the publication of his scholarship on the ancient cities. He therefore wrote his very substantial manuscript entirely in code and left it to Hamilton at his death in 1795, but it

[13] Winckelmann, *Briefe* III.317, 328. D'Hancarville revisits this volcanic theory and expands it in his fourth volume (*AEGR* IV. 82–5).

was subsequently lost by the Society of Dilettanti.[14] His eight volumes of notes and sketches on Vesuvius survive, however, and Hamilton drew on them extensively, with acknowledgment, in at least two of his later publications. In a note to the engraving of Piaggio's last sketch of the mountain, included in the *Philosophical Transactions* for 1795, Hamilton explains that it was left unfinished because friends arrived to carry the 80-year-old monk down the mountain amidst a hail of ash and red-hot pumice stones.[15]

Hamilton's last major collaborator, Wilhelm Tischbein, was neither antiquary nor natural historian, but from 1789 director of the Royal Academy of Art at Naples. Hamilton thus chose a professional when he appointed Tischbein to supervise the publication of his second vase collection, another four folio volumes that appeared from 1793 to 1803 (imprint dates 1791–95). Publication of the last two volumes was severely disrupted by the threat of French invasion, which culminated in the short-lived Parthenopean Republic of 1799. The same cause nearly prevented Tischbein from publishing his own collection of engravings, but after a narrow escape these plates gradually appeared as *Homer nach Antiken gezeichnet* (9 vols, 1801–21). This incident illustrates the common tendency of Hamilton's collaborators to use their connection to advance their own careers. Many of Tischbein's images were drawn from Hamilton's vases while they were also being recorded for the *Collection of Engravings*. Among many other differences, the second vase collection was a more academic enterprise than the first, with Tischbein's students contributing most of the drawings and engravings, and it was produced more cheaply for accessibility. Among the most obvious differences are the contrast between the outline style of engraving here and the coloured prints of *AEGR*, and Hamilton's more active role in writing some of the later text.[16]

Hamilton's increased authorial presence reflects a number of changes in the conditions and assumptions under which antiquities were collected as the century waned. It became legal to excavate around Naples and less difficult to export antiquities. Archaeology was gradually becoming a more self-consciously scientific domain in which Hamilton's Baconian method of inquiry, honed in his study of volcanoes, became a more relevant qualification. One of his tasks in the introductory matter to the *Collection of Engravings* was to settle the conflict over whether any of the red- and black-figure vases

[14] Jenkins details the loss of Piaggio's archaeological manuscripts, quoting the reference to his 'miserable' pension from a letter of Charles Townley (*V & V* 44–5).

[15] 'An Account of the Late Eruption' 115. I take up Piaggio's Vesuvian diary in the sixth section of this chapter.

[16] See the illuminating discussion of this stylistic shift in Brylowe, 'Two Kinds of Collections' 43. Although Hamilton wrote some text for the second collection, it remains the case that he relied on 'hireling scholars' for his antiquarian publications (*V & V* 41), while using his own words to compete in a new literary market for volcanoes—a decision that may support the emphasis on incipient specialization in Hamblyn, 'Private Cabinets' 202.

were truly of Etruscan, rather than Greek, origin. Hamilton restores the Grecians' bragging rights, but without otherwise challenging d'Hancarville's theory. D'Hancarville himself was well aware that these vases were called 'Etruscan' only by convention, but his investment in a pre-Hellenic origin for art led him to hold out the possibility that genuine Etruscan vases might yet be found (*AEGR* II.134, 136). Part of the interest of Hamilton's lengthy preface to the *Collection of Engravings* is historical. He recalls the very different circumstances of the first collection twenty-five years earlier, when the government had not yet slackened its control over the nation's antiquities, not even for Hamilton. He recalls his excavation of a tomb at Trebbia in 1765, recorded in the first collection, despite the risk of exposure, in a dramatic engraving by Carlo Nolli.[17] According to an anecdote recorded by Carl Justi, Hamilton disguised himself and lodged incognito with peasants in order to carry out some of these excavations. In his article on Pompeii he noted the enterprise of the local people who traded in artefacts before the official excavations began there, and undoubtedly he took some cues from them.

In 1790 the circumstances were entirely different, as illustrated by a parallel engraving that served as the frontispiece to the second collection (Figure 2.1), showing the now legal and very public excavation of a tomb at Nola in that year.[18] This caption to this engraving, taken by Tischbein's student A. Clener from a drawing by Christoph Heinrich Kniep, begins with a Latin translation of 'Society of Antiquaries of London', to whom the book is dedicated. A very much idealized Hamilton is shown at the corner of the sarcophagus with another antiquary behind him, and two labourers behind them continuing the excavation. Another worker in the foreground has put down his tools to admire the vessels that have come to light. The most striking new face in this image is that of Emma, the second Lady Hamilton, who seems to be directing this worker. The composition also suggests a contrast between admiration, as expressed by this pair, and the more critical examination being carried on by the two antiquaries behind them.

Lady Hamilton herself has often been thematized by scholars (and by satirists including James Gillray) as the choice 'piece' in Hamilton's collection.

[17] *Collection of Engravings from Ancient Vases* I.24. For the image see Figure 5.4 in Chapter 5, below.

[18] Hamilton himself refers to 'His Sicilian Majesty's having lately taken off the Prohibition to search for Antiquities' (*Collection* 2). He does not mention the issue of export, which had been expressly prohibited by Neapolitan law, as Ramage shows in her 'Sir William Hamilton as Collector'. Ramage infers that Ferdinand must have 'decided to turn a blind eye' to Hamilton's trafficking (472) but also notes that Hamilton had some legal troubles, as reported by Carl Justi. Justi's fascinating sketch of Hamilton's collecting practice is partly based on Winckelmann's *History of Ancient Art*, but also apparently draws on legal documents that Justi fails to cite explicitly. See his *Winckelmann und seine Zeitgenossen* III.437–8. Winckelmann's version is conveniently quoted in Jenkins and Sloan (*V & V* 141–2). For Hamilton's own brief mention of small rural collector-dealers, see his 'Account of the Discoveries at Pompeii' 162.

Fig. 2.1. Engraving by A. Clener from a drawing by Christoph Heinrich Kniep. Sir William Hamilton, *Collection of Engravings from Ancient Vases*, Vol. 1 (1791 [1793]), Frontispiece (Dedication to the Society of Antiquaries). A tomb at Nola. Courtesy of Margaret M. Bridwell Art Library, University of Louisville.

Recalling that one of the most widely engraved George Romney portraits of Emma depicted her as a bacchante, it is tempting—and, it seems to me, more apt—to include her with the Neapolitan rustics and other modern primitives mentioned above as supposed evidence of modern continuity with most remote antiquity.[19] As much as d'Hancarville or any of Hamilton's collaborators—or for that matter the local population, which occasionally turned threatening—Emma had her own agenda, which by now may be more widely known than Hamilton's own. Most famously, she cultivated a modern primitive persona herself in the performance of her Attitudes. Later I will touch briefly on a less well-known performance of hers, an eruption narrative that illuminates Emma Hamilton's role in the network of inquirers and unruly objects that instigated knowledge work while dismantling neoclassical consensus and circumventing teleological histories of nature and art.

2.3 WORD AND IMAGE IN *CAMPI PHLEGRAEI*

So far I have focused on Hamilton's relatively early *Remarks upon the Nature of the Soil of Naples*, roughly contemporaneous with d'Hancarville's first two volumes. Hamilton's landmark book on volcanoes, *Campi Phlegraei*, coincides with d'Hancarville's last two volumes, which continue to show the influence of natural history, and of Vesuvius in particular. Hamilton's earlier publications on volcanoes have archaeological moments as well, but *Campi Phlegraei* more closely resembles the *Collection of Etruscan, Greek, and Roman Antiquities* because both feature colour plates and present themselves as fine art books. Carlo Knight observes that the beautiful hand-coloured prints by Pietro Fabris 'overshadowed the text', 'with the consequence that artistic values prevailed over any others'. This consensus was also shared by historians of science, but more recently Karen Wood has made an impressive case for the organic continuity between *Campi Phlegraei* and the earlier stages of its text, reprinted more or less intact from the Royal Society's *Philosophical Transactions*. Wood stresses the continuity between phases of these five letters, from composition to oral presentation (in absentia) at Royal Society meetings to print and beyond. Drawing on Simon Schaffer and Steven Shapin's paradigm of scientific witnessing, she refers to each of these phases as 'witnessing tools' aimed at a variety of overlapping audiences. *Campi Phlegraei* restores the physical specimens and the element of visual drama—lost in the transition from oral

[19] The term 'piece' originates with Hamilton's nephew (and Emma's previous lover) Charles Greville. See Constantine, *Fields of Fire* 128, 137. Emma sat for Romney many times, but his *Emma Hamilton as a Bacchante* (1785) circulated widely as an engraving. Elisabeth Vigée-Lebrun painted her *Lady Hamilton as a Bacchante* in 1792.

presentation to print—in order to 'achieve a more complete witnessing experience', Wood argues.[20]

These visual components, along with several other new features, lend *Campi Phlegraei* an antiquarian aspect as well. On the one hand, the book resituates the 'vigilant eyewitness' of natural phenomena in a visual context (and a book market) associated with the Grand Tour, which helps to explain Knight's view that 'artistic values prevailed'. On the other hand, the increased visual and verbal attention to antiquities necessarily extends to the material traces of volcanic activity as well. *Campi Phlegraei* thus deploys two cognate sciences of antiquity, yet exploits the tension between them to delimit new fields of inquiry.

In their final form as a plate book, Hamilton's observations more fully demonstrate the broad geographic range and remote antiquity of volcanism. *Campi Phlegraei* includes a new prefatory letter addressed to the President of the Royal Society in the style of the five that had appeared in the *Philosophical Transactions*. Wood rightly points out the importance of Hamilton's decision to follow the printed texts of his letters, which 'encouraged readers to credit this independent publication with the status of a learned journal'.[21] Nonetheless, his prefatory letter imposes an ambitious new framework on the whole, which not only attains the 'narrative power' of an epistolary novel in its collected form, as Wood observes, but also argues an overarching thesis supported by other new features of the volume. Hamilton systematically argues for the volcanic origin of the entire region between Vesuvius and Ischia, placing many of the fifty-four plates and his accompanying explanations directly at the service of this thesis. In addition to the prefatory letter and the fifty-four often substantial explanations written to accompany each image, there are minor changes to the letters that reflect ongoing volcanic activity. The prefatory letter incorporates a formidable 'witness' in the Genevan naturalist H. B. de Saussure, who is quoted at length.

Many of the explanations contain extensive references to the classics, further expanding the role of classical antiquity established in the letters. The colour plates themselves also link Hamilton's observations more strongly to the science of antiquity and to tourism. The plates include close-up depictions of volcanic specimens and many views, some of which are devoted to volcanic craters or their immediate surroundings, and some of which are panoramic and include classical ruins as well as sites familiar from classical texts. In some cases, most obviously in Fabris's view of the Temple of Isis at Pompeii, the choice of volcanic scenery is certainly conditioned by the classical

[20] Knight, 'Sir William Hamilton's "Campi Phlegraei"' 195; Wood, 'Making and Circulating Knowledge' 90. Knight cites Martin Rudwick and other historians of science in support of his view.

[21] Wood, 'Making and Circulating' 89.

resonances of a particular site. The explanations alone contain more than a dozen specific references to classical texts (some with quotations) and a larger number of allusions to ancient authors or history.[22]

Another important addition is ethnographic in nature. I mentioned above Hamilton's reference to the contemporary practice of medicinal bathing in volcanic hot springs, and other social uses of the landscape feature in the plates and the new text, which draws upon Fabris's expertise in customs and manners to present the history of geological causes operating in the present (actualism) as somehow symmetrical to traces of antiquity in present customs. This elaborate symmetry captures the essential insight fostered by the stronger antiquarian emphasis of this collaborative work. To speak in modern disciplinary terms, *Campi Phlegraei* sharply juxtaposes human and geological time.

By today's standards even the oldest of the craters and other volcanic remains that mark the 'fields of fire'—a region along the northern tip of the Bay of Naples, some 30km west of Vesuvius—are very young. By eighteenth-century standards, however, 37,000 BP would be an antiquity so remote as to be hardly conceivable, and Hamilton's new thesis forcefully articulates the divide between recorded antiquity and the prehistoric (if not pre-human) past suggested by the different lavas he observed: 'If the circuit I have described, can be fairly proved to owe its very existence to Volcanick explosions, at various, and in some parts at very remote periods', he declares, 'I flatter myself [that] I shall have open'd a new field for observation on this curious subject' (*CP* I.3). The word 'circuit' signals his debt to tourism, and indirectly the strong affinities between this book, with its beautifully composed plates and meticulous explications, and other topographical and antiquarian volumes.[23] Some of the Fabris paintings that were used for this work predate his acquaintance with Hamilton and were probably painted as *vedute* for the tourist trade.

In some ways the sequence of plates and explications itself resembles an itinerary. The first volume is devoted to the text of the letters. In the second, larger volume, which contains the plates and explications, the most obvious formal division falls between plates XLI and XLII: the first group of plates are views and the last thirteen all depict specimens of volcanic matter arranged for display. The first, larger sequence begins with views in the vicinity of Naples,

[22] Many of the original letters already cite classical authors, and there are some substantial Latin quotations (especially in Letter V), but these tend to be from historians (Tacitus, Pliny, Strabo, others), whereas poets tend to be more prominent in the explanations of Fabris's prints. The breadth of these references increases along with their frequency in *Campi Phlegraei*. While some notes were added for this volume (e.g. vol. 1, p. 64n.), many had been added when Hamilton provided the text for the first collected edition of the letters, *Observations on Mount Vesuvius, Mount Etna and Other Volcanoes* (1772).

[23] The English painter Paul Sandby produced several such volumes, including one that featured some engravings after Fabris, *Views in and near Naples* (1777–82).

including several of Vesuvius (I–XV), and then moves gradually westward via Posillipo and Pozzuoli to the island of Ischia (XVI–XXXII) before returning to Vesuvius (XXXIII). The last nine images suggest a less systematic movement among various islands, including Sicily, but five plates feature aspects of Vesuvius, and the sequence culminates with the excavation of the Temple of Isis (Plate 1). The itinerary form serves Hamilton's argument, as each new view provides further evidence, explicitly pointed out on almost every plate, that the region 'owes its very existence to Volcanick explosions'. Although he cannot claim, as Banks did concerning marine fauna, that 'so large a field of natural history has remained almost untrod', Hamilton can plausibly suggest that the very well-trod 'fields of fire' are rendered 'a new field for observation' by his antiquarian natural history.[24]

Fabris's justly famous depiction of the excavation in Pompeii forcefully juxtaposes classical antiquity and natural history. Hamilton's explanation, as indicated by the title of this plate, is more concerned with natural history: 'View of the first discovery of the Temple of ISIS at Pompeii, which City was buried by showers of pumice stones and other volcanick matter, during the great eruption of mount Vesuvius in the year 79' (*CP* XLI). His commentary addresses the antiquities as well, but foregrounds geo-history in declaring that 'the intention of this plate is to shew the exact similitude of the strata of rapilli and erupted matter that cover this city, [and whose volcanick origin cannot be disputed] to the soil of many parts of the neighbourhood of Naples. [see P. XL.]' Hamilton's cross-reference to the previous plate, which depicts a road-cut roughly 30km to the west, deploys the plate-book form to bolster his case. The main difference between the two sites is that the ancient city in this case makes it possible to date a fixed point within the stratigraphic column: 'Under the foundations of these Buildings are other strata of lava... intermixed with strata of vegetable earth, proving beyond any doubt diverse eruptions of Vesuvius, long before that which destroy'd the City, and which is the first recorded by History'. Confirming this relationship between buildings and strata, Hamilton provides a key to the objects numbered on the plate, including the modern soil (4) and a series of lavas (3) as well as the two main structures (1, 2). This known case of volcanic stratigraphy provides crucial support for his thesis concerning the volcanic origins of the whole 'circuit', and yet the site is hardly typical and does not yield readily to Hamilton's suggestion that its importance is primarily geo-historical.

Though strategically underemphasized in the commentary, the emergence of classical antiquity in material form provides the visual dynamic of Fabris's composition and enables Hamilton's geo-historical conjecture. D. T. Moore suggests that the excavations were what prompted his interest in volcanoes to

[24] Beaglehole (ed.), Endeavour *Journal of Joseph Banks* I.157.

begin with, and notes that his geological achievements are closely tied to these sites. Hamilton, Moore points out, observed that the ancient cities were initially covered by pumice rather than lava and that 'what a present-day geologist would recognize as palaeosoils are widespread'.[25] Hamilton's earlier *Remarks*, which includes the anecdote of the statue with its impression in the tufa, is explicitly addressed to 'the curious in antiquity' (4), but this same essay is also his most geological to date, focusing on stratigraphy rather than specific modern eruptions. Natural history and antiquity converge here because we have a 'very accurate' classical source, Pliny the Younger, for the firm date (79 CE) that marks the division between the strata overlying and those underlying the ancient towns (*Remarks* 3). Although the foundation of Pompeii is more secure than anything in d'Hancarville, Hamilton uses essentially the same antiquarian strategy as d'Hancarville to open a vista on prehistory, a prehistory of very uncertain extent.

Neoclassicism thus produces a pre-classical past, the human dimension of which appears in the living scene as depicted by Fabris. There is a difference between the original builders and the labourers reconstructing the temple through excavation, but this difference vanishes in the face of the volcano's continuing action. In the same way, volcanic action conjures up the prospect of even earlier populations. The genteel spectators, including Hamilton and Fabris themselves to the right of the column (Fabris is seated, sketching), contemplate these human strata through the present. The details of Neapolitan labouring-class costume and tools belong to the repertoire that Fabris established in his *Collection of Various Costumes and Trades* (1773), which he dedicated to Hamilton. This excavation plate includes a larger number of figures than any of the others, and the figures are larger relative to the landscape as well, highlighting Fabris's ethnographic interest.

The plate and text together promote the neoclassical convention of comparing ancient and modern, but they also create a new context for such reflection by introducing material concerns with the landscape, the state and preservation of the classical remains, and the economies of publishing and tourism. On a smaller scale, the specimen trade similarly mingles the archaeological and geological pasts, as when the fragment of a wall from Herculaneum appears in a collection of volcanic specimens because of a mineral concretion attached to it: 'There can not be a more satisfactory proof of the volcanick origin of this kind of stone', Hamilton notes in this case (*CP* XLV). The court of Ferdinand IV was notorious for restricting access to the archaeological sites and even to the museum at Portici where most of the finds were kept. Even to the elite readership of Hamilton's volume it would have been news that 'this temple and many other edifices of the ancient city of Pompeii, have been

[25] Moore, 'Sir William Hamilton's Volcanology' 171, 182.

already cleared by order of His Sicilian Majesty, and the beautifull Monuments of Antiquity that have been found there, will appear in the sequel of the truly magnificent and Royal publication of the antiquities of Herculaneum' (XLI). In his first letter Hamilton observed that part of the museum's collection had to be recovered from a lava flow generated by the 1767 eruption (19–20). In this context the 'Royal publication', *Le Antichità di Ercolano Esposte* (a plate book like Hamilton's own), may well appear more stable than the museum as a technology for preservation, though it too was subject to strictly limited circulation.[26] Tacitly at least, Hamilton recognizes the agency of Vesuvius in creating the most stable preservation environment of all, and Fabris celebrates the freshness of these remains by using vivid colours for the image, especially on the pediment of the lower building. This plate provides good evidence that antiquarian subjects are not inimical to geological explanation, encapsulating the method of the work as well as the general features that make this scene a 'microcosm' of the history of archaeology. In Alain Schnapp's words: 'Since ancient times Naples had been the terrain on which human history encountered natural history, and the excavations of Herculaneum and Pompeii represented the fruit of such interaction.'[27]

Hamilton's argument for the volcanic origin of the region depends on the unrecognized antiquity of the volcanic landforms as well as on learned support and his own authority as a connoisseur and naturalist. The lava underlying the foundations of Pompeii provides one dramatic and datable example. The uncertainty of this territory motivates his frequent repetition of these claims as well as many of the formal innovations intended to render *Campi Phlegraei* more accurate and authoritative in its representations of nature. In his prefatory letter, Hamilton begins with the ubiquity of volcanism, adding that not only the 'circuit' around Naples but also 'Pausilipo, whose charming scenes inspired Virgil's muse' (4), and even Rome itself owe their 'existence to Volcanick explosions'. Hamilton proceeds with the authentication of these claims: 'I speak with greater confidence, since I have had the pleasure of accompanying Monsieur de Saussure, Professor of natural History at Geneva, and shewing him every one of my observations, on the spots themselves, when that Learned Philosopher . . . agreed perfectly with me in opinion' (4–5). Hamilton had not had the benefit of such a formidable witness in his previous letters, and he quotes Saussure's letters to him at length, focusing particularly on observations that confirm his own views on the volcanic origin of the Roman campagna (8–10). Saussure notes telling details of structure that suggest a much earlier date for the volcanic topography than for the city itself. He adds, however, that 'these observations fix no precise date, and give us only vague ideas of a most remote antiquity' (10). It is worth noting that, in

[26] See Shell, 'Publishing Pompeii'.

[27] Schnapp, 'Introduction: Neapolitan Effervescence' 162.

translating Saussure's French, Hamilton recycles his own phrase, 'most remote antiquity' (cp. *Remarks* 13). In Wood's view, Hamilton is downright 'polemical' in asserting his authority in this prefatory letter, and he does not merely rest his claims on Saussure's authority. He points out that his own observations have been made in the course of more than ten years and further underscores his claim when he charges a competitor, Johann Jakob Ferber, with plagiarism (*CP* XLIII). Hamilton claims credit as well for employing Fabris, though he seems to recognize that he owes the new authority derived from this visual documentation primarily to the painter's eye.

The new visual and verbal resources of *Campi Phlegraei* enhance the reader's ability, already invoked in earlier versions, to infer 'the great changes our globe suffers, and the probability of its great antiquity' from a 'little stone'.[28] This formulation provides a clue to the importance of the thirteen plates presenting specimens. If we accept that these specimens typify rocks occurring over a large region, then this geographical scope in turn supports the longer geological time scale. Along with range, depth plays a significant role in the analysis. As early as 1770 Hamilton observed lavas more than 200 feet thick (*Remarks* 36), and in the new plates and explanations he adduces more evidence of this kind. Canon Giuseppe Recupero, Hamilton's guide on Etna, used the same source of evidence to advance more specific claims concerning the age of the earth. Hamilton does not include this detail in his own account of Recupero, but another British traveller made a point of Recupero's radical views on the age of the earth and provoked great controversy, as Dennis Dean has shown. Wood argues that Hamilton recognized his own views as 'radical', deliberately using 'a common language of curiosity' to make them palatable to the Royal Society and other readers.[29] But Hamilton's increasing conviction in his series of volcanic publications suggests that he was aware of the power of his own reputation to gain acceptance for novel ideas. In a later essay on Calabria Hamilton sounds even more like the controversial James Hutton: 'Nature is ever active; but her actions are, in general, carried on so very slowly, as scarcely to be perceived by mortal eye, or recorded in the very short space of what we call history, let it be it ever so ancient.' The exhaustive search for evidence of volcanism that structures the narrative of *Campi Phlegraei*, then, also sets the agenda for future research, as indicated by this preoccupation with origins even amidst the wreckage left by the Calabrian earthquake.[30]

[28] Letter VI, not included in *Campi Phlegraei*; see Hamilton, *Observations on Mount Vesuvius* 178–9.

[29] Dean, *Romantic Landscapes* 139–43; Wood, 'Making and Circulating' 80; cp. her description of his 'polemical' letter (89).

[30] Hamilton, 'Of the Earthquakes which Happened in Italy' 207. 'My object', he explains, 'was not to visit ruins, but the greater phenomena produced by the [1783] earthquakes' (179). These greater phenomena turn out to include (he hypothesizes) the birth of a new volcanic island

2.4 PIETRO FABRIS AND THE NEAPOLITAN CONTEXT

These broad claims about natural forces may be correlated with the increasing empirical rigour of Hamilton's accounts, which derives not only from his own data and Fabris's pictures, but also from the large body of evidence gathered by Italian naturalists. In his letters Hamilton referred to Italian eyewitness accounts of eruptions dating back as far as 1538, and the explanations written for *Campi Phlegraei* add many new references to both antiquaries and naturalists, with particular emphasis on new and forthcoming works by Neapolitan scholars such as Nicola Andria, Antonio Minasi, and Domenico Cirillo. Though at times condescending to the Italians, Hamilton takes pains to acknowledge his debt to their research, a debt ignored by much of the Anglophone scholarship on Hamilton. The visual richness and apparent originality of *Campi Phlegraei* account for its great influence in the historical literature on volcanology, but its form and content both derive from an Italian tradition that scholars are now engaged in recovering. Maria Toscano has put this relationship in perspective by describing *Campi Phlegraei* as 'a work which is to be properly viewed as the culmination, not the point of departure, for this type of scholarship in Italy'.[31]

Hamilton depended on Fabris's Neapolitan expertise. However, perhaps out of concern for his reputation as a patron of British art, Hamilton initially dwells on the artist's Anglo-Italian identity, introducing him as 'a most ingenious and able artist, a native of Great Britain, [employed] to take Drawings of every interesting spot, described in my letters, in which each stratum is represented in its proper colours... likewise the different specimens of Volcanick matter, such as lava's, tufa's, pumice stones, ashes, sulphurs, salts &c., of which the whole country I have described, is evidently composed' (*CP* 5). Referring presumably to the engraver and/or the colourists who coloured the plates, Hamilton adds that these do 'honour' as well to 'the able artists of this country', i.e. the Kingdom of Naples (6). Fabris himself was unusually mobile, working in both London and Naples in the 1750s and 1760s before settling in Naples in the 1770s. Though he may have been born in England to immigrant parents (*V & V* 167–68), Hamilton's debt to him derives from the artist's intimate knowledge of the Neapolitan landscape. It is of a piece with Hamilton's debt to Italian fieldwork.

The collaboration with Fabris gave Hamilton wider scope in *Campi Phlegraei* to engage with natural history as well as antiquities, supplying a rich

(206–7). In *Campi Phlegraei* the emphasis on volcanic origin, though ubiquitous, is especially strong in the commentary on Plates III, VIII, XI, XV, XVIII, XXII, XXVIII, XL, XLI, XLIV, XLV.

[31] Toscano, 'Figure of the Naturalist-Antiquary' 228. Italian naturalists and/or antiquaries cited in the explanations added for *Campi Phlegraei* also include Andria, Gaetano de Bottis, Padre Paoli, Scipione Mazzella, and Ignazio Sorrentino.

visual context for the history of the wider landscape. Invoking the Royal Society's motto, *Nullius in Verba*, Hamilton invites his audience to 'examine with attention', assuring distant readers that they 'may safely trust to the representations' of these sites as confirming the truth of his 'assertions' (*CP* 12). He assures us further that Fabris has 'compleated this collection under my eye, and by my direction, with the utmost fidelity' (5). Wood comments that 'each image . . . embodied the skilled labour of artists, engravers, and colourists and of Hamilton's management of these workers'. If *Campi Phlegraei* constitutes 'a more complete witnessing experience', then the independent expertise of these workers deserves equal attention. The creative direction provided by Fabris's eye is most obvious in the case of the images that Hamilton included even though they predated his acquaintance with the artist: Plates V (1754), X (1756), XII (1760), and possibly others. The last of these inspired a pair of 'before and after' views, one recording the eruption of 1760 in process and another entitled 'View of the present state of the little mountains raised by explosion in the year 1760'.[32] Hamilton provides a cross-reference to encourage the comparison, and many other added cross-references correlate particular strata with those appearing in another view, or with particular specimens from the last group of plates. The frequent deployment of 'sections', a recent addition to the emerging visual language of geology, also testifies to Fabris's influence.[33]

Though recognizing Fabris's importance from the start, Hamilton shows some of the ambivalence associated with collaborations that crossed the bounds of social class. His remarks in the text and in private letters confer on Fabris the ambiguous status of the knowledge worker—he is by turns a crucial source of accuracy and authority, an object of patronage, and a commodity. His first mentions the project in a letter to his nephew Charles Greville, one of his most frequent correspondents, who genuinely shared his geological interests. Writing about a year before the publication, Hamilton promised that by the time of Greville's next visit, the 'new edition

[32] Wood, 'Making and Circulating' 90; *CP* XIII. This pair of before-and-after plates (XII–XIII) typifies a significant formal feature of the work. Plate XII was re-engraved for a didactic volume by J. F. Bertuch, possibly from an earlier version by Fabris but more likely from *Campi Phlegraei*, which was disseminated widely in various recycled forms. See *Bertuch's Bilderbuch für Kinder* VI, Plate 52, Fig. 1. Wood (p. 86) richly documents the wide dissemination of Hamilton's texts via extracts and unacknowledged quotations. *CP* V, X, and XII carry pre-1764 dates, but many of the plates are undated and some of these probably also predate Fabris's acquaintance with Hamilton.

[33] The term is used in its geometrical sense to describe cross-sections of the cones of active and extinct volcanoes (XV, XIX). It is also used in the explanations to classify views of roadcuts and ravines as 'sections' of the strata (XVI), and Hamilton even presents a section in tabular form to supply updated information (from a new well) for one of his letters (*CP* 64n.). On the emergence of geological sections, see Rudwick, 'Emergence of a Visual Language for Geological Science'.

of my volcanick letters, with about forty colour'd prints by Fabris, & giving the very clearest idea of every stratum in this country of all the craters will be finished, & I will with great pleasure accompany you with the book to the spots themselves. I wish every book of natural history was executed with such fidelity'.[34] Two recurring themes emerge in subsequent letters: rising costs ('Fabris's demands keep me very short') and a growing number of plates, reported over the next six months as fifty and even sixty. Revisiting some of the themes of his correspondence with d'Hancarville, and applying what he had learned about the management of knowledge work, Hamilton represents the process to Greville as a kind of addiction: 'You cannot conceive how interesting the work is grown by the additional drawings, all of which have been taken on the spot'. He reports taking Greville's sound advice to 'secure the original drawings', though *Campi Phlegraei* never was a profit-making enterprise like the *Etruscan, Greek, and Roman Antiquities*. The context in which Hamilton introduces the project, however, reinforces the sense of Fabris as a commodity, because he compares Fabris to one Moreni, a Veronese 'fossil man' who is 'ready to sell himself and his collection'.[35] In print, Hamilton characterized his patronage as both philanthropic and philosophical: he chose to publish, he says, so 'that the ingenious artist himself, might at the same time reap a moderate and constant benefit from his labours, particularly as he is unfortunately in a declining state of health' (*CP* 6).

Fabris shows himself at work in six of the plates. The written record of his life is even scantier than the biographical record of Parkinson, but these images prove at least that he was robust enough in health to make numerous climbs and excursions with and without Hamilton. Three plates show him working by himself; the other three, including the Temple of Isis, show him working 'by [Hamilton's] direction' as promised. This image of Fabris sketching a 'hollow way ... on the flancks of Mount Vesuvius' (Plate 2) stands out because he places himself in the foreground. Sam Smiles's rubric of 'visual documentation' seems especially apt for this deliberately unspectacular image, which shows neither an eruption, nor spectators, nor a sweeping panorama (unlike the other, more spectacular plates on which Fabris chose to show himself working). The artist concentrates intently on the strata, which are carefully delineated by colour, a form of attention that Hamilton stresses in his prefatory letter. Although defined by its scale as a landscape, the image also presents us with a specimen in the form of the boulder on which the artist is seated. As often the strata are numbered and a key provided, though Hamilton

[34] *H & N* I.38, and cp. 91–93, 107, for Greville's mineralogical interests.

[35] *H & N* I.41, 43, 44, 17. The first reference to Moreni reads, in full: 'he is ready to sell himself and his collection, which is admirable, & the British Museum would do well to take him into their service to arrange their Natural History' (17). Moreni is mentioned in conjunction with Fabris in the letter of 6 June 1775 (I.38).

stops short of using numbers to correlate strata on the opposing side of the gorge.[36] It seems plausible to attribute to Fabris the choice of a perch that clearly shows the stratification of the near side, corresponding to the compacted soil overlying one of the younger lavas on the upper left. The nearer view of this 'specimen' shows the similarity in structure between these dark strata and the most recent lava, which ran down the middle of this gorge in 1767. The gorge, called Fossa Grande, was 'cut by torrents of rain water' during an earlier quiescent period; Hamilton observes that 'it is probable that a very great space of time elapsed, between the running of the lavas 4. & 5., to have allowed of the accumulation of the thick stratum between them'. Referring to the specimen section, he also notes that one set of stones with gems and crystallizations (LIV) was found here after a heavy rain.

This self-portrait captures an important aspect of Hamilton's debt to Fabris. The painter shows his ability to focus on natural history fieldwork and provides an image of himself witnessing to support Hamilton's claim that readers 'may safely trust to [his] representations'. Other images in the work further represent Fabris's attention to the sequence and particularly the colours of the strata, such as a splendid view of Capo dell'Arco on the island of Ventotiene (XXXIV). In this view Hamilton singles out a 'Stratum of Tufa, under which is a stratum of Rapilli, mixed with dark Volcanick ashes, and under that is a Stratum of deep Orange colour'd Tufa, which curious Stratum runs over both Islands' (i.e. Ventotiene and Ischia). Fabris (and his colourist) carefully separated the *rapilli* and ashes into three distinctly coloured layers, the lowest of which is purple, complementing the burnt orange of the 'curious' tufa. In retrospect, it is reasonable to imagine Fabris placing his palette at the service of science, as Thomas Webster was to do more consciously forty years later in *Description of the Isle of Wight*, which imitates *Campi Phlegraei*.

More than Parkinson and other artist/naturalists such as Schnebbelie and Webster, Fabris creates a complex visual record of knowledge work in practice. The view of Ventotiene includes three fishing boats at work and a gaudy pleasure boat, with six liveried rowers, that has just discharged several tourist/naturalists. Here Fabris depicts two kinds of labour and seems to conflate natural history fieldwork with ostentatious tourism; elsewhere he includes both working and idle peasants, superstitious city-dwellers, and naturalists more clearly distinguished from casual tourists by their activity. The tiny figures on the far right in the view of Fossa Grande (Plate 2) are too indistinct to be correlated with any of these social strata. The next plate, depicting a very similar setting, substitutes two travelling farm labourers in

[36] The numbers are nearly impossible to make out in my reproduction. The original pages are 39 × 21.5 cm, with relatively narrow borders around the prints. Strata 1–5 represent 'lava… mixed with Volcanic cinders and Rapilli', and 6 (at the bottom of the gorge) is lava from the 1767 eruption.

the position of the artist. Fabris uses these figures to create a narrative arc linking both scenes (XXXIX, XL) to the Temple of Isis (XLI). He strengthens this continuity by highlighting two strata of *rapilli* ('loose white pumice stones' mixed with ash and lava) that Hamilton flags as 'exactly similar' to those covering Pompeii. In the excavation scene we recognize Hamilton supervising Fabris because their positions echo those in the best-known version of this artist-patron motif, a sweeping panorama of the crater-shaped harbour on Nisida that shows their two figures in the foreground (XXII).

If Hamilton's natural history owes something to Fabris's careful study of the landscape, Fabris owed his commission in part to the Italian convention of illustrated scholarly works on antiquities and natural history, as Toscano has shown. I would argue that Fabris used this convention to his advantage, engraving several plates himself, as indicated by his decorative signature, and colouring some copies himself in gouache to create a 'de luxe' edition.[37] Working in the tradition that Toscano describes, Fabris would have been well aware that these 'naturalist-antiquary' images use visual conventions to credential the artists as not only recorders but reliable witnesses in their own right. His exposure to these works may have influenced his witnessing presence in individual paintings done before the book project was conceived, such as the famous night view of an eruption attended by Hamilton with the King and Queen in 1771 (XXXVIII). Hamilton would have recognized the relationship between these conventions and the expectations of his Royal Society audience, and might have seen (perhaps with encouragement from Fabris) an opportunity to expand his readership. Toscano describes artists such as Fabris and the 'even more exemplary' Willem Fortuyn, employed by the naturalist-antiquary Antonio Minasi, as professionals: 'The scientific method of the naturalist-antiquaries was based on the chemical analysis of rocks (always collected in person) and on drawings made on the spot (often the work of professionals).' Toscano adds that from about 1735 Neapolitan intellectuals such as Minasi visited a given site 'with the aim of reconstructing its vicissitudes (whether human or geological), reaching as far back in time as possible', a preoccupation that is also central to *Campi Phlegraei*. She notes the naturalist-antiquaries' explicit preference for gouache or pen and ink over the less immediate genre of easel painting and relates this preference to the convention of depicting the investigators at work: 'This custom originated largely to give the viewer an idea of the proportions of the scene, but it came to testify, with pride, that the artist had worked from real life and that the

[37] Plates XVII, XXV, XXXVI, and XL carry this signature, which according to Jenkins and Sloan is demonstrably etched into the plate (*V & V* 167). They also note that a comparison reveals the involvement of colourists whose skill levels varied, and point out that the 'deluxe' editions have gouache and egg white applied in addition to or instead of watercolor. Many of the original gouaches are reproduced by Knight in *Hamilton a Napoli*, a splendid book in its own right.

naturalist had actually supervised the artist in his work.'[38] It is all the more remarkable in this context that Fabris depicts himself as unsupervised in three of the images.

Though biographical materials are scarce, *Campi Phlegraei* itself allows us to trace Fabris's collaboration with Hamilton in surprising detail. Many of the plates are undated, including a depiction of Fabris sketching inside the caldera of Vesuvius, among perambulating tourists—possibly including Hamilton—during the eruption of 1767 (IX).[39] The earliest firm date comes from the text, in which Hamilton mentions that Fabris made the journey to Sicily with him in 1768, producing one image included in the volume (XXXVII). No other images from this tour survive, but Hamilton began collecting landscapes by Fabris not long afterward. According to one account, he 'clearly saw this artist as the one who could best fulfill his requirements for images of the Campi Phlegraei' (*V & V* 168). The book may have been a joint project from the start. Certainly Fabris was moving confidently in the homes of visiting British nobility by 1771, as attested by an elegant pair of oil paintings of *Lord Fortrose at Home in Naples*. One of these includes Hamilton playing the violin and a detailed representation of the artist with his face turned toward the viewer.

Jenkins and Sloan, who reproduce these paintings (*V & V* 128–29), also give the most precise account of the little that is known about Fabris's life. They give his dates as 'fl. 1756–84' and speculate that he might have been born in London to French Huguenot parents. They also note that he exhibited in London in the 1760s and record possible evidence for his collaboration with Paul Sandby, whose 'bodycolour drawings from the 1760s are the closest English works' to those of Fabris (167). Jenkins and Sloan argue that the use of gouache or bodycolour in landscape was an English innovation that Fabris brought to Naples and popularized there (168). It is hard to be sure, as Fabris was forgotten so thoroughly that it took a concerted research effort by Carlo Knight to secure the attribution of *Campi Phlegraei* to him in 1984; traditionally it had been attributed to Philip Hackert.[40]

[38] Toscano, 'Figure of the Naturalist-Antiquary' 227, 225, 227. Toscano argues that the naturalist-antiquaries 'based their research on Bacon's inductive method, but also on Newton's experimentalism and the technique of digging down into the archives of the world, *les archives du monde*, advocated by Buffon' (227), which would help to explain their affinities with the British Royal Society. But she also emphasizes the independent Neapolitan tradition (from circa 1750) of illustrated studies of Vesuvius in this vein, stating that the 'naturalist-antiquary mentality developed in this southern Kingdom relatively early, and not out of passive acceptance of the English model' (228).

[39] One of the spectators wears the unusually long red coat that is Hamilton's signature article of dress in many other images. The red coat, which becomes more firmly associated with Hamilton in plates dating from the 1770s, may have been associated with his regalia as Knight of the Bath. Cp. the portrait by David Allan (*V & V* 107).

[40] Knight, 'Sir William Hamilton's "Campi Phlegraei"' and *Hamilton a Napoli* 145–51. Jenkins and Sloan's dates may be amended slightly, since Hamilton indicates that one of the paintings dates from 1754 (*CP* V). They also note that Fabris exhibited at the Society of Arts and

Fabris's earlier *Collection of Various Costumes and Trades of the Kingdom of Naples* (1773) helps to contextualize the human elements of *Campi Phlegraei.* The first plate shows *commedia dell'arte* figures gawking at a monumental inscription (in Italian) that records a dedication to Hamilton, the only published text by Fabris himself. The last lines tell us that the collection was drawn and etched by Hamilton's 'humble servant P. Fabris, Englishman [*Inglese*]'. The colourful customs and manners of the common people were equally prominent in *vedute* with Vesuvius and antiquities, contributing a picturesque element. Fabris's volume makes it clear that he studied these as carefully as volcanic phenomena, and like his gouaches for Hamilton these etchings privilege objectivity and accuracy as much as romantic effect.[41] These etchings, together with his numerous watercolours and engravings on this theme, depict traditional male and female costume, often elaborate, from numerous small towns and several islands in the Bay of Naples. They show popular customs, including the tarantella (a dance that Hamilton was known for having mastered), the consumption of macaroni (by hand), and tradesmen with their equipment, especially street vendors. Franco Mancini notes the early modern genre of 'street cries' as an important precursor for this artistic interest in popular customs and manners. There is no letterpress in the volume, but Fabris's ethnographic attitude comes across in some of his longer titles, one of which also sheds light on a widely reproduced plate from the *Supplement to Campi Phlegraei.* In 'Devotions of the Ordinary Neapolitan People to St. Anthony', a barefoot woman peers into a box containing an image of the saint offered to her by a sacred type of hawker with a bell. One of the terrified spectators of the 1779 eruption, as depicted by Fabris in the *Supplement* (II), holds up a similar sacred image in supplication, and another plate in the volume also features popular devotion—a subject of perennial interest to foreign tourists.[42]

the Free Society. The artist looks about fifty in the 1771 painting of Lord Fortrose's concert. Moore gives a brief account of the little that is known of Fabris's later life ('Sir William Hamilton's Volcanology' 184).

[41] In his reprint of this volume, Mancini groups it with a third, ethnographic strand emerging slightly later than the realistic depiction of natural history and antiquities that flourished in the cosmopolitan art scene of late eighteenth-century Naples. In his introduction Mancini notes that this 'growing sea of etchings and engravings dedicated to customs and dress', avoiding 'social satire', strictly 'limited itself to describing with microscopic accuracy the infinite facets of quotidian life' (Fabris, *Raccolta di varii vestimenti* 9). For the picturesque use of these elements in large paintings, see the examples in *V & V* 136–7, 130–1.

[42] For specimens of this interest see the treatment of St Januarius in Hamilton's first letter to the Royal Society (*Observations* 11) and Emma Hamilton's letter of 4 August 1787 (*H & N* I.133). It is notable that two of the four full-size plates in the 1779 *Supplement to Campi Phlegraei* (Plates II–III in the Berlin copy) explore this theme of popular devotion, which is not addressed in *Campi Phlegraei* proper.

Though an independent strand of artistic practice, the interest in popular customs and manners is linked in important ways to the study of natural history and antiquities, and this configuration plays a major role in d'Hancarville's volumes as well. Some of the details of costume in *Campi Phlegraei* contribute to the overall performance of visual documentation. The labourers engaged in excavation at Pompeii (XLI) wear white knitted caps identical to those of a pair of Neapolitan vintners presented in an ethnographic study by Fabris, indicating his close attention to every aspect of the scene. A similar cap on the third spectator, in an image that prominently features Hamilton instructing Fabris (XXII), serves to distinguish him socially from both the artist and patron in their three-cornered hats. In one of the sections of strata in a gorge, the close attention to the burden of one of the travelling farm labourers—he carries his basket on a stick, like a peasant featured in the costume volume, along with a curious wooden frame on his back and another basket, all filled with goods for the market—mirrors the artist's close attention to the strata.[43] One form of attention reinforces the other. Smiles has written that 'accurate observation' in the art of this period 'was not orientated entirely to landscape and natural phenomena but also toward the social and material world'. Expressed as visual documentation, this form of attention, Smiles argues, produced 'data-rich' images of phenomena 'below the threshold of traditional scrutiny', including neglected aspects of antiquities and natural history as well as 'labouring-class occupations'.[44]

By depicting quarrying and other rural occupations in *Campi Phlegraei*, Fabris links the ancient and modern populations through their use of natural resources. Hamilton repeatedly notes the use of volcanic hot springs by 'the common people of Naples' (XXI). Fabris's plate in this case includes a naked bather and Hamilton reports that he has witnessed the 'learned physician and... excellent naturalist' Cirillo's experiments on this water. In a gloss on one of the specimen plates, Hamilton describes a 'piece of lava incrusted with pure white salts, and such as the Peasants, who inhabit the neighbourhood of Mount Vesuvius, make use of at their tables' (XLVI). These examples and the quarry carefully depicted by Fabris (VIII), along with the tourist trade on the mountain, provide examples of economic geology—the use of resources that links the ancient and modern populations. After the return of Cook and Banks on the *Endeavour*, Hamilton was able to add a note to his 1770 letter citing the artefacts they collected for geological evidence: 'all the implements of stone brought by Mess. Banks and Solander from the new-discovered islands in the South-Seas, are evidently of such a nature as are only produced by Volcanos'

[43] For the basket hung over the shoulder, see Fabris, *Raccolta di varii vestimenti* 16; the white caps are featured in a separate engraving titled *Il vinaio* (*Raccolta* [Introduction] 8).

[44] *Eye Witness* 10, 7, 10. Smiles devotes one chapter of this study to costume books like Fabris's (77–110), though he focuses on Great Britain.

(84n.). In an analogous case described by Toscano, the naturalist-antiquary Giuseppe Giovene used the Polynesian stone tools to elucidate a much more remote era of human antiquity when he compared them to Neolithic jadeite axes uncovered in a saltpeter mine in the eastern part of the kingdom.[45] In all these cases, natural history and antiquity converge in ethnography.

The interest in Neapolitan customs may be traced to the revolutionary experience of the excavations there, which uncovered the quotidian realities of an ancient civilization known for so long only by its monuments. As the comparisons to South Seas artefacts attest, the conflation of spatial with temporal distance was an established trope in the literature of exploration, but on the classic ground of Naples it was purely a question of the lapse of time. At the Temple of Isis, the distance between ancients and moderns vanishes in the face of ongoing geological activity, and this recognition strengthened the supposition that present-day Neapolitans were connected by tradition not only to Rome but to a more remote antiquity. Hamilton's particular attention to the shop fronts and the volcanic paving stones of Pompeii in his essay for *Archaeologia* (also illustrated by Fabris), should be seen in this light. His pioneering actualism in natural history—the geological method (as it later came to be called) of making inferences about the past from present processes—owes something to the scrutiny of Neapolitan customs in light of archaeological discovery, exemplified in the costume book by Fabris and his continued attention to the issue in *Campi Phlegraei.*

This ethnographic actualism informs geological actualism most visibly in a study of the quarrying operation (Plate 3) at Resina (now Ercolano), which prompts further reflections by Hamilton on volcanic paving stones—this time as used in modern Naples. In this plate, the quarrymen are at work on a comparatively modern lava, numbered as usual on the plate: '1) Representation of a thick Stratum of Lava, that ran into the sea from Mount Vesuvius in the terrible eruption of 1631. It is at present the principal Quarry from whence are taken the Stones for the Pavement of Naples'. Fabris accentuates the division of labour, displaying the several stages of cutting, finishing, and transfer (via the cart shown in the centre) into boats bound for Naples. Hamilton points out that because the lava ran into the sea, quarrying may now be done only fifty paces from the shore. The geographic context is clear from the previous plate, which shows a bit of the quarry in the foreground but

[45] 'When I saw the hatchets belonging to the Tahiti islanders in the museum of Signor Poli in Naples', Giovene remarked, 'I was surprised to find that they resembled exactly those from Pulo at Molfetta' (quoted in Toscano, 'Figure of the Naturalist-Antiquary' 231–2). The latter axes were identified as Neolithic much later. In his note on Pacific implements, Hamilton states that 'these gentlemen [Banks and Solander] have assured me, that no other kind of stone is to be met with in the islands' and so concludes, 'I am induced to think, that these islands (at so great a distance from any continent) may have likewise been pushed up from the bottom of the sea by like explosions' (84n.).

also includes the arc of the bay, two coastal villas, and Vesuvius in the background, with travellers carefully marked by their costume as belonging to different classes. Although the figures are rendered smaller by scale in the quarry scene, Fabris makes the most of regional costume to set off the figures against the stark colours of the strata. The tiny strollers at the top of the column add a contrasting hint of leisure and figure the lapse of time in relation to the depth of the strata beneath them.

The production of paving stone, the production of ceramic clay (XXXII) and even 'the small works of Moles in a field' all provide materials for 'comparison' with 'the great operations of volcanick fires' (XXVIII). Fabris contributed much of the detailed attention to contemporary Neapolitan society that enlivens Hamilton's project of comparing the small with the great and the present with the past. Fabris also contributed the knowledge of landforms that he gained as an artist-naturalist beginning at least ten years before Hamilton's arrival in Naples.

2.5 CLASSICAL SITES AND MODERN MEDIA

Even as they document the connections between economic and historical geology, Hamilton's explanations of the plates also draw deeply from the mine of classical literature, locating the Elysian Fields, Lake Avernus, the Sibyl's Cave, the tombs of Virgil and Scipio, the Gates of Cumae, and more, in Fabris's views.[46] These allusions recast the volcanic 'circuit' as Grand Tour material. In relation to the more philosophical versions of his letters printed in the *Philosophical Transactions*, it seems fair to say that Hamilton uses the explanations of the plates to recolonize the landscape with classical associations. There are important differences here from the classical allusions in narratives of the *Endeavour* voyage (see Chapter 1). The material site itself is classic ground, linking modern Neapolitans in some sense genetically to ancient Greeks, though culturally they may be almost as removed as modern Tahitians are. It is therefore less clear how or whether illustrated works on Naples are contaminated by neoclassical presuppositions. Parkinson was praised because his human figures were less marred by neoclassical anatomical conventions than the work of others, and his writing eschewed the classical allusions so present in Banks and Hawkesworth. We have no text from Fabris, but his images necessarily incorporate a classical repertoire due to his location as well as the nature of the market. Moreover, the presence of academic conventions in landscape, considered an inferior genre in any case, does not

[46] See, respectively, Plates XVII, XXVIII, XXIX, XVI, and XXVIII (for both Scipio and Cumae).

obviously compromise his ability to document volcanic activity. Finally, although attention to Neapolitan customs was certainly stimulated by the discoveries of Herculaneum and Pompeii, the study of ancient customs itself took an anthropological turn as well, showing a reciprocal influence between ancient and modern ethnography.[47] As familiar as they are, then, ancient sites take on a new look in the pages of *Campi Phlegraei.*

Twelve of the forty-one views in the volume are sweeping panoramas that necessarily include numerous ancient sites, ranging from Roman ruins to natural features incorporated in ancient myth. In Plate 4, Fabris takes one of these sites as his point of view: 'View taken from the spot N. 1 near Puzzoli, on which anciently was Cicero's Villa call'd the Accademia. There are still some remains of the famous Portico' (XXVI). Typically for these views, both cultural and natural objects are identified by numbers in the image: these include, from left to right, the Cape of Misenum, Bauli, the Castle of Baia, Baia, Nero's Baths, Monte Nuovo, Monte Gauro, and in the left foreground, the Temple of Serapis, with its three columns, at the edge of Pozzuoli. As in the case of Pompeii (XLI), Hamilton frames the motivation of the image in terms of natural history rather than antiquity: 'the intention of this Plate is to shew . . . the great change that has been made in this country, in the short space of 48 hours'. He refers here to Monte Nuovo, formed of matter that erupted from level ground in a very short time, most of it in forty-eight hours, in 1538; the flanks are still bare in this image. Nonetheless, most of his verbal attention is given to the 'curious remains of antiquity', including additional facts about the site and a quasi-promotional reference to recent publications on the area by Padre Paoli and the engraver Filipo Morghen. Monte Nuovo quite literally erupted onto this storied scene, altering the coastline and jutting out against the curve of the bay and the older, forested hills, frowning across the water at the orderly ancient-modern town.

The 'great change' suddenly wrought on this scene by a much older and ongoing natural process reveals the painterly atmosphere of classical permanence as an illusion. The two gazing figures certainly mirror this tension between opposing ways of viewing the scene. The rustic lounging figure is horizontal, impassive, perhaps caught up in the illusion or even furthering a programme of class distinction between enlightenment and superstitious ignorance, which Wood points out in certain other plates. The aristocratic spectator stands erect, commanding, and above all vigilant, alert to the 'great change' that a casual viewer of the landscape—and even more so the print—would fail to see. Again,

[47] Smiles points out such a reciprocal influence between costume books and antiquarian studies: 'What distinguished costume studies from fashion plates was their coverage of other times, other classes and other cultures, which, in turn, encouraged the development of more recondite studies of remoter periods and little-known peoples' (*Eye Witness* 80). Smiles provides ample evidence from the period of a perception that modern improvements in illustration were greatly advancing the knowledge both of antiquity and of 'foreign and remote nations' (*Eye Witness* 9).

the distance between ancients and moderns is diminished, as Hamilton, the 'modern Pliny', stations himself opposite Cape Misenum, 'from whence Pliny discover'd the eruption of Mount Vesuvius, which proved fatal to him' in 79 CE (XVII). Monte Nuovo renewed this fatal activity, which had seemed at a safe distance from modernity, and it continues in the Baths of Nero, still hot enough to boil an egg (XXVI). These continuities reassert the classical identity of the landscape even as it is undermined by Hamilton's naturalistic temporality. Classical ruins such as the Temple of Serapis, the Temple of Isis, and ruins mentioned elsewhere (XXIX, XXXIV, XXXVI), provide an indispensable context for the new volcanology.[48] Classical place names and their attendant myths—such as Forum Volcani, Lake Avernus, and 'the Sybil's Grotto as described by Virgil' (XXIX)—also serve Hamilton's thesis of volcanic origin well.

Classical texts are an even more constant presence than ruins in *Campi Phlegraei*. Although Hamilton himself was diffident about his classical scholarship, his not infrequent references and allusions, added to those already present in his early modern sources, signal the humanistic legacy informing his study of volcanoes. Although my main concern in this book is to examine cases in which natural history and antiquarianism uncovered non-classical forms of antiquity, both of these disciplines were themselves strands of the humanistic legacy that produced the European system of classical education. Much as Monte Nuovo represents a geologically recent 'great change' to the landscape, Hamilton's detailed discussion of the event illustrates a great change in scholarship since the mountain's appearance in 1538. His sources for this discussion, such as Marco Antonio delli Falconi, are still remembered by volcanologists as the first modern eyewitness accounts of volcanic eruption. In the *Remarks* Hamilton quotes Falconi and another period source at great length (in his own English translation), revealing at least two techniques that shaped his own Enlightened study of volcanoes. The first is ekphrastic response, taking the eruption as an aesthetic experience and comparing his sensations to those captured by ancient authors. Falconi, according to Hamilton's translation, felt as though the Titans were rising up again in revolt: 'methought I saw those torrents of burning smoke that Pindar describes in an eruption of Etna, . . . in imitation of which, as some say, Virgil wrote these lines: Ipse sed horrificis juxta tonat Aetna ruinis' (22–23). Aesthetic response is a common feature of all Hamilton's writings on volcanoes, though he also draws on the more recent materials available to him, particularly industrial metaphors and the natural sublime.[49] The second technique that he shares

[48] In the previous page of commentary (XXV), Hamilton notes that a thin layer of volcanic matter had adhered to the top of the Temple of Serapis during some past eruption, but point of view prevents us from seeing it in the plate (XXVI).

[49] Falconi quotes *Aeneid* III.571: 'Aetna, just beyond, rumbled and flashed/Formidable in eruption' (Virgil, *Aeneid* 86 [lines 760–61 in Fitzgerald's translation]). For an account of

with these earlier accounts is the analogy to humoral medicine. Among other examples, Hamilton's definition of lava as 'foul vitrified matter' may be compared with Falconi's idea that 'the eruption returned periodically, like fits of ague or gout' (25).

The rise of humanistic education in England also underlies the custom of the Grand Tour, of which *Campi Phlegraei* is a complex souvenir. Far from being a mere participant in that custom, Hamilton did much to promote the place of Naples on the itinerary, and his own Palazzo Sessa became a highly desirable station for Grand Tourists in Naples. By 1776 he had guided many distinguished visitors around his circuit, most of whom were interested in antiquities as well as volcanoes.[50] More than any other publication, *Campi Phlegraei* reflects and mimics his activity as a cicerone, reproducing the sites that he favoured and the guidance he supplied. Hamilton's hospitality provided a significant part of the foundation for his European reputation, and for some the book was literally the souvenir of a visit to Palazzo Sessa. Some of his guests were included among the necessarily small number of buyers for the book, along with those who hoped to pay him a visit and others for whom the book was an acceptable substitute.[51] Hamilton was himself a passionate collector of souvenirs, as vividly imagined by Susan Sontag in *The Volcano Lover,* and the structure of the book reflects this passion. The volumes are presented physically as part of a larger repertoire of souvenirs, ranging in price from mineral specimens and inexpensive views produced for the tourist trade up to Old Master paintings and ancient artefacts, including vases and sculpture. As both a real and virtual guide, Hamilton contributed to the tourist economy. He also depended on that economy as a collector and connoisseur, and *Campi Phlegraei* demonstrates some of the lessons that he learned from his own guides, including Fabris and Bartolomeo Pumo, whom he dubbed 'my cicerone of the mountain'. The prints by Fabris were, for many, the main attraction, serving as a collection superior to most collections that one might make on the spot in Naples. Hamilton's numbered keys and commentaries made the engravings souvenirs for those who had visited the sites.

Hamilton is mapping classic ground onto a new and larger physical geography of volcanic activity and insisting that this activity also requires new temporal co-ordinates 'out of the reach of history', beyond the classical memory with which this ground is charged. Yet his mapping depends on the tourist economy largely predicated on the prestige of cultural memory.

eighteenth-century aesthetic experience of volcanoes, see Heringman, 'Style of Natural Catastrophes'.

[50] For an overview of this aspect of Hamilton's career, see Carlo Knight, 'Art of Going through Life' (*V & V* 11–23).

[51] Walpole, for example, commented: 'I shall desire to be a subscriber to your Vesuvius; but I wish you had not exchanged your taste in painting and Antiquity for Phenomena' (*H & N* I.46–7).

The final section of the book offers souvenirs that are specimens of antiquity in Hamilton's characteristic double sense. The last thirteen plates show collections of specimens, many of them brightly coloured and represented in close-up *trompe-l'oeil* views that achieve an uncanny vividness comparable to photorealism. Groups of eight to ten specimens are viewed from above arranged on table tops or on shelves, as they would appear in a collector's cabinet. Structurally, this section concludes the work by offering us souvenirs to take home once we have completed the 'circuit' or itinerary of the preceding views. Each virtual 'little stone' records the memory of 'the great changes our globe suffers', but Hamilton acknowledges that virtual specimens are inadequate to this purpose. He therefore includes what amounts to an advertisement for 'complete collections of all the natural productions of Mount Vesuvius... now made and sold by Tomaso Valentiani at Naples opposite the King's Theatre' (XLIX).[52] The featured specimens include colourful minerals transformed by sulfuric gases at Solfaterra (XLIV, XLVI, LIII), seashells embedded in tuff (XLVII), 'curious stones' suitable for making snuff boxes (complete with a snuff box [XLVIII]), and 'large honey comb pumice, such as the Ancients made great use of in... their Buildings' (LI). Other specimens with strongly classical connotations include a piece identified as 'the true Lapis Obsidianus of the Ancients' (LII) and 'crystals commonly call'd Gems of Mount Vesuvius' (LIV).

Hamilton's commentary on the specimen plates provides unexpected glimpses of his close involvement with the tourist trade at all levels. Persistent scholarly suspicion of Hamilton's motives is anticipated, and partly explained, by Goethe's widely cited suggestion that he may not have obtained all his specimens by legitimate means.[53] Hamilton's explorations of antiquity were far from disinterested, yet he did promote the advancement of archaeological and geological knowledge, and it is a kind of historical fallacy to insist on the contradictions between the selfish motives of the collector and the scientific agenda of the naturalist-antiquary. The specimen of tufa attached to a fragment of painted wall, from the theatre of Herculaneum, shows in a small way how these motives can coincide, as Hamilton notes: 'As the labourers in the Quarrys are now attentive to preserve such bodies, as they meet with in the tufa, some further curious discoveries may be made hereafter' (XLV).

[52] Although he goes on to say that Valentiani includes more specimens than are strictly necessary (L), ironically these collections must have been considerably cheaper than Hamilton's virtual one, as the original market price for *Campi Phlegraei* is estimated at 12 guineas (Wood, 'Making and Circulating' 91). Three of the specimen plates are reproduced in Krafft, *Volcanoes* 3. Moore notes that although 122 specimens sent by Hamilton are preserved in the British Museum, none of them match those depicted on the plates (186). Sleep points to these specimen plates to illustrate some of Hamilton's lasting contributions to geology, especially to terminology and classification ('Sir William Hamilton' 333).

[53] *Italian Journey* 310–11.

Hamilton was in an unusually good position to capitalize on 'curious discoveries', enjoying a diplomatic latitude further extended by the Earl of Elgin when he acquired most of the sculpture from the Acropolis twenty-five years later. In this reference to the labourers, however, Hamilton shows a characteristic willingness to publicize avenues of discovery. At the same time, this reference is symptomatic of his willingness to cross class boundaries in the name of knowledge. There is nothing especially democratic in the Royal Society's long-established practice of delegating fact-gathering activities, though Hamilton pursued this policy more systematically than many inquirers did. In this context, however, the excavators in Fabris's view of the Temple of Isis deserve to be re-examined as potential knowledge workers in a restricted sense: these labourers are, after all, more likely than the spectators to have found specimens both volcanic and antiquarian.

2.6 ON THE MOUNTAIN: BARTOLOMEO PUMO AND ANTONIO PIAGGIO

The career of Bartolomeo Pumo, the 'Cyclops of Vesuvius', vividly illustrates the connections between the tourist trade and the conduct of knowledge work in eighteenth-century Naples. A self-educated guide who came (as far as we know) from the streets of Naples, Pumo carved out a subject position for himself in the consciousness of at least some aristocratic connoisseurs who were all too apt to objectify and exoticize the members of his class as a whole. He also suffered the contempt of the learned and even physical abuse from angry customers, though not (as far as the record shows) from Hamilton. Fabris, like Pumo, left no written record, though his many incorporated self-portraits visually record his aspirations to gentility, particularly by way of costume. He also left a visual record of Pumo, which has not, as far as I can gather, been noticed by modern scholarship. The plate, which depicts an unusually tall man in working-class dress instructing Hamilton (Plate 5), dramatically turns the tables on the usual scene of instruction and may help to explain Pumo's stage name as well. The Cyclopes were giants and, like other giants and mythological beings, literally a part of nature.

Pumo's epithet may therefore allude to the perception that the common people of Naples somehow shared the substance of the mountain, with their at times explosive and ungovernable activity. Along with Pumo, Piaggio, Emma Hamilton, and Wilhelm Tischbein, the Neapolitan crowd is a major actor on the increasingly volatile scene of Hamilton's later career. Hamilton meditated on the 'irregularities of [the] lower class' and eventually fled Naples along with the court in 1799. Piaggio had a good deal to say about the 'rustic' Pumo in particular, while Lady Hamilton and Tischbein, though collaborators in very different senses, both negotiated the ambiguity of their own class positions by engaging with the Neapolitan crowd and popular superstition.

The small figures indicate the scale of this 'View... of the inside of the cone of the mountain of Somma' (Plate 5), but Fabris invests the two central figures with a number of significant details. Hamilton may be identified, as usual, by the long red coat. The figure facing Hamilton and the viewer is marked as labouring-class by his clothing, particularly the knitted white cap that is also worn by the labourers at the Temple of Isis. Unlike the many rustic and labouring-class figures throughout the volume, however, this worker is mentally active, speaking and gesturing broadly in what must be interpreted as a scene of instruction. In this sense Pumo is Hamilton's double, a 'cicerone' who surely taught him something about volcanoes as well as the nature of the guide's role. We more commonly see Hamilton himself in this role, as when he is instructing Fabris or guiding the royal entourage in two particularly well-known plates (XXII, XXXVIII). The identification of this figure with Pumo is warranted, in my view, because he is pointing out a geological feature that was of particular interest to Hamilton (as we know from several other plates in the volume), the vertical lava flows formerly surrounded by channels that have eroded away.[54] These are first mentioned in the explanation of Plate V, a striking view of lava flowing down the snow-covered sides of the mountain, taken by Fabris in 1754. In this instance Hamilton points out the analogy between the vertical channels cut in the snow and the vertical channels on Somma—the remnants of the cone associated with the eruption of 79 CE, adjacent to the modern Vesuvius. The melting snow suggested the idea of more substantial channels eroding over time. This idea supports Hamilton's inference that Somma too was once an active volcano (V). The 'perpendicular strata' that remain behind also appear in other views, together with the 'lava like ropes' presented here in the right foreground.[55] These stark volcanic features dominate this plate, giving it a more technical aspect than the other interior view of Somma, which is bustling with tourists (XXXIII).

Hamilton mentions Pumo twice in print. The first instance occurs in the *Supplement to Campi Phlegraei* when he credits Pumo with the discovery that certain lava flows may be safely crossed on foot (3). This lesson makes for a dramatic scene of instruction:

> The wind changing, when we were close to this gentle stream of Lava, which might be about fifty or sixty feet in breadth, incommoded us so much with its heat and smoke, that we must have returned without having satisfied our curiosity,

[54] On Hamilton's geological contributions see Moore, 'Sir William Hamilton's Volcanology' 182, 187, and Sleep, 'Sir William Hamilton (1730–1803)'.

[55] The latter, also featured elsewhere in the book, runs under the scoriae and is 'forced out of a small passage' (XXXIII). The Hawaiian term *pahoehoe* has since been adopted for this type of lava.

> had not our Guide proposed the expedient of walking across it, which, to our astonishment, he instantly put in execution, and with so little difficulty, that we follow'd him without hesitation, having felt no other inconveniency than what proceeded from the violence of the heat on our legs and feet.

This is a good example of the topos memorably glossed by Kevin Nevers as 'the problematization of focal hyperbole by peripheral meiosis'. The figure is certainly characteristic of Hamilton's narrative technique, though it does not prove that he is chiefly motivated by a drive to 'subdue and subsume nature', as Nevers argues.[56] Whatever his conscious or unconscious motives may have been, he could not have accomplished them without Pumo, as he acknowledges in a note to the word 'guide': 'BARTOLOMEO, the Cyclops of Vesuvius, who has attended me on all my expeditions to the Mountain and who is an Excellent Guide'. Though only his first name is given, the printer of the *Philosophical Transactions* (whose text Hamilton adopts) follows the convention of capitalizing any personal name cited in the notes. A second, shorter reference in Hamilton's last article supplies the last name and another epithet, 'my old Cicerone of the mountain'. By this point, in 1794, they had made sixty-eight journeys to the summit over thirty years, ten of them between 1779 and 1794; in the earlier instance Hamilton estimates that he had also made over two hundred partial ascents.[57]

Hamilton gives Pumo some of the credit he deserves, but other references to the Cyclops of Vesuvius are more slighting. Hamilton at least makes clear that Pumo understood when 'the crust of the Lava was so tough' that one could walk upon it (*Supplement* 31–4). Fabris's memorable image of Pumo pointing out volcanic features suggests that he conveyed a great deal more to his patron over the years. Winckelmann's disparaging reference in the narrative of his ascent in 1767 could mean that Pumo was still young and inexperienced at this time, or even perhaps that Hamilton recommended a different guide in this case. The guide remains unnamed here and Winckelmann instead credits d'Hancarville with the leadership, if not exactly with expertise: 'in order to arrive at the mouth, we had to cross the burning hot lava, which our guide refused to do; since no other means of persuasion was effective, we coerced him with the stick, and [d'Hancarville] went before with a torch and we followed with scorched shoes'.[58] Another such beating is recorded by the diarist Nathaniel

[56] Nevers, 'Immovable Objects' 32. Nevers divides the 'narrative technique' that Hamilton 'perfected' by the end of his career into three aspects: 'in literary terms, the problematization of focal hyperbole by peripheral meiosis; in contextual terms, the ironization of natural catastrophe by human proximity; in lexical terms, the subversion of a rhetoric of astonishment by the discourse of science'.

[57] 'Account of the Late Eruption' 94.

[58] *Briefe* III.317. Winckelmann essentially personifies the stick in describing its disciplinary action, an interesting nuance that is lost by my use of 'we' in the translation: 'da kein Mittel war ihn zu bewegen, nöthigte ihn der Stock'. If the guide was indeed a youthful Pumo—he had met

Wraxall, and in this case the guide is explicitly identified as Pumo. The aggressor was an imperial personage perhaps especially accustomed to obedience, the 'irascible' Austrian emperor Joseph II, visiting the court of his sister and brother-in-law in Naples. Wraxall presents the anecdote as a direct quotation from Hamilton, whom he visited in 1779: 'I accompanied him to the summit of Vesuvius, and with concern saw him break his cane over the shoulders of the Guide, Bartolomeo, for some slight offence which he had given his Imperial Majesty'.[59]

The most substantial source on Pumo is the Vesuvian diary of Padre Antonio Piaggio. As a fellow student of the mountain, Piaggio takes Pumo much more seriously than tourists such as Emperor Joseph and Winckelmann, who treated their guides like cattle. Piaggio's treatment, however, is hardly flattering, since he regarded Pumo as a dangerous rival. Piaggio's magnificent eight-volume diary is still awaiting publication, and his equally voluminous treatise on Herculaneum, written in cipher, was lost before it could ever be published. Nonetheless, he wrote with a view toward publication and engaged in spirited debates with published works even for his audience of one—Hamilton, who commissioned the diary.

Understandably, Piaggio developed a kind of paranoia with regard to other scholars who were more successful in publishing their work on natural history and antiquities. One of these scholars, whose volcanology he refutes in a 252-page essay that concludes the diary, is identified only as NN. Another rival was the Abbate Domenico Tata, who dedicated a book on the extinct volcano Monte Vulture to Hamilton in 1778. In Piaggio's view, Tata was in league with Pumo, and another long interpolated essay of 1787 is dedicated to refuting 'the system of him and of Bartolomeo concerning these subterranean channels [*cammini coperti*] and the surges of lava that they asserted came from the summit of the mountain'. (Tata and Pumo argued, incorrectly, that lava erupting from new apertures had to come by way of the summit.) Most of this second essay is concerned with Tata, but Piaggio has specific objections to Pumo and evidently the debate originated between him and Don Vincenzo Merli in the presence of a local nobleman, Piaggio, and others. In Piaggio's view any evidence presented by Pumo should be discredited because of his low status, and he charges Tata with stretching the republic of letters too far by sharing authourity with 'this rustic guide, even with Bartolomeo, who has nothing, ... even though he frequently has the privilege of conducting visitors

Hamilton by this time—Winckelmann might have been pleased to note that the beating had its pedagogical effect. Goethe had better luck with his guide on Vesuvius in 1787, when he was initiated into the secret of walking on the lava flow, though he found it challenging (*Italian Journey* 206).

[59] Wraxall, *Historical Memoirs* I.248.

to the mountain, and who claims that a new saint, for the sake of four candles that he lights for him, has taken him under his protection'.[60]

Piaggio touches on several themes here that are prominent in the *Supplement to Campi Phlegraei*, including popular superstition and the danger of political subversion. Hamilton also mentions the names of Pumo and Piaggio for the first time in this slim third volume, bound to match the two volumes of *Campi Phlegraei* by collectors when it came out three years later. The occasion was the 1779 eruption of Vesuvius, one of three major eruptions during Hamilton's tenure as envoy (1767, 1779, 1794). This sequence of eruptions (five between 5 and 11 August) was accompanied by social unrest, which plays a more prominent role in this narrative than in the earlier ones, particularly in the account of the third eruption (evening of 8 August):

> Numerous processions were formed in the streets, and women and children, with dishevelled heads, filled the air with their cries, insisting loudly upon the relics of St. Januarius being immediately opposed to the fury of the mountain: in short, the populace of this great city began to display its usual extravagant mixture of riot and bigotry, and if some speedy and well-timed precautions had not been taken, Naples would, perhaps, have been in more danger of suffering from the irregularities of its lower class of inhabitants than from the angry volcano.

We do not learn what the precautions were, because Hamilton hastens, as he says, 'to return to [his] subject', but he does note that when the relics were brought out the next day, during the fourth 'fever-fit' of the mountain, the people were 'well convinced, that to this ceremony alone Naples may attribute its happy escape'.[61] Fabris echoes this concern with popular devotion in the two plates mentioned above (*Supplement* II, III), which include sacred images but not the relics themselves.

Though Hamilton returns to many of the same themes in this supplement, including the sublime effects of the eruption and the possible recurrence of a disaster like that of Pompeii, he places more emphasis than before on the process of the eruption, specifying many variations in the phenomena accompanying each eruptive event. The title page of the *Supplement* mirrors this emphasis on process, including four night views of eruptions and lava flows, in contrast to the vignettes on the first title page, which show volcanic products such as volcanic islands and dormant craters. Hamilton's careful gloss on these

[60] *Abbozzi di Fenomeni del Vesuvio* [Royal Society MSS 2–9], RS 5, f. 408, 402–3. According to Piaggio, 'the Duke... had imposed silence on this controversy', and he feels hurt that Tata has renewed it in a published text that evidently refers to Piaggio and misrepresents his position. Carlo Knight gives an account of this episode that is sympathetic to Piaggio, noting that he had often witnessed Pumo's 'cynical' manipulation of tourists. See Knight, 'Un Inedito di Padre Piaggio' 120; the last section of this article (116–31) helpfully contextualizes the controversy.

[61] 'Account of an Eruption' 59, 64; the *Supplement* text is based on this article. Hamilton typically veers away from social commentary after a few sentences, as 'foreign to [his] purpose' (71), but nonetheless returns to it frequently throughout the essay.

vignettes points out that one of the night views is based on a drawing by a new contributor, Piaggio.

Hamilton introduces Piaggio here as an artist—even advertising a 'magnificent copper plate' from his drawing, on sale in Naples by October—but Piaggio became especially significant as a diarist, keeping a daily log from Resina (Ercolano) over the next fifteen years. Hamilton and Piaggio came to terms within a month of the eruption, as indicated by the date of Piaggio's first entry, 12 September.[62] As Hamilton's letter to the Royal Society is dated 1 October, he probably had access to the first batch of entries from Piaggio's diary. The mountain continued to be active for some time, sending up great clouds of steam and occasional 'flashes of fire' (*vampa di fuoco*). One such cloud drifting near the city had caused the panic described by Hamilton on 8 August. As often in the diary Piaggio begins his explanation as a commentary on popular terminology: the peasants (*i Paesani*) use the term 'pine' (*pigno*) to refer to high volcanic clouds that are 'extraordinarily menacing or dark', though Piaggio objects that these clouds do not 'truly resemble' the pine's shape so much as the peasants' 'idea' of the tree.

Piaggio continued to make many sketches and drawings, knowing the value of visual documentation from his work on the Herculaneum papyri, and in these early entries he calls on the authority of a witness (one Ropoli) who confirmed the accuracy of his sketches of five differently shaped 'jets' [*butti*] of fire that were seen on 12 September. The entry for 1 November includes a remarkable series of five sketches of a 'terrible black cloud', 'impregnated' with spherical cinders, that formed a vortex whose rapid fractal development Piaggio captures precisely in the midst of his astonishment. The eddies of this cloud, he writes, formed themselves into *Tortiglione* (spirals or twists), which he attempts to represent as linked units (labelled *A-E* on the drawing): 'while I am in the act of drawing . . . it has already grown from *B* to *C*, I cannot depict its change from the first state because it is changing every minute in a thousand different attributes on the part of the whole machine . . . *A* is the first smoke which continues to eddy without confounding itself with the others'. The 'machine' eventually reached a height six times that of the mountain (then about 3,600 feet) and emitted a 'shudder' of wind that could be felt miles away.[63]

It seems fair to say that Hamilton developed a procedure of data-mining in his work with Fabris that he extended and made permanent in his fifteen-year relationship with Piaggio. Hamilton consulted Piaggio's diary regularly and made use of it in subsequent publications on volcanoes (crediting Piaggio, as we have seen). Fabris brought primarily visual skills to the collaboration, but

[62] According to the covering letter prefixed to the manuscript, Piaggio 'was unworthily treated at Naples and his labours ill rewarded, and the poor man would have been in real want if Sir William had not allowed him about 20 £ per annum to keep this Diary or Journal' (RS 2).

[63] For the narrative of 12 September, see *Abbozzi*, RS 2 [vol. 1], f. 3–4; for 1 November, including the first image, see f. 140–1 in the same volume.

Piaggio was known to Hamilton and his contemporaries primarily as an antiquary. According to Moore, they may have been introduced by Winckelmann in 1765, when Piaggio was a member of the royal antiquary's staff employed on the papyri and other finds from Herculaneum.[64] While all three of them participated in what Toscano calls the 'naturalist-antiquary mentality', the balance in Piaggio's case inclined toward classical antiquity, and he continued to work on his Herculaneum manuscript, even though his many attempts to publish it were rebuffed. According to Jenkins, Hamilton cherished the idea of writing his own book on the buried cities and planned to use Piaggio's manuscript as a data-mine for that project as well. In Hamilton's view Piaggio was a better antiquary than 'that wicked and ignorant rascal' Camillo Paderni, the royal antiquary, who compromised the antiquities by his 'barbarian spoiling'. Only Piaggio's drawings, Hamilton wrote, gave a sense of the antiquities 'as they came out of the Earth'.[65]

Unlike Piaggio's antiquarian writings, his substantial Vesuvian diary survives and has been diagnosed as both eccentric and deeply knowledgeable. The diary was almost forgotten before Carlo Knight published a substantial study (1990), which includes numerous reproductions of Piaggio's illustrations. Knight reports that Domenico Bassi, director of the Office of Papyri (Naples) from 1906 to 1927, published a small selection of Piaggio's papers with the caveat that the prolific diarist 'was affected by an acute and somewhat dangerous form of graphomania'. Knight points out that Piaggio was sixty-six when he began his career as a diarist, already somewhat infirm and prone to anxiety, which contributed to his scholarly defensiveness and paranoia. Knight also emphasizes that Piaggio was correct in his dispute with Pumo and Tata (which Hamilton recognized, though apparently he was unable to intervene effectively in the dispute) and concludes that, by the end of his career, Piaggio 'knew Vesuvius better than all other "experts"'.[66]

For Hamilton and for foreign tourists who visited him on the mountain, however, Piaggio appeared in the guise of the 'good old father' rather than the anxious scholar.[67] Although subject to a range of social and national

[64] Moore, 'Sir William Hamilton's Volcanology' 170. Hamilton's first reference to Piaggio in print includes a synopsis of his findings to that point and already indicates his intention to donate the volumes to the Royal Society. See Hamilton, 'Some Particulars' 366–7, and cp. 'Account of the Late Eruption' 74–5.

[65] These quotations appear in an 1803 letter from Hamilton to Charles Townley discovered and quoted more fully by Jenkins (*V & V* 44).

[66] Bassi, quoted in Knight, 'Un Inedito' 63; Knight, 'Un Inedito' 124, 129. In his systematic description of the eight volumes, Knight notes that the first 200 pages are devoted to what is technically known as the second phase of the eruption of 1779, which lasted from 12 September–31 December (66). He also states that Hamilton originally intended to write a sequel to *Campi Phlegraei* and employed Piaggio with that project in mind (71).

[67] RS 9, title page. Hamilton added a note to the title page of the eighth and final volume of Piaggio's diary to indicate that he had bound a sheaf of 'the first sketches of the drawings' with this volume 'to show the spirited touch of the good old father when he drew from Nature'.

prejudices, Piaggio took conscious pride in his contributions to knowledge. So too did Pumo, if Fabris's portrait (Plate 5) is any indication. Though hardly free of prejudice himself, Hamilton acknowledges their role as knowledge workers in a variety of ways.

2.7 EMMA HAMILTON AND WILHELM TISCHBEIN: KNOWLEDGE WORK AND SOCIAL CHANGE

Emma Hart, when she was not yet Lady Hamilton, enthusiastically witnessed the same 1787 eruption that spawned the controversy between Piaggio and Pumo. She mentions Piaggio in her account, which occurs in a long journal-letter typically cited in biographical scholarship for other reasons, primarily because it attests to Hamilton's growing fascination with her. Hart had been in Naples over a year and declared herself 'never... happier' in this letter to Greville, whose mistress she had been. She reports that they had watched the eruption initially from Sorrento: 'After breakfast we vew'd the lava running down 3 miles of Vesuvua, and every now and then black clouds of smoak, rising into the air, had the most magnificent apearance in the world. I have made some drawings from it, for I am so used to draw now, it is as easy as ABC.' Two days later she describes an ascent of the mountain, using language that is both familiar and unfamiliar at the same time:

> We was last night up Vesuvus at twelve a clock, and in my life I never saw so fine a sight. The lava runs about five mile down from the top; for the mountain is not burst, as ignorant people say it is. But, when we got to the Hermitage, there was the finest fountain of liquid fire falling down a great precipice, and as it run down it sett fire to the trees and brushwood, so that the mountain looked like one entire mountain of fire. We saw the lava surround the poor hermit's house, and take possession of the chapel, notwithstanding it was covered with pictures of Saints and other religios preservitaves against the fury of nature. For me, I was enraptured. (*H & N* I.130-31)

Like many other tourists, including Hamilton himself, Hart responds to Italian popular piety as an almost organic part of the spectacle of eruption. Italian 'superstition' is useful as a wedge for Enlightenment critique and documentation, for Hart as for Hamilton, and although she refers only briefly to Piaggio, she seems to enlist him on the side of Enlightenment. Though introduced as 'Padre Antonio, an old priest who lives on the mountain', Piaggio seems to be exempt from Hart's general condemnation of Italian superstition and 'ignorance' when she takes seriously his prediction of new volcanic activity 'on the Portici side'. Most of the Italians, from various clergy to the tutors of the crown prince to her own maid, are brought forward as examples of ignorance and bigotry. One virtue of her account is that she does not veer away from these observations as 'foreign' to her 'purpose', as Hamilton does. In the course of the same eruption narrative, she performs an experiment on her maid Teresa: 'I took last night one of my maids, who is a great biggot, to the top of the house, and I shewd her the mountain. But, when she saw the great fire, she fell down on her knees, and cried out, "O Januario mio, Antonio mio," So I fell down on my knees and cried aloud, "O Saint Loola mia, Loola mia"' (*H & N* I.133). The maid then upbraids her, understandably, for not believing in the saint and being ungrateful to God.

From a strictly formal point of view, Emma Hart is merely ringing the changes on a set of standard northern European tropes concerning southern Italy. The same themes are present in Hamilton, Goethe, Tischbein, and many others. Viewed more sociologically, however, this letter is remarkable for the energy with which Hart made these ideas her own. Thanks to the recent archaeological discoveries and its own renewed activity, Vesuvius embodied the presence of the past, and not only the classical past. Like the Enlightenment philosophers and artists who paid her homage, Hart was inclined to see Neapolitan popular culture likewise as an eruption of the deep past. From an English point of view, this is the feudal, Catholic past, though there is often a hint of the evolutionary past—in the sense promoted by stadial theories of the history of society—in these descriptions as well.

In one respect at least Hart resembled the ordinary Neapolitans more closely than the philosophers, or perhaps she took her cue from them when she famously adopted the role of the modern primitive herself. She cultivated this role as the star of a one-woman show, her famous 'Attitudes', in which she brought to life a variety of roles and images from antiquity, improvising under Sir William's occasional direction. The identification was reinforced by disparaging comments on her own lack of education, such as Frederick Augustus Hervey's description of her speech as 'Emma's Doric' (*H & N* II.vi). Emma's objectification by these performances is re-inscribed by much modern scholarship, though some scholarly and popular literature gives her more of the credit she deserves as inventor and co-producer of her Attitudes.[68]

Hart's eruption narrative has a place within the larger network that allowed Hamilton, Fabris, Pumo, Piaggio, and others to work out new possibilities for intellectual and social agency. In the letter that contains this narrative, Hart observes herself as an object of numerous gazes: as an artist's model, often represented as a nymph or some other personification of Nature; as a quasi-religious icon to the common people; and as a sexual object to Hamilton and his guests. Her creative response puts to the test Bruno Latour's contention that 'objects too have agency', if not quite in the sense he intends. Hart's own

[68] See, again, Nolta, 'Body of the Collector' 111–12. For contrasting examples, see Sontag, *Volcano Lover*, and Cox, *Poetry and Politics in the Cockney School* 178–86.

narrative invites the analogy between her intellectual involvement with vases and volcanoes, as a unique form of social mobility, and the catastrophic force of the volcano itself, a variation on the common analogy between the mountain and the Neapolitan crowd. I have argued elsewhere that Hamilton elaborated the figure of telluric agency from a long and rich English literary tradition of representing foreign earthquakes and eruptions, a tradition developed further by his own enthusiastic readers, including Humphry Davy.[69] Here I am concerned to show how that figure operates as a sociopolitical rubric, especially for observers working in close proximity to the volcano itself.

The analogy between eruption and revolution was not new in 1789, when the French Revolution made it especially popular. In 1778 Hervey (later Earl of Bristol), one of Hamilton's more freethinking correspondents, expressed his political dissatisfaction by declaring: 'when I reflect on our desperate situation, I almost wish for a Vesuvius, or rather an Etna, in the midst of Great Brittain' (*H & N* I.57). For the Hamiltons and their circle in the 1790s, very much invested in the status quo in Naples, the superstitious and violent eruptions of the Neapolitans were more troubling—much more so than the volcano itself. But at the same time, the volcano and its legacies stimulated unusual social alliances, new forms of patronage and knowledge work (like those explored by Hamilton and his collaborators), and new philosophy. In this sense the telluric agency of the volcano provides an apt metaphor as well as a real environmental cause—not only for the raucous, unpredictable agency of the Neapolitan crowd but also for the new forms of social agency explored by gifted individuals such as Emma Hart, d'Hancarville, Piaggio, and many others connected with Hamilton. D'Hancarville, in fact, began his literary career with a radical egalitarian tract on political arithmetic, in which he used the volcano as a figure not for revolution but for despotism (*Antichità* 36). In his writing on ancient vases, he uses the Neapolitan crowd as a lens on the deep past, reconfiguring their 'superstition' as something closer to ancient religion. Vesuvius in this context has metonymic force, imbuing the notoriously vulgar King Ferdinand, and at times even Hamilton himself, with the glow of raw nature. Wraxall, for example, makes connections between the modern tarantella and the ancient worship of Dionysus, much as d'Hancarville does—except that Hamilton is the dancer in this case: 'I have seen him, not more than two years before his decease, perform the "Tarantella,"... which, as it is undoubtedly a copy of the Bacchant amusements of Antiquity, demanded no slender portion of animal strength and spirits'.[70]

69 Heringman, 'Style of Natural Catastrophes'.

70 Wraxall, *Historical Memoirs* I.229. Wraxall refers to Ferdinand as 'a rustic, elevated by fortune or accident, to a Crown' (233). He quotes Hamilton as saying that the king's 'ordinary Italian is a Neapolitan dialect, such as the lowest of his subjects, the *Lazaroni*, speak in their intercourse with each other' (235). Hamilton participated enthusiastically in Ferdinand's hunting parties, in which sometimes thousands of animals were killed. In one letter Greville

The painter Wilhelm Tischbein, Hamilton's most significant collaborator in the 1790s, connected ancients and moderns in his own way, linking the Neapolitans not so much to the fiery stones of the mountain as to the *pierres parlantes* more familiar from neoclassicism: classical sculpture. Implicitly contradicting Winckelmann's argument that the nature imitated by the Greeks survived only in classical sculpture, Tischbein provides a whole catalogue of statuesque subjects that he found among the common people of Naples and used for paintings of classical subjects. He dubs one chimney sweep 'the most beautiful man in Naples', incorporating his legs into a painting of Achilles. Emma Hamilton, though not a native, belongs to this category as well. 'No mouth more beautiful [than Emma's] may be found in all antiquity', Tischbein declares, and he deploys her in a painting of Orestes and Iphigenia, in addition to portraits commissioned by Hamilton.[71] Tischbein explains the classical beauty of the Neapolitans as an expression of the superabundance that characterizes their natural world. He also offers an extended eruption narrative that parallels those of the Hamiltons. Tischbein finds not only superstition, but downright fanaticism in the panic of the crowd when he attempts to reason with them: 'I was more alarmed by the unrest of the people, who are fanatical in the highest degree, than by Vesuvius itself and decided that the time was not right to quiet them with reason.'[72]

A later passage balances this parallel between demotic and demonic agency with a more rational view of the volcano's agency as curatorial. Tischbein's narrative of his collaboration with Hamilton begins with this curatorial anecdote of art treasures preserved by geology, two newly uncovered vases that belong with 'the most precious works of art that the Earth has conserved [aufbewahrt] for us'.[73] Hamilton celebrated this discovery by announcing to Greville in March 1790 that 'a treasure of Greek, commonly called Etruscan, Vases have been discovered within these 12 months, the choice of which are in my possession, tho' at a considerable expence' (*H & N* I.142). This letter also announced his intention to publish the vases in what would be his third multivolume folio publication. It shows Tischbein's influence in the hypothesis mooted here that 'many of the subjects [are] from Homer'.

While Tischbein was supervising his students at the Neapolitan Academy in the drawing and engraving of scenes from these vases for Hamilton, he was simultaneously pursuing his own publication project, *Homer nach Antiken gezeichnet*. Hamilton reportedly laughed at Tischbein's ambition to collect all

takes him to task for, as it were, going native in this manner: 'it is with regret I hear you boast of your barbarity, and leave only one half line to the antiquities or virtu' (*H & N* I.68).

[71] Tischbein, *Aus meinem Leben* 246–7. Winckelmann, by contrast, is much stricter in rejecting all modern bodies categorically for their 'individual wrinkles', their 'emaciated tensions and depressions', which were lacking in the Greeks. See his *Reflections* 17–19.

[72] *Aus meinem Leben* 263.

[73] *Aus meinem Leben* 289.

available 'Homeric' motifs—from bronzes, gems, and wall painting as well as vases. To add to the excitement, the threat of a French invasion from the mid-1790s moved Tischbein to have his engraver commit the designs directly to copper, rather than working from drawings as he did for Hamilton's *Collection of Engravings*. He succeeded in getting all these plates onboard a ship bound for Leghorn as Naples was 'teeming with Jacobins' in 1799, and the eventual publication of his work proves that they arrived safely in Germany. As Director of the Neapolitan Academy, Tischbein associated himself resolutely with high art, and his views of the Neapolitan crowd, as well as his taste for literary subjects, should be viewed in this light. For the same set of reasons, *Homer nach Antiken gezeichnet* (with text by Christian Heyne) is a more doctrinaire piece of neoclassicism than *The Collection of Engravings*, his collaborative publication with Hamilton.

The Collection of Engravings participates more fully in the revisionary process by which vases, gems, and other minor forms are elevated to the status of 'speaking stones', like sculpture. Like the volcano, which promoted an increasing mixture of high and low in the conduct of knowledge work (if not directly in social structure), the 'mine of vases' disseminated by Hamilton and Tischbein brought 'low' design into contact with 'high' art in new and dramatic ways. The most famous case of such contact, Josiah Wedgwood's Etruria ceramic works, was the immediate result of the first vase publication in 1768, and Wedgwood was one of several 'practical men' who went on to publish scholarly works of their own as a result. Another commercial protégé of Hamilton's, Henry Clay, applied 'Etruscan' design to snuff boxes—a neat complement to the volcanic snuff boxes depicted in *Campi Phlegraei* (XLVIII)—and later asked permission to publish a letter of Hamilton's as an advertisement for his new business as 'japanner to the king'. Greville purged the advertisement of Clay's 'awkward' prose and assured him that 'it could not be displeasing to [Hamilton] to appear from a distance as superintending the Arts in this country' (*H & N* I.41). Thora Brylowe has argued that one of the publications inspired by the *Collection of Engravings*, Thomas Kirk's *Outlines* (1804), further domesticated Tischbein's renditions of the vase paintings so as to 'protect the delicate English reader from indecency', in keeping with middle-class taste.[74]

[74] Brylowe, 'Two Kinds of Collections' 50. The outline style of engraving practised by Kirk, John Flaxman, and others, which arguably 'sanitized' neoclassicism, is related to the design industry, though not identical with it. (Flaxman, for example, created designs for Wedgwood.) Other publications related to Hamilton's collection by 'practical men' associated with the design industry or the art market include George Richardson, *A New Collection of Chimney Pieces Ornamented in the Style of Etruscan, Greek, and Roman Architecture*, modelled on a work by Piranesi; Wedgwood, *Account of the Barberini, Now Portland Vase*, which recounts Hamilton's acquisition of the vase; and James Christie, *A Disquisition upon Etruscan Vases*, which responds to d'Hancarville. On commerce and neoclassicism, see Coltman, *Fabricating the Antique*.

My study of Hamilton's collaborations tends to support the broad scholarly consensus that he effectively sponsored a revolution in taste. This revolution, however, stimulated broader public access to classical taste through market forces rather than violent eruption. Many scholars have pointed out that Hamilton's self-interest guided his 'supervision of the arts' to a large extent—some commentators seem to feel that it did so exclusively.[75] My point here has been that Hamilton's philosophical commitment to the deep past, however mixed his motives, led him to collaborate with a large number of artists and intellectuals, respectable and otherwise, who put their own stamp on the knowledge produced under Hamilton's auspices. The three folio publications certainly promoted Hamilton's reputation, if not his purse, but they also promoted the careers of new kinds of experts and new divisions of knowledge, including non-classical antiquities and romantic anticlassicism. Brylowe presents Kirk's sanitized neoclassicism as a romantic preference for idealizing depiction over flawed ancient objects. D'Hancarville mounted up from ancient objects to a syncretic apprehension of Western as congruent with Eastern antiquity, a revision that was equally disruptive to neoclassicism. In the process, Etruria and Magna Graecia emerged as different historical and geographic domains, while classical and human antiquity increasingly had to be resituated in relation to geological time. This last shift also transformed the affiliation between antiquities and natural history, as I have argued in this chapter. That affiliation was the precondition, however, for the collaborative accumulation of verbal and visual materials that characterized Neapolitan research in Hamilton's time. Bringing the vases and volcanoes back together also brings the dynamic social relations and environmental history of this moment back into view.

Naples truly represented a mine of antiquity for the naturalist-antiquaries of the eighteenth century. The finds in Herculaneum and Pompeii presented a copious 'mine of literature', as Hervey recognized, and Hamilton was delighted to find a fresh 'mine of vases' so late as 1790, when his collaboration with Tischbein began.[76] The images of quarrying and excavation in *Campi Phlegraei* vividly illustrate the complex division of labour involved in any mining operation, and Hamilton fully embraced the collaborative possibilities of mining work, helping to create networks in diverse areas of natural history and antiquity. My concern has been to show the extent of these networks and the dynamic processes that they set in motion, particularly in the case of *Campi Phlegraei*. Forcefully juxtaposing the temporalities of nature and of

[75] See, again, the publications by Vickers, especially 'Value and Simplicity'.

[76] *H & N* I.55, 144. Hervey proposed in 1778 that he and Hamilton form a company like the East India Company to work 'this mine of Litterature like any other mine whatever either of Gold, Silver, or Alumn' (55).

history, this work accelerated the fragmentation of antiquity into multiple domains.

Hamilton's pioneering emphasis on volcanic action as the larger context for the evolution of this cultural landscape depended on a multitude of actors, beginning with Vesuvius. The unexpectedly violent eruption of 1779 set in motion new developments in the careers of Fabris, Pumo, and Piaggio, among others. The *Supplement* occasioned by that eruption provides a glimpse of these developments and registers a growing interest in the energies of ordinary Neapolitans, both religious and otherwise. The revolutionary power of these energies became apparent in the wake of the Parisian eruption of 1789, with the establishment of the short-lived Parthenopean Republic ten years later. One of Hamilton's naturalist-antiquary colleagues, Domenico Cirillo, was executed in 1799 for his role in the republic; today he is commemorated as a political martyr. Hamilton's intense scrutiny of the incendiary past also promoted new social combinations, but these were conceived within the more quiescent framework of the Republic of Letters. We now turn from volcanoes back to vases and to Hamilton's collaboration with d'Hancarville—a collaboration that was far more difficult to manage than the others, yet also uniquely rich in possibilities.

Part II

Greek Vases and Deep Time in Naples

3

Baron d'Hancarville, Sir William Hamilton, and the Collaborative Production of Antiquities

It may be . . . that [this work] will attract the attention of a Posterity curious to know what the antiquaries of our centuries did and thought. In that case I would like to inform them (which may seem extraordinary) that the ancient authors of 1769 were grateful to the public.

(D'Hancarville, *AEGR* II.2)

3.1 ANTIQUARIAN CONDITIONS OF PRODUCTION

There is no definitive account of all the artists and artisans who collaborated on the production of the *Collection of Etruscan, Greek, and Roman Antiquities* (*AEGR*). The first and most detailed list is given by the author, Baron d'Hancarville, in a letter to his patron, Sir William Hamilton, who was still largely unfamiliar with the details of book production. Early in 1767 d'Hancarville writes of himself: 'For nine and a half months he has provided sustenance for seven people who eat three times a day. He has had two additional domestics in his house; he has provided lodging for the artists, for besides those who are in his house he has paid for Bracci's lodging and the majority of Raimond's and that of Cardon and Joseppino'.[1] Later he adds, 'my wife is working night and day to make our books' (*Antichità* 131). In his final, desperate letters from Florence, he claims to have been feeding fifteen people (136) and employing eight families on the book (139).

[1] *Antichità* 129. All d'Hancarville's letters to Hamilton are printed from manuscript in Griener's *Antichità* as Appendice 1 (118–44). This chapter depends heavily on Griener's fine collection of these letters, which has been used very little in English-language scholarship.

D'Hancarville's initial tally, when he drew up the first of many budgets for Hamilton, had called for only four artists and two 'domestiques', one to operate the press and one to apply acid (122). In an important essay Nancy Ramage identifies six printmakers and three designers whose signatures appear on the plates, and three more artists are named in d'Hancarville's correspondence. Ian Jenkins identifies another possible contributor in George Redmund, described by Hamilton as a 'good sober creature' who 'drew all the figures' in 'the Etruscan work'.[2] Then there is the printer, Francesco Morelli, who set the text (presumably in his own workshop, though using d'Hancarville's characters) and applied for the imprimatur. And there are the translators who provided the parallel English text.

All these contributors are well worth tracing as individuals, to the extent that the record allows. Ramage's essay on Giuseppe Bracci is still the only scholarship of this kind. My more modest aim in this chapter is to examine d'Hancarville's accounts of these collaborators and their mode of working together. Unlike Sydney Parkinson, whose artistic contribution was deliberately minimized in the official publication of the *Endeavour* voyage, Bracci receives credit explicitly in d'Hancarville's text, and artists' names appear prominently on many of the plates in *AEGR*. Moreover, d'Hancarville's letters reveal that an elaborate division of labour was devised by him as a crucial piece of the economic plan for the work.

This thorough documentation of the production process, along with unique formal features of the work itself, make this particular scholarly folio publication the touchstone of my argument that the knowledge produced in natural history and antiquities depended on a complex and explicit division of labour. My other chapters approach book production in the context of subject matter, relying on available scholarship (such as Wood's study of *Campi Phlegraei*) to account for the material context of book production. In this case, however, I have devoted a chapter solely to the history of a book: the especially full correspondence and other documents associated with *AEGR* provide the grounds for inference about the production of other similar works, and close attention to technical detail is also warranted by the content, d'Hancarville's ambitious synoptic account of human antiquity, which will occupy our attention in the next two chapters. The correspondence also shows that d'Hancarville continually sought to enlist Hamilton as an equal partner—intellectually, fiscally, even socially—while the latter maintained the stance of a patron.

The number of knowledge workers involved in *AEGR*—at least eighteen—exceeded by far the number involved in Hamilton's two later folio

[2] *V & V* 48. On the other artists, see Ramage, 'Initial Letters'. In addition to Bracci, Ramage lists the designers Beaulieu and Tierce and the engravers Pignatari, Cardon, Nolli, de Grado, Piroli, and Lamberti (446). Jenkins points out that Redmund, who also conducted d'Hancarville's wife to Florence after his flight, might be the same artist referred to as 'Raimond' by d'Hancarville (*V & V* 63n76). No such name appears on any of the plates, nor do the names Joseppino and La Marra, mentioned only in d'Hancarville's letters (*Antichità* 129, 136).

publications.[3] With the exception of d'Hancarville, more is known about Hamilton's later collaborators, such as Fabris, Piaggio, and Tischbein. *Campi Phlegraei* and the *Collection of Engravings* are in some other ways less complicated objects of study. Hamilton contributed as an author to both of them and supervised their production more closely. Fabris worked with a small number of colourists and assistants and Tischbein reports delegating work to a chosen few students and a single engraver.[4] Hamilton relied on a single guide, Bartolomeo Pumo, for most of his work on Vesuvius, as opposed to the numerous guides, collectors, and dealers who contributed to the first vase collection documented in *AEGR*. The second vase collection was assembled more quickly and without legal impediments. Tischbein's more efficient operation also focused exclusively on vases collected by Hamilton, as opposed to *AEGR*, which expanded to include objects from many other collections. And Hamilton himself wrote a substantial portion of the *Collection's* English text, whereas a brief but tantalizing footnote by d'Hancarville himself tells all we know concerning the translators who rendered his French prose—with some apparent difficulty—into English for *AEGR*. It is instructive to begin with these shadowy figures, since d'Hancarville bothers to mention them and above all because the reception of the work owes so much to them.

Translators even of freestanding editions of scholarly works often remained invisible in the eighteenth century, so d'Hancarville's acknowledgment of his English translator(s) seems remarkable. Their absence becomes remarkable, too, with the sudden disappearance of the parallel English text in Volumes III–IV. The prefaces, the only English passages in these two volumes, are so badly written that they could be taken as evidence for George Rousseau's contention that d'Hancarville 'could not write English'.[5] All the more reason for his gratitude toward the 'persons' who translated the initial volumes 'from regard to me', and who 'have attended more to Clearness and Precision than to Elegance of expression . . . [since] they imagined, that intended for the advancement of the Arts which require being treated simply, it was sufficient

[3] Based on the primary sources and the findings of Ramage and Jenkins, at least eighteen people must have made significant contributions to *AEGR* during the most active phase of the project (1766–1770), including d'Hancarville, his wife, his translator, the printer, twelve artists, and the two 'domestics'. It may be stretching the term 'knowledge worker' too much to apply it to the labouring-class artisans responsible for the acid and the mechanics of the press, or to d'Hancarville's wife, only mentioned once in this context. My figure of eighteen tallies with d'Hancarville's later figure of fifteen, assuming that he is excluding himself, his wife, the translator, and/or Morelli. (This does leave out Hamilton and any additional translators—d'Hancarville makes them plural.) To identify the 'huit familles à cet ouvrage' (*Antichità* 139) seems more challenging.

[4] On Fabris and his colourists, see *V & V* 167; on Tischbein's production process, see *Aus meinem Leben* 292.

[5] Rousseau, 'Sorrows of Priapus' 116.

to render the Ideas of the Text in the most litteral manner' (*AEGR* I.xxii). D'Hancarville himself has aimed, he says, at being 'usefull' rather than 'brilliant'—and although the latter claim is suspect, it does reinforce his compliment to the translators and exonerates them. I see no obvious reason to distrust his characterization of the translators, from which it follows that there were at least two and that they did not do it for money. While the tenor of d'Hancarville's correspondence with Hamilton, among other factors, rules out the possibility that Hamilton did any of the translating, it is conceivable that the first Lady Hamilton might have had a hand in it. D'Hancarville credits her with two important ideas for the development of the work and her correspondence shows that she passed many lonely hours and days while Hamilton was ensconced in the royal hunting lodge with Ferdinand IV.[6] D'Hancarville's own wife may also have served as his translator, which would justify the remark about his wife's hard work.

There is much in the translation to suggest that it is not the work of a native speaker—the lack of a subject for 'intended' in my last quotation will stand as an initial example. D'Hancarville's only other reference to the translation also suggests that he supervised it closely and had constant contact with the translator: his footnotes, he tells us, appeared 'too tedious to deserve a translation' (II.156)—unfortunately for readers who lack French, he announces this decision just before launching into a nine-page double-column footnote that provides the only sustained explanation of the paintings on Hamilton's vases, supposedly the main subject of the work. (Volumes III–IV offer more explications, again in French.) Hamilton and his fellow virtuosi were fluent in French, of course, but d'Hancarville wanted the work to compete primarily with English-language folios such as Stuart and Revett's *Antiquities of Athens*, and at least some members of the artisanal readership cultivated by Hamilton did lack French.[7] D'Hancarville acknowledges his translators in a postscript to the ambitious preface to his first volume, perhaps the most nakedly autobiographical paragraph in the whole work.

What begins as a defence of his translators ends up as an apology for his own style and even his authorship, sketching very concisely the network of

[6] D'Hancarville attributes the expansion of the work from three to four volumes to Lady Catherine's suggestion (*Antichità* 127) and also credits her with a plan that he laments not having followed, which was to finish the production in London (140). Constantine quotes Lady Hamilton's very poignant last letters to Sir William in *Fields of Fire* 112–13.

[7] In one of his early proposals d'Hancarville details the features of five recent folios with which, he says, his and Hamilton's volume will compete successfully: two English works besides Stuart and Revett (1762), Robert Wood's *Ruins of Palmyra* (1753) and Robert Adam's *Ruins of the Palace . . . at Spalatro* (1764), as well as two Italian works: the first volume of the state-sponsored *Antichità di Ercolano* and the *Musaeum florentinum* (*Antichità* 123–4). Elsewhere in the letters, and of course in the work itself, he refers to numerous other antiquarian folios, perhaps most notably the Comte de Caylus's *Recueil d'antiquités égyptiennes, étrusques, grècques, romaines et gauloises* (1752–1755), from which d'Hancarville seems to have borrowed his title.

economic dependency that links the translators, the author, the engravers, the designers—even the Venetian foundry that cast the type—and the public of the *Antiquities*. His unorthodox emphasis on subject matter over literary brilliance, he explains, proceeds from two causes: 'his taste for antiquity, and since it must be own'd, the situation of his fortune . . . therefore he may be pardon'd for being the Author of four Volumes in Folio, since he has been forced into it' (xxii–iv). This confession of economic necessity is more truly unorthodox, it seems to me, than d'Hancarville's prose style, and it triggers another, more audacious confession that has already (and deservedly) been noticed by several commentators: 'When he saw, the hundred thousand Characters cast on purpose to compose these present Sheets come from Venice, he was frighten'd to think of the few good things and the immense quantity of bad ones, which these Characters might bring to light' (xxiv). But he consoles himself and us, concluding, 'My Designers, and Engravers, to whom I have already great obligations, will see those Obligations increas'd, should their talents cause the follys of the Author to be forgotten.' He then draws our attention to the work as a whole, 'which, considering the number of Plates it contains, and the manner in which they are executed, is certainly very cheap'.

D'Hancarville presents himself as both patron and client of his collaborators and of the public, who, he claims, have effectively financed the *Antiquities* through their loyal purchase of his previous publication. John Boydell's ambitious project for refining public taste and employing worthy artists while reaping a profit, the Boydell Shakespeare Gallery, earned him a title in the 1790s that would have flattered d'Hancarville too: 'the commercial Maecenas'.

Evidently only one copy survives of the earlier publication, a series of engravings of scenery and antiquities around the Bay of Naples aimed primarily at Grand Tourists.[8] Since there was little or no letterpress in this work, d'Hancarville owed even more of his success to the skill of his engravers, and his generous praise of the artists in the text of *AEGR* as well his letters must reflect this continuing debt. Many of the same artists were involved in both projects, including Antonio Cardon and Giuseppe Bracci.

[8] Griener refers to this work as 'un bellissimo libro illustrato sul golfo di Napoli', entitled *Golfo di Napoli in Campagna felice* and first published in 1765 (*Antichità* 50–1). Lissarrague and Reed refer to the same work as 'a set of four maps with views and antiquities' in 'The Collector's Books' 276. To confuse matters further, they also speculate that when d'Hancarville refers to the 'kind reception' given by the 'public' to his last work, he means the 'ancient and modern history of Sicily' rather than the former work; this is surely a misreading, as the context of this passage (*AEGR* I.xxii) makes clear that he is proposing the work on Sicily for the future as an enticement to the public to buy the present one (*AEGR*). Further study of the earlier work is needed in light of d'Hancarville's very frequent references to 'mon golfe' in the letters.

Bracci gets star billing in *AEGR*. Throughout the four volumes, many of the in-text plates are signed by him as designer, and his designs were lavishly engraved by Cardon, Carmine Pignatari, and other talented artists. D'Hancarville credits him in the text with two major technical achievements. 'It is not to us', meaning himself, that readers 'owe the discovery of the manner in which the plates are printed', he writes, 'but to Mr Joseph Bracci, a most able and ingenious Artist; it is to him also that we are indebted, for . . . the method of making all sorts of Vases upon a given form, without altering their system' (I.172). The first of these techniques, for reproducing the vase paintings in full colour, has been widely recognized as the book's greatest innovation.[9] The second technique was arguably even more crucial to d'Hancarville's authorial project. Time and again he refers to the 'rules' and the 'sistem' he has found that will enable artists to produce originals on 'Etruscan' principles: to the extent that Wedgwood and other producers were able to imitate the models in the collection, they owed their success more to the minutely calibrated sections and perspective views of the vases than to any of d'Hancarville's lengthy deductions concerning the methods of ancient vase painters (e.g. II.56–152). D'Hancarville's claim to have found the 'sistem the Ancients followed' (I.vi), as well as his claim to be advancing Hamilton's project of influencing the ceramics industry and popular taste, therefore both rest on Bracci's authority.

Bracci's relationship with d'Hancarville vividly illustrates the interdependence of book production, cultural capital, and neoclassical theory that this chapter is concerned to demonstrate. Their dynamic relationship activated a much larger network, bringing the agencies of book production to bear on the natural agency of the eruptions that buried the vases; on the workmanship, preservation, and fluctuating currency of the vases themselves; and even on the currency of the copper plates on which the two men with their collaborators caused the images of these vases to be engraved. As early as 1767 d'Hancarville speaks of an attempt to pawn the plates of his work on the Bay of Naples to fund his work for Hamilton (*Antichità* 129). He made sure to carry the very substantial number of plates for *AEGR* with him to Florence in 1770. His success in pawning them there in 1772—apparently for the value of the copper alone—secured what little leverage he had and materially affected the history of the work, even beyond its completion in 1776 to the final release of Hamilton's copies in 1780 (52).

[9] See, for example, Lissarrague and Reed, 'The Collector's Books' 275. It is worth noting that Hamilton echoes d'Hancarville's language about Bracci when he calls Pietro Fabris 'a most ingenious and able artist' in his preface to *Campi Phlegraei* in 1776 (1.5). It might be argued that Bracci saved the project money, since according to d'Hancarville's reckoning he only paid half as much for the coloured (yellow) as for the black ink (*Antichità* 125).

3.2 THE FISCAL LOGIC OF VIRTÙ

In d'Hancarville's expansive letters to Hamilton about the project, Bracci plays a leading role in both the colourful narratives and the copious financial accounting that these letters contain. According to Ramage, Bracci left his post of many years as a designer at the royal Neapolitan tapestry works to take part in d'Hancarville's project, and this experience even more than his earlier efforts in topography and antiquities would have qualified him to supervise the engraving of the sections and perspective views. He is the only artist named as having a monthly salary, and his salary was certainly the highest at 56 écus, roughly £5 10s. (*Antichità* 136). He is the only artist to be paid at all during one of d'Hancarville's many episodes of bankruptcy (135), and another time d'Hancarville notes that Bracci, though up to his neck in work, will make an exception and execute a *veduta* for one of Hamilton's friends—a view of Vesuvius in eruption (130).

D'Hancarville's patronage of Bracci and the other artists is socially complicated and far from disinterested, but he takes unusual care to acknowledge their contributions, in the text as well as the prominent signatures visible in most of the in-text illustrations. (These include the frontispieces, most of the vignettes, and even many of the ornamented capitals—some 120 images—but almost none of the 400-odd full-size plates following the text, including those in colour.) He also foregrounds his close relationship with the artists in many of his letters to Hamilton, striking a note sure to please the ambassador so widely known (and so copiously acknowledged) for his patronage of artists. This emphasis provides an important clue to the nature of the relationship between Hamilton—who saw himself primarily as d'Hancarville's patron—and d'Hancarville, who worked hard to portray himself as a collaborator and social equal rather than a client. Almost every scholar who has published on their joint project has been mystified by Hamilton's decision to trust d'Hancarville. Recently Thora Brylowe has reiterated the lack of any evidence revealing how the two men established contact: 'many of the details of Hamilton and d'Hancarville's relationship . . . have been lost to time'.[10]

D'Hancarville's second tally of seven collaborators confirms that Hamilton accepted his proposal to lodge and feed the artists while they worked on the project, along with most other aspects of the plan. The idea must have appealed to Hamilton for its fiscal logic, which seemed sound enough initially, but also because d'Hancarville staged a scenario that Hamilton would have enjoyed supervising himself: a troop of artists working away at designs for his publication, the objects drawn from his collection and even the many decorative designs

[10] Brylowe, 'Two Kinds of Collections' 35. See also Haskell, 'The Baron' 30, and Constantine, 'Winckelmann and Sir William Hamilton' 60.

adapted from books and prints in his own library.[11] The scene is somewhat reminiscent of the cultural evenings with which Hamilton enriched the social life of Naples, according to Goethe and many other visitors who recorded their experiences. Hamilton would have stopped short of hosting the production process in one of his own houses, but it is easy to see how d'Hancarville—already notorious across Europe for his aristocratic impersonations and other frauds—established an identification with Hamilton that was attractive to both of them. Hamilton must have known something of his reputation, but d'Hancarville's successful identification with Hamilton's own ambitions bound them together during the long and tempestuous course of their relationship. This identification rests on d'Hancarville's relationship to the artists and on his sure grasp—being an habitual debtor himself—of Hamilton's financial anxieties.

Debt is the focal point not only of the division of labour but also of the complex affinity between the two men, which becomes less mysterious in this light. D'Hancarville persuaded Hamilton in 1766 not only that he could produce a lavish and successful book on his vase collection, but that Hamilton's own 50 per cent share of the profits would be sufficient within a few years to pay off a debt of £6,000 that Hamilton evidently confessed to him in a moment of indiscretion. Establishing new collaborative relationships and a new method of book production under his immediate supervision—in a large house probably rented for the purpose—was the centerpiece of this plan. D'Hancarville was not only a lover of antiquity, as he tells us in his preface, but a great connoisseur of books on antiquity, particularly the genre of large illustrated books in which he hoped that *AEGR* would make its mark. In proposing the project to Hamilton, he offers a numeric table comparing the price, number of plates, pages of letterpress, and other features of the proposed work to five recent antiquarian books and then gives extremely detailed estimates of his costs to justify the competitive price that he assigns to the proposed work (*Antichità* 123–24). This accounting gains credibility from d'Hancarville's intimate knowledge of book production. He will buy the copper in bulk, he writes, he has a good friend in the paper industry, and he knows how to make sure that the acid used on the plates has not been diluted by the vendor (121). The most important cost-saving measure, though, is his plan to lodge and feed the artists, saving about a third of the standard daily wage.[12]

[11] See further Ramage, 'Initial Letters' 455, and Lissarrague and Reed, 'Collector's Books' Appendix 1. Lissarrague and Reed demonstrate that fully a third of the 180 colour plates of vase paintings were taken from vases in other collections, and hence largely from earlier published engravings of those vases (283–6). This article documents an astonishing range of visual borrowings from previously published plate books on antiquities, most of which were in Hamilton's library.

[12] The passage is admittedly problematic: 'Les artistes se payent à un prix aussi bas que celui indiqué au-dessus parce que Mr d'Hancarville les loge, les nourrit, les entretient à 5 carlins par jour ce qui n'est guère que la moitié de ce qu'il dépense' (*Antichità* 122). The 'price indicated

The artists responded well to this treatment. A large majority of the plates were finished within a year, by early 1767, and the number of artists involved increased significantly from the initial estimate of four. They must have been satisfied with their accommodation and stimulated by the access to books and antiquities and by the resulting intellectual exchange. None of this would have been possible in an ordinary printer's or engraver's workshop, and the end product reflects the excitement that the project must have generated in its early years. Hamilton's friend and fellow naturalist-antiquary Joseph Banks, as we have seen, accommodated the artists for his *Florilegium* project in his own house from 1771–84 and supervised them there—a plan that Hamilton may have regretted not following himself. D'Hancarville attributes the ardour of his collaborators—who worked eagerly, he says, from before dawn until late—to his efforts to pay them well and on time, though this would eventually become impossible as the project dragged on (127). Here d'Hancarville speaks as an entrepreneur, recurring to the fiscal logic of bigger is better—'the more rich the work, the more we shall profit' (130)—that guides all his decisions until the project goes irretrievably into the red.

Although these difficulties inevitably overshadow all subsequent views of the project, it is important to bear in mind two points concerning its early history. First, d'Hancarville and 'his' artists really were extremely productive in the initial phase. Second, some disappointment was inevitable given the outlandish projections that d'Hancarville made and that Hamilton clearly accepted. An undated letter assigned by Pascal Griener to the first half of 1767—less than a year after d'Hancarville received Hamilton's initial investment of 420 guineas (*Antichità* 118)—states that 494 usable plates have already been engraved; 150 sets are printed up to Plate 82; and 400 copies of the letterpress are ready for Volume I (124–25). Winckelmann's letters confirm that proofs of the images had reached Rome by early April and at least some of the text by June of 1767.[13] By 1768, d'Hancarville's difficulties completing the accompanying text and paying for the printing began to

above' is 16 carlins a day; if, however, the 5 carlins room and board is 'barely half' of what he spends per artist, then the artists must be receiving 5–6 carlins in wages, bringing d'Hancarville's total expenditure to 10–11 carlins, a substantial savings over the 16 carlins/day indicated in his previous paragraph as an artist's standard daily wage. This explanation does not quite tally with d'Hancarville's prose ('aussi bas que' rather than 'plus bas que'), and it may be that he is 'blinding' Hamilton 'with science', to paraphrase Jenkins's view of d'Hancarville's rhetorical style (*V & V* 97). On the five competing volumes see n. 7 above.

[13] See the letters to P. Usteri of 8 April and 27 June, *Briefe* III.247, 278. Winckelmann also anticipates that the total number of plates in the work will exceed the promised number of 468 by an even wider margin, approaching 600 by the end (III.366). In various early letters to Hamilton, d'Hancarville promises that his work will have more pages per volume than Palladio's volumes on architecture (*Antichità* 123); be 'both relatively and absolutely cheaper' than Caylus's *Receuil d'Antiquités* (121); and have more plates than all comparable volumes except the *Musaeum florentinum* (124).

cause long delays. His letters, therefore, draw attention to the fact that he has delivered and even exceeded the promised number of plates (468), a performance that impressed Winckelmann as well. But he also projected astronomical sales of 4,500 or more copies of the work (121), though conceding that they would need to sell only 1,335 copies, not counting subscriptions, to get Hamilton clear of his debt (125). If nothing else, the unstable nature of copper as a medium makes these figures unrealistic: both Hamilton and d'Hancarville must have known that the quality of the images would begin to deteriorate after 500 impressions or so.[14] Did they expect to engrave new plates for subsequent print runs?

This technical blind spot in d'Hancarville's calculations has gone unnoticed in the scholarship, and Winckelmann's correspondence suggests that Hamilton was unfazed by it too: 'd'Hancarville . . . will make a profit near £20,000 on [the work], as Hamilton has written to me'.[15] No doubt 'Winckelmann exaggerates', as Griener puts it (50), but he is exaggerating the expected profits (*Gewinn*), and not Hamilton's costs. Griener's tendency to dismiss all d'Hancarville's calculations as 'fantastical' leads him to misread this letter as well as one of the entries in his own collection of d'Hancarville's manuscripts, in which d'Hancarville asks Hamilton to cede his rights to all profits above £8,000 (133)—not by any means, as both Griener and Jenkins assume, because Hamilton has invested that much by 1769, but rather because d'Hancarville still wishes them both to believe that Hamilton's debt will be paid, and then some.[16] The patron's debt is the work's raison d'être: in 1767 d'Hancarville reminds Hamilton that he proposed the work as 'a certain means toward paying his debt' (126), and more than two years later is still advising Hamilton

[14] Antony Griffiths, prints curator at the British Museum, points out (pers. comm.) that although plates deteriorated after 500 impressions it might be possible to get as many as 2,000 impressions 'from a plate of this kind, perhaps more with retouching where needed'. If such high demand had existed, then perhaps technology could have been marshalled to meet it—but nonetheless it seems to me that both d'Hancarville and Hamilton overlook the factor of wear.

[15] Here is the original sentence from Winckelmann's letter to C. Heyne of 19 March 1767: 'Der Unternehmer dieses Werks, ein bekannter Avanturier von grossen Talenten, welcher sich izo d'Hancarville nennet, gewinnet auf dasselbe, wie mir Hamilton schreibt, an 20,000 P. Sterling' (*Briefe* III.242). Hamilton's letter is lost, but if he wrote anything like this it not only suggests that he believed d'Hancarville's projections but also confirms that he wished to conceal his own participation in these expected profits. D'Hancarville's sales projections, though seemingly ignorant of limitations on production, are supported by impressive demographic analysis that recalls the bent for social science displayed in his first publication, *Essai de politique et de morale caculée* (1759). He breaks down the potential market into four large segments, including artists and artisans, libraries, print collections, and academies/universities (*Antichità* 120–1).

[16] At this point d'Hancarville is not anticipating a further investment from Hamilton beyond the total of £1,350 recorded in his letter of 30 January 1770 (136). By the time he is finished, he claims, that sum will amount to 'less than one-sixth' of the total invested (133), as against the 50% or more share of profits implied by the upper limit of £8,000. Yet Griener cites this letter of 1769 to support an estimate that Hamilton had invested £8,000 (50), which seems more plausible if the purchase price of the vases is included, as it is by Constantine (*Fields* 65).

to represent the plates to his banker, Ross, as a 'fund' guaranteed to produce annual payments (134). I believe that previous scholarship, swayed in part by d'Hancarville's bad reputation, has exaggerated his deceptions and Hamilton's costs, of which only £2,770 can be documented.[17] I have attempted to sum up the fiscal state of the case as I see it, though I will have occasion to return to those reckonings that seem to me to shed the most light on this unorthodox patron-client relationship.

The aesthetic pleasures of the workshop can also be gleaned from the testimony of Winckelmann, who visited d'Hancarville in 1767 and agreed to contribute text for the second volume, though he did not deliver it before he was killed the following spring. Winckelmann, too, lodged with d'Hancarville. He reports in a letter that his room has been 'decorated with so-called Etruscan vases that belong exclusively to me, and [d'Hancarville] is daily adding to their number'.[18] Early reactions, including Winckelmann's detailed and enthusiastic letters, do not single out Bracci or any other artist by name, but their importance for the success of the project appears later in a letter from d'Aix, one of the financial backers d'Hancarville found to continue the project after he fled to Florence. D'Aix informs Hamilton that he has brought the same team of Neapolitan artists who worked on the first two volumes to Florence to complete the final two volumes 'in the same form and character as the two others' (143).

Winckelmann's testimony suggests that d'Hancarville gave his collaborators the ability to imagine themselves in the role of the collector, and he sought to give the same ability to his readers. Moreover, he anticipated that the work would find an avid readership and one-ninth of its purchasers among artists and artisans, including designers of jewelry, textiles, stucco ornaments, and many other goods besides earthenware (120). In a sense, then, his production process was at the same time a marketing experiment that surfaces in explicit appeals to a commercial and/or professional readership (e.g. *AEGR* I.xx). When d'Hancarville claims that his book will enable 'the Artist who would invent in the same stile' as the ancient vase painters to 'do so with as

[17] Based on exhaustive study of manuscript sources by both Griener and Jenkins, a total expenditure of £2,770 by Hamilton is all that can be documented, although it is certainly possible that some expenses went undocumented. The three main sums include £1,350 between 1766 and 1768, as documented in d'Hancarville's accounting of 30 January 1770 (*Antichità* 136), which is meticulous and also consistent with the earlier letters; another £1,000 in December 1772 (51); and then a final payment of £420 to the creditors in 1778 or 1779 for the release of Hamilton's copies (*V & V 63, 79*). This is not counting subscriptions. By the middle of 1766 d'Hancarville had received 420 guineas from Hamilton and was 'assured of' 179 subscriptions from London and another 60 from Naples (*Antichità* 118–19). In 1779 Hamilton wrote to his bookseller, Thomas Cadell, that d'Hancarville 'had all the subscription money' (*V & V* 49).

[18] 'In Neapel hingegen habe ich bey einem der größten Avanturiers eine eigene Kammer, die mit sogenannten Hetrurischen Gefäßen, welche mir eigenthümlich gehören, ausgezieret ist, und von demselben für mich vermehret werden' (*Briefe* III.366).

much truth and precision, as if he had the Originals themselves in his possession', the simile suggests both physical ownership of the collection and a privileged ability to recreate the objects faithfully (I.vi).

Several commercial designers, most famously Josiah Wedgwood, read the *Antiquities* in this sense, and though it was Hamilton's ambition to disseminate neoclassical taste through commercial manufacturing, d'Hancarville deserves some credit for making this happen. His assumed baronial identity marks him as a 'social climber', in Brylowe's words, who identified with the aristocracy, but his letters express tremendous satisfaction in his ability to organize and labour in the production of his own book.[19] In a certain sense d'Hancarville sacrifices his aristocratic pretensions in favour of his authorial ones, selling or pawning all his valuable property, including even his cherished Raphael (*Antichità* 129), to fund the ongoing work. By the end of 1769, with the second volume two years overdue, he stakes his moral authority on his responsibility to his crew of artists in a gorgeous poetic simile: 'I see nothing around me but suffering faces . . . [I am] like a pilot whose vessel is overwhelmed [*abîmé*] by a storm, . . . the waves draw him back in just as he is about to land.'[20]

In his more defiant letters to Hamilton, d'Hancarville uses his relationship to the artists to assume a quasi-professional identity that gives him a strong autonomous position, if not equal social status, vis-à-vis his nominal patron. Paradoxically, his identification with Hamilton works not through his aristocratic pretensions but through his artistic and intellectual credentials, especially through his antiquarianism, a distinction to which Hamilton equally aspired. This tone is more pronounced in the earlier letters leading up to and immediately following the success of the first volume, in which he emphasizes Hamilton's great debt to him for his authorial and managerial efforts.[21] From late 1769 the gloomier tone captured by the nautical metaphor predominates, but d'Hancarville's sense of himself as Hamilton's partner resurfaces dramatically in a 1773 letter, written when the project was finally moving forward again after three and a half years of catastrophic delay in Florence. Here he writes as a colleague, no longer a supplicant but a fellow *savant* who wants nothing more than to exchange ideas and please his favourite reader. As

[19] Brylowe assumes too readily, it seems to me, that 'the social climber d'Hancarville was completely uninterested in the world of manufactures' ('Two Kinds of Collections' 33).

[20] '[J]e ne vois autour de moi que des physionomies qui souffrent . . . Semblable à un pilote dont le vaisseau abîmé par la tempête, vient de s'engouffrer dans la mer, en vain il tient encore le gouvernail pour s'aider à gagner le rivage auquel il est prêt à aborder, les flots de la mer le retirent au moment qu'il est prêt à toucher la côte' (*Antichità* 135).

[21] 'Vous pouvez à présent observer le changement de scène', he writes, in one case: 'c'est que par le passé j'ai toujours été votre débiteur et par la nature de la chose vous aller toujours être le mien pour l'avenir' (126); elsewhere he asserts that Hamilton's total investment—£1,350 as of January 1770 (136)—'ne sera pas *la sixième partie* de ce que j'y ai placé jusqu'à présent et j'aurai compté pour rien le manuscrit, la peine, l'industrie et l'intelligence qui ont créé cet ouvrage' (133).

Rosemary Sweet has observed of eighteenth-century antiquarian correspondence, 'antiquarianism was not class neutral, but . . . it did provide a language within which people from very different backgrounds could communicate and exchange information' (*Antiquaries* 60). First d'Hancarville curses his book because, he says, it almost came between them (*Antichità* 140). For all his language of passionate friendship, this is surely a kind of flattery. But the scholarly content of the letter, which challenges a thesis of Winckelmann's and asks Hamilton to confirm a piece of evidence in the Duke of Noia's collection in Naples, suggests that antiquarianism *was* class-neutral for at least one party in these exchanges, or at least that the illusion of the republic of letters had real social consequences.

D'Hancarville was right when he insisted in his earlier, more defiant letters that he was not merely Hamilton's protégé or client, and in this letter he reminds him of the reason. On the brink of renewed success, d'Hancarville writes: 'I now understand very well the affliction in which I once found you at your country house, reading a letter from your banker. One must have passed through these trials to know them. You are now clear of them, and I will be soon' (140). Like Hamilton, then, he is struggling to get out of debt, and like Hamilton he will succeed. This conclusion may be less naive than it sounds, since Hamilton always hoped to receive government compensation for his long diplomatic service, only to be bitterly disappointed in this hope less than two years before his death. D'Hancarville understood debt as an enabling condition of virtu at any social level and probably had no such illusions. Since Hamilton and his correspondents by this point (1773) routinely refer to d'Hancarville as a 'rogue' and a 'good-for-nothing fellow', the Baron's identification with him as a fellow debtor must have struck him as one more instance of presumption.

D'Hancarville had, however, made Hamilton his dependent in more ways than one, holding hostage by delaying and delaying a work that Hamilton counted upon to increase the reputation of his collection. When it became impossible to maintain the fiction of massive profits, d'Hancarville substituted intellectual currency, calling his new work 'the most singular piece of writing onancient art, which alone explains everything' (141). It has often been noted that Hamilton employed d'Hancarville partly because he felt that his own classical scholarship was inadequate to the task. In this ebullient letter d'Hancarville finally claims that his eagerness for Hamilton's friendship has made him write 'a very good book under the pretext of writing a catalogue', a book amounting to a '*métaphysique de l'art*' (141).[22] This is more than mere effrontery. Hamilton

[22] Original passages from this section: 1) 'Je conçois bien à présent l'affliction où je vous trouvais une fois en votre maison de campagne, sur une lettre de votre banquier. Il faut avoir passé par ces peines-là pour les connaître. Vous en êtes dehors, et j'en sortirai'; 2) 'la meilleure chose que j'aie jamais faite, et peut-être la plus singulière que l'on ait écrite sur l'art des anciens, qui seule explique tout'; 3) 'un très bon livre sous prétexte d'écrire un catalogue'. D'Hancarville's last letters no longer refer to the work as 'un moyen assuré de payer sa dette'

had more scholarly ambition than most of his fellow Dilettanti, and d'Hancarville's braggadocio seems calculated to remind him or at least insinuate that he coveted the role of the flamboyant scholar as much as d'Hancarville coveted his patron's role as aristocratic collector. If nothing else, this letter certainly reminds Hamilton that he initially entrusted d'Hancarville with the publication because he persuaded Hamilton that it would get him out of debt. In the event, the £8,400 Hamilton received for the actual vases from the British Museum barely reimbursed him for the cost of acquiring and publishing them.[23]

3.3 DESIGNING ART HISTORY

As François Lissarrague and Marcia Reed have shown, Bracci, Beaulieu, and the graphic artists working with d'Hancarville made extensive use of existing visual materials in their compositions. Many of the vignettes and ornamented capitals included with the text were based on designs in other antiquarian publications, especially those of Piranesi, and in the later volumes many of the vase paintings too were taken from existing engravings, as the scope of the publication expanded far beyond the Hamilton collection itself.

In this context it is worthwhile to focus on designs that appear to be original, or to show in some other way the imprint of d'Hancarville's art-historical concepts or of Bracci's unique abilities. I have already noted the philosophical, conjectural aspect of d'Hancarville's art history, the interest in 'sistem' or first principles that led him, as he puts it perhaps smugly, to write a good book on the pretext of writing a catalogue. From the moment Wedgwood hit on the name Etruria for his factory, d'Hancarville's Etruscan emphasis has been cited as an influence, but this was a self-conscious choice on d'Hancarville's part and he drops the emphasis when it becomes inconvenient in Volumes III and IV. He himself was well aware that Hamilton's vases were not Etruscan, but Etruria made a good brand name and it also made a convenient vehicle for his quasi-anthropological discourse about first principles in the first two volumes.

(126). Often he is replying to letters by Hamilton that would certainly, if found, shed more light on their relationship and specifically on whether or to what extent Hamilton shared the belief that the work could get him out of debt. On d'Hancarville as rogue, see below Section 3.4. On Hamilton's diffidence concerning his own classical scholarship, see Constantine, *Fields* 33–4.

[23] As noted above, Griener puts the cost of publication alone at £8,000, but I think on insufficient evidence (*Antichità* 50). Constantine adds the original cost of the vases to reach this figure (*Fields* 65), and two other commentators agree on a figure of £6,000 for the publication cost (see the summary in Brylowe, 'Two Kinds of Collections' 53). Although Hamilton surely spent more than the documented total of about £2,800 on the publication, it would take more work to support these higher estimates.

Fig. 3.1. Unsigned engraving. *Collection of Etruscan, Greek, and Roman Antiquities* [*AEGR*], Vol. I (1766 [1768]), Frontispiece (Dedication to George III). Courtesy of Margaret M. Bridwell Art Library, University of Louisville.

The collaboration with Bracci reinforces this association between Etruria and first principles in several striking examples. The frontispiece to Volume I (Figure 3.1), one of the most highly finished engravings in the work, forges powerful connections among three apparently discrepant objects: George III, Etruria, and painted vases. In Volume II, d'Hancarville admits candidly that no painted Etruscan vase has yet been found (*AEGR* II.134–36). But he as well as Bracci surely had exposure to the telluric nationalism that informed depictions of the landscape of Etruria as the cradle of Italian civilization, as indicated here by the River Clanis (foreground) and the rocky Appenines (right background). In his explication of this vignette d'Hancarville also makes national symbols of the fragment of entablature, the ornamented border of the inscription, and the fasces leaning against it, all supposed to commemorate Etruscan inventions. Of the three, only the fasces is still attributed to the Etruscans by modern archaeology. The most important association forged by this image is that between Etruria and the origin or first principles of the arts, a connection reinforced by the topographical elements of the river and the cave. Not only Bracci but several of the engravers, including Antonio Cardon and Carlo Nolli, had experience with views and scenery.[24]

D'Hancarville and Bracci take a very different approach in one of the title vignettes to the first chapter of text (Figure 5.1, in Chapter 5). Bracci restrains his sense of landscape here, though the picturesque motif of ruins overgrown with vegetation appears clearly both in the vignette and in the initial capital L. Bracci designed twenty-six of the forty ornamented capitals used in the four volumes, as Ramage has noted, and this is one of six for which no prior pictorial source has been found. The reason in this case is that d'Hancarville, probably with Bracci alongside, made these observations in the field. Both the inscription and the Temple of Neptune are from Paestum, where 'they search[ed] with Mr Hamilton for every circumstance that might instruct us, as to the plan and size of this town' (I.96). D'Hancarville's affinity for the authorial 'we' makes it hard to be sure, but it seems likely that he is referring to himself and an artist when he reports their chance finding of the inscription, which appeared on the interior of a wall as workers were knocking it down for building stone. The added French text here (Figure 5.1) alludes to this incident, even though the narrative of the discovery comes much later in Chapter 3. The specimen holds at least a double significance as both an empirical confirmation of the antiquity of Etruscan letters (the subject of the first chapter) and the sign of an archaeological approach that is more rigorous than the fanciful composition of the frontispiece.

Several of the vignettes designed by Bracci combine these two approaches. The river motif in the tailpiece to the Volume I text (Figure 3.2) echoes the allegorical depiction of the Clanis or Liri in the frontispiece to this volume.

[24] See Lissarrague and Reed, 'Collector's Books' 275–6, and Ramage, 'Initial Letters', on the artists' résumés. The vase painting depicted in this image is reproduced as a colour plate near the end of the volume (Plate 122); for details on this vase, an Attic *pelike*, as well as a reading of the dedication, see Jenkins and Sloan (*V & V* 146–7). D'Hancarville's own explication of the painting is given in Volume II (166). For an account of the Etruscan origins of the fasces, see MacKendrick, *Mute Stones Speak* 40–1. For a good account of the role of telluric nationalism among antiquaries in both northern and southern Italy, see Lyons, 'Neapolitan Context'.

Fig. 3.2. Engraving by Carlo Nolli. *Collection of Etruscan, Greek, and Roman Antiquities* [*AEGR*], Vol. I (1766 [1768]), p. 173 (tailpiece to text section). Courtesy of Margaret M. Bridwell Art Library, University of Louisville.

The unspecified spring flowing downhill toward the viewer once again allegorizes the first principles of the arts in a happy synthesis of d'Hancarville's argument with the picturesque repertoire of Bracci and Nolli. The cliffs suggest a metamorphosis of the architectural ruins that serve as the backdrop or matrix of similar artefacts in the other vignettes, ruins echoed here by the fragment of a wall on the left. The expanded topographical scale of this design also permits an allusion to the original use of the two larger vessels as containers for water. For d'Hancarville some reference to utility is a crucial feature of all design, one of the first principles that he cites throughout the work. In his explication of the image—another record of his own involvement in these designs—d'Hancarville refers to these artefacts as 'different *morceaux* taken from the cabinet of Mr Hamilton' (177), indicating that the collection comprises much more than just the painted vases featured on the folio plates and provides a kind of ethnographic context for the vases themselves.

Another group of vignettes borrow Piranesi's trick of incorporating tiny rustic figures that further monumentalize the scale of the artefacts and inscriptions featured in the images. An unsigned pair in this style head the preliminary discourse on painting in Volume II, and the first preface concludes with a monumental inscription set in a landscape (Figure 3.3). Like the vases at the fountain of the arts, the artefacts in this landscape, again referred to as 'morceaux' from Hamilton's collection, presuppose an elaborate *mise-en-scène*

Fig. 3.3. Engraving by Carlo Nolli (possibly from a design by Giuseppe Bracci). *Collection of Etruscan, Greek, and Roman Antiquities* [*AEGR*], Vol. I (1766 [1768]), p. xxv (tailpiece to preface). Courtesy of Margaret M. Bridwell Art Library, University of Louisville.

that relocates the collection from the cabinet back to the countryside. We see them again in situ, not as they were originally discovered but rather reorganized as components of a *paysage moralisé*, with the figures suggesting a roughly modern date. The inscription is crumbling and overgrown but clearly legible, and the arms have been completely cleaned of rust and debris. D'Hancarville's explication of this vignette concentrates largely on the text of the inscription and its implications for military history as well as pre-Roman religion.[25] Although it is related to the arms and the legionary ensign, an eagle, all in Hamilton's collection, d'Hancarville seems to have inserted the inscription, from another collection, as an occasion to display his learning. Like the previous engraving, this one is signed only by the engraver, Nolli, but the style suggests that Bracci also had a hand in this composition. A widely reproduced pair of vignettes exhibiting the silvan origins of candelabra and columns and the religious origins of sculpture, among other first principles,

[25] He is exclusively interested in the larger inscription for both cultural and philological reasons: it proves, he says, that Roman soldiers were allowed to carry their women with them on campaigns, at the time this was made; and the spelling of Poeninus with the oe diphthong proves that he is not a Carthaginian deity (*AEGR* I.175–76). Patricia Crown suggests (pers. comm.) that d'Hancarville's conjugal thesis may be a muted echo of the erotic history recorded in monuments created for eighteenth-century landscape gardens.

again bears no signatures at all (Figures 5.2–3, Chapter 5 below); an able draftsman, d'Hancarville may have composed them himself.

Each volume of *AEGR* carries a separate dedication, and the most spectacular of these features d'Hancarville himself as the miniaturized figure establishing the scale of the monument (Figure 3.4). D'Hancarville commemorates his friend,

Fig. 3.4. Engraving by Carmine Pignatari from a drawing by Giuseppe Bracci. *Collection of Etruscan, Greek, and Roman Antiquities* [*AEGR*], Vol. II (1767 [1769]), Frontispiece (Dedication to Johann Joachim Winckelmann). Courtesy of Margaret M. Bridwell Art Library, University of Louisville.

Fig. 3.5. Engraving by Carmine Pignatari from a drawing by Giuseppe Bracci. *Collection of Etruscan, Greek, and Roman Antiquities* [*AEGR*], Vol. III (1767 [1776]), facing p. 4 (tailpiece to preface). Research Library, The Getty Research Institute, Los Angeles (90-B2863 v.3)

sometime collaborator, and—in his own mind, perhaps—arch-rival, Winckelmann, who was killed while Volume II was in progress. The inscription prompts us to read the mourning figure as d'Hancarville humbly offering his text at the feet of the great deceased antiquary, an attitude that he also assumes in his writing. The magnificent rendering of a Roman columbarium attests to Bracci's close study of Piranesi as well as his own gift for architectural draftsmanship. Both this and the next image (Figure 3.5) are signed by Bracci and the engraver Carmine Pignatari, the team responsible for the archaeological illustrations of Paestum. The monument here—a much less magnificent funerary altar set in a desolate landscape—may be read as one of d'Hancarville's myriad inflections of the modesty topos. The inscription, however, echoes the inscription on Winckelmann's sarcophagus, and though there are no mourners present d'Hancarville isn't really dead either.[26]

[26] The inscription reads as follows: D M/P Vict Hancarville/ANNVM AGENS PLVS MINVS XXXXIIII/SIBI FECIT ORCO PEREGRINO/VIXIT PARTHENOPE TENET NVNC FLORENTIA/SI QUID OBIECTAMENTI APVD VOS EST/MANES MOESTAM REFICITE ANIMAM/TV QUI LEGIS/HAVE ET VALE. I am indebted to Ray Marks and Jonas Cope for the following translation: 'To the spirits of the dead [Dis Manibus], P[ierre] Vict [François?] D'Hancarville, about forty-four years old, made this [epitaph] for himself, for when he should be a pilgrim in the underworld. He lived [died?] at Naples. Florence now holds [his body]. If there is any objection among you, restore, spirits [of the dead], his sad soul. You who read [this], hail and farewell.'

PLATE 1 *View of the first discovery of the Temple of ISIS at Pompeii.* Hand-colored etching from a gouache by Pietro Fabris. *Campi Phlegraei: Observations on the Volcanos of the Two Sicilies* (1776), Plate XLI. By permission of Staatsbibliothek zu Berlin-Preußischer Kulturbesitz, Abteilung Historische Drucke.

PLATE 2 *Interiour view of one of the deepest hollow ways cut by the torrents of rain water, on the flancks of Mount Vesuvius, call'd the Fossa Grande.* Hand-colored etching from a gouache by Pietro Fabris. *Campi Phlegraei: Observations on the Volcanos of the Two Sicilies* (1776), Plate XXXIX. By permission of Staatsbibliothek zu Berlin-Preußischer Kulturbesitz, Abteilung Historische Drucke.

PLATE 3 *Representation of a thick Stratum of Lava, that ran into the sea from Mount Vesuvius in the terrible eruption of 1631.* Hand-colored etching from a gouache by Pietro Fabris. *Campi Phlegraei: Observations on the Volcanos of the Two Sicilies* (1776), Plate VIII. By permission of Staatsbibliothek zu Berlin-Preußischer Kulturbesitz, Abteilung Historische Drucke.

PLATE 4 *View taken from the spot N. 1 near Puzzoli, on which anciently was Cicero's Villa call'd the Accademia.* Hand-colored etching from a gouache by Pietro Fabris. *Campi Phlegraei: Observations on the Volcanos of the Two Sicilies (1776),* Plate XXVI. By permission of Staatsbibliothek zu Berlin-Preußischer Kulturbesitz, Abteilung Historische Drucke.

PLATE 5 *View of a part of the inside of the cone of the mountain of Somma.* Hand-colored etching from a gouache by Pietro Fabris. *Campi Phlegraei: Observations on the Volcanos of the Two Sicilies* (1776), Plate XXXV. The figure at the right is Bartolomeo Pumo. By permission of Staatsbibliothek zu Berlin-Preußischer Kulturbesitz, Abteilung Historische Drucke

PLATE 6A (*Left*) *Unsigned engraving. Collection of Etruscan, Greek, and Roman Antiquities* [*AEGR*], Vol. IV, Plate 81. [Eros and three women]. Vase untraced. Research Library, The Getty Research Institute, Los Angeles (90-B2863 v.4).

PLATE 6B (*Below*) *Unsigned engraving. Collection of Etruscan, Greek, and Roman Antiquities* [*AEGR*], Vol. I, Plate 24 (upper half). Corinthian column krater (boar hunt). Now British Museum Vases B37 (1772,0320.6). Courtesy of Margaret M. Bridwell Art Library, University of Louisville.

PLATE 7 Unsigned engraving. *Collection of Etruscan, Greek, and Roman Antiquities* [*AEGR*], Vol. IV, Plate 69. Apulian *kylix* (interior: medallion); now British Museum Vases F456 (1772,0320.695). According to d'Hancarville, the scene represents votive *boetiles* being removed from a *cassette mystique*. The more prosaic modern rendering is "Eros, holding box" (Gisler-Huwiler et al., *Collection* 477). Research Library, The Getty Research Institute, Los Angeles (90-B2863 v.4).

PLATE 8 Unsigned engraving. *Collection of Etruscan, Greek, and Roman Antiquities* [*AEGR*], Vol. I, Plate 45. Apulian *podanipter* (foot-bath). Tondo: woman seated on an Ionic capital. Now British Museum Vases F462 (1772,0320.393). Courtesy of Margaret M. Bridwell Art Library, University of Louisville.

3.4 THE AUTHOR AS PRODUCER

D'Hancarville stops short of dedicating the third volume to his own memory. His memorial appears as the tailpiece to the audacious preface, capping a final piece of effrontery. In abandoning the arguments of his first two volumes, including the thesis of Etruscan origins, for a new approach, d'Hancarville declares the author of those volumes dead. His justification for doing so is somewhat enigmatic, though his letters suggest that he may be addressing the naysayers who predicted—in the course of a five-year lapse between the second and third volumes—that his work would never be finished. 'The Publick', he writes simply, 'does not forget, that a right has been given me, to look upon myself as an author that does not exist for some years past; not being at a loss, if needs must, to name the places where I have been burry'd, and the People who have instigated my burryal, I think myself impowered to hold my former opinions, to be them of a man dead this long time' (*AEGR* III.3). He informs us that his goal is still to describe 'the infancy of the Arts' (4), but he is not so vain as to cling to his own exploded opinions, the greatest weakness of which (he says) was simply the weight allowed in them to the views of other scholars—always an impediment to progress. There is no testament here to his collaborators, as in the preface to Volume I, no acknowledgment of the complex network of dependency involving Hamilton, the artists, and the public. In 1992 Griener discovered what may be the largest absence of all: d'Hancarville also fails to mention Octavien de Guasco, whose *De l'usage des Statues chez les Anciens* (1768) in fact provided some of the new 'opinions' for which he here claims such absolute originality. Guasco is suppressed or 'cancelled by the adventurer's "historiography"', for d'Hancarville took the 'Machiavellian' measure of having the false date 1767 printed on the title pages of his third and fourth volumes (*Antichità* 82).[27] This plagiarism notwithstanding, and despite the fact that he wrote these volumes in isolation—in debtor's prison, or from his refuge in the convent of Santo Agostino—the argument of the new volumes is in some ways more sophisticated. But these later volumes lack the enthusiastic record of collaboration that preserves for us the innovative and—initially at least—superbly effective production process that made the work possible.

It is hard to reconstruct Hamilton's own view of this production process, or that of his contemporaries. D'Aix, the new investor who brought the Neapolitan artists to Florence in 1773, typifies a continuing tradition of flattery when

[27] For a fuller treatment of the plagiarism issue see Chapter 5 herein. It seems more likely to me that the wrong imprint date was engraved on the title pages simply because they themselves were engraved in 1767. The text of the work took another nine years to complete, but the plates really were finished by 1768, and there is good evidence that d'Hancarville genuinely expected the text to follow much more quickly.

he writes to Hamilton of the 'four volumes of which you are, Monsieur, the sole creator [*l'unique créateur*]' (143). Hamilton's contemporaries typically did give him most of the credit for the publication, and rightly so, insofar as he spent a great deal of money and showed remarkable perseverance over a period of some fifteen years (before his copies of III–IV were finally released). As a consequence modern scholars are tempted to assign to Hamilton a somewhat exaggerated supervisory role in the production process. But Hamilton left very few traces of his feelings about d'Hancarville in his letters—scholars cite the same few passages again and again—and did not describe the production process. D'Hancarville's work on the project certainly improved his contemporary reputation and the near-unravelling of the project in its later years did not reverse those gains entirely. He portrays himself throughout the published text as a neglected and even persecuted genius; in his correspondence with Hamilton, he also portrays himself as a fellow antiquary, a solid business partner, and a faithful if unjustly suspected friend.

Hamilton's remarks, suggesting a gradual erosion of trust, can be quickly summarized. In the absence of any surviving letters to d'Hancarville, the best evidence of Hamilton's continued trust in him appears in a letter to Sir Horace Mann: 'His knowledge of the Antiquities of this country & the perfection of his Etruscan work . . . is I believe the real cause of his disgrace'. Biographers are quick to add that the reason actually cited by Ferdinand's government for d'Hancarville's 'disgrace' and banishment from Naples was that he published pornography.[28] One of his earlier letters, however, also lends some support to Hamilton's suggestion here, because it shows that he did conceive of the project as competing with the Neapolitan Academy's official publication. Expanding *AEGR* to four volumes, he writes, will 'render my book competitive [*concurrent*] with the Herculaneum' [i.e. with the Academy's multi-volume *Antichità di Ercolano*] (127). David Constantine, Ramage, and others cite a 1773 letter to Wedgwood and Bentley that shows how disillusioned Hamilton became over the next three years: 'I found the rogue I employed in prison at Florence, and the plates pawned' (*H & N* I.19). There appear to be no more substantive references to d'Hancarville in Hamilton's letters.[29] He resurfaces

[28] Constantine, *Fields* 62–3, quoting State Papers (PRO) 105/319, f. 730 (cp. *V & V* 48). As Hamilton's diplomatic counterpart in Florence, Mann helped to broker his dealings with d'Hancarville and his creditors throughout the 1770s.

[29] As far as I can tell, the large selection printed by Alfred Morrison in *H & N* includes only this letter to Wedgwood and Bentley of 2 March 1773. Cp. Constantine, *Fields* 72, and Ramage, 'Publication Dates'. This and the previous passage are the only two first-hand accounts of d'Hancarville by Hamilton that occur widely in the scholarship. Jenkins cites a second letter to Mann in the State Papers (*V & V* 63, nn. 74, 76). Apart from these few brief references by Hamilton himself, the most canonical source on d'Hancarville's character is Mann, who gives a detailed assessment of this 'rogue' and 'good-for-nothing fellow' in letters of 1772–73. See *Walpole's Correspondence* XXIII.447–9, 456–7. A later letter from Mann (18 November 1777)

briefly in a letter to Hamilton from his nephew, Charles Greville: 'D'Hankerville is again *sur le pavé*, his creditors have taken his hotel at Paris, in which I am told he had some fine furniture, & even his pot de ______ was Sevre china; he disappeared without giving the least hint to anyone; it is a matter of curiosity to me where he will find an asylum, & what country has not had full demonstration of his talents' (*H & N* I.149–50). A reference by another correspondent at this time, the sculptor John Flaxman, is more flattering. Flaxman describes his newest work in progress to Hamilton, saying that he owes the idea 'to the ingenuity & learning of Mr D'Ankerville, & the advantage of his acquaintance to your introduction' (*H & N* I.166). Hamilton must have introduced them in 1777 when he was on leave in London (*V & V* 184), where d'Hancarville finally proceeded to compile a true catalogue of Hamilton's collection—which had been in the British Museum since 1772—before he acquired a new patron in Charles Townley.

D'Hancarville's letters make it clear enough that Hamilton grew weary of his requests for more money. As early as 1767 d'Hancarville began reciting long lists of other expedients he had tried (pawning his Raphael among them). He complains that his enemies, M. Turbin and Mme. Trestondam, have conspired to ruin his reputation with lenders and potential subscribers to the work. He refers to this conspiracy as 'a meal for cannibals, which resembled that of Pelops and in which I had to be eaten alive' (*Antichità* 128). As begging letters go, d'Hancarville's are quite evocative, but they also provide insight into the state of the work. In November 1768 he reports on the long-delayed second volume, noting that the plates are finished and the printer has reached page 115 of the letterpress with sixty-five pages left to go—but funds have dried up, and 'the workers (and myself) must have something to live on tomorrow' (133). Nonetheless, he writes, Hamilton can see the finished manuscript for himself: 'I beg Mr Hamilton to read this with attention and observe that that these are neither words, nor reasonings, nor promises, but facts' (132).

Hamilton was evidently embarrassed at having to subsidize the work at all. Writing from Florence in January 1770, d'Hancarville reminds him that he had pledged to keep Hamilton's financial involvement a secret, but indicates that he may be forced to reveal it at last if Hamilton joins the ranks of his creditors (136). Like many of his letters, this one is largely a *compte rendu*, a particularly succinct and forceful one that also gives the clearest account to-date of Hamilton's involvement: he advanced £1,350 between 1766 and March 1768 with no more subsidies since that time. If Hamilton attempts to seize the plates and other stock in Florence, d'Hancarville's creditors will be justified in assuming that he has cheated Hamilton, whereas in fact he is simply the

details the arrangements he made for Hamilton with d'Hancarville's Florentine creditors concerning the publication of Volumes 3 and 4 (*Walpole's Correspondence* XXIV.319 n.3).

partner in an investment to which he himself has contributed the lion's share, according to the detailed accounting in this letter. D'Hancarville concludes this poignant letter by lamenting the distress of his wife, whom he has left behind in Naples. He carried this point (Hamilton helped her get to Florence) and eventually the fiscal one as well, since Hamilton invested another £1,000 in 1772). In addition to the large expenses of boarding fifteen collaborators for 'almost five years', rent on the house he hired for the purpose, and Bracci's salary, d'Hancarville cites a host of obviously relevant smaller expenses (two-thirds of which, again, he claims to have borne himself): 'I have paid the draftsmen's fees; Mr Raimond, Cardon, La Marra; the engravers; the printers; [for] the plates; the finishing [*brunissage*]; the varnishes; the acids; the printing costs; paper; ink; colours; machines; instruments; oil and thirty more articles that I omit here' (*Antichità* 136).

Hamilton must have learned something about book production from these detailed accounts, but his impatience is understandable. His desire for secrecy is more surprising. Banks lavishly (and openly) subsidized the production of his *Florilegium* and Richard Gough invested a great deal in the drawings and engravings for his five-volume *Sepulchral Monuments*, to cite two examples treated at length elsewhere in this book. Hamilton must still have hoped that his folios would make substantial profits and really, as d'Hancarville had promised, help him pay his debts.

Winckelmann's letters, in giving a sense of the atmosphere in d'Hancarville's home workshop, confirm that d'Hancarville at least took care of (and returned) the vases he was illustrating. They also give a sense of the scholarly abilities that must have influenced Hamilton's choice of him to begin with. In the 1767/68 letter quoted above, Winckelmann is clearly excited to be involved with 'the greatest adventurer' and to sleep surrounded by Etruscan vases, even if only 'so-called'. His extravagant praise for *AEGR* in this letter—he calls it 'the most magnificent work the world has seen'—echoes the more obvious flattery characteristic of his letters to Hamilton, whom he treated more or less as a patron. But it attests to his admiration for the work's 'author and editor' as well, and he praises the adventurer's scholarship explicitly in a number of letters.[30]

Constantine has reconstructed the negotiations that led Winckelmann to promise to contribute explications of the vase paintings engraved in *AEGR*. This expectation may be one reason, as Constantine suggests, that no explications at all appear until the very end of the letterpress in the second volume (*AEGR* II.154–68). Constantine speculates engagingly that these explications are in fact largely Winckelmann's unacknowledged contribution. As indicated

[30] The Winckelmann letter quoted above (n. 18) and again here is dated 6 February 1768, but actually describes his final visit to Naples of September–October 1767 (Winckelmann, *Briefe* III.366).

by the beautiful dedicatory plate, Winckelmann was killed while this volume was in progress, and d'Hancarville does quote a manuscript letter from Winckelmann to provide an alternative explanation for a particular vase painting (164). Constantine calls this 'an eerie survival of what we know to have been a substantial correspondence'—a correspondence presumably continued after Winckelmann had studied the vases in d'Hancarville's workshop and returned to Rome.[31] Since it post-dates the visit, Winckelmann's enthusiastic letter strongly suggests that he intended to return to 'his' room in d'Hancarville's house and valued the connection. He was willing to acknowledge d'Hancarville as author, as d'Aix was not, and his praise for this 'man of great talents' undoubtedly enhanced d'Hancarville's reputation.[32]

D'Hancarville went to London in 1777, shortly after fulfilling his obligations to the new proprietors of *AEGR*. This London period is relatively well documented in letters by Townley and Richard Payne Knight, among others, and of course in the work he produced under their patronage, *Recherches sur l'origine, l'esprit et le progrès des arts de la Grèce* (1785), as well as reviews of that work. Although Townley claimed that he was 'sculking about in the City' at first, before he acquired Townley's patronage, Flaxman was evidently impressed by him then and Thomas Bentley (of Wedgwood and Bentley) also wrote in respectful terms of his expertise after d'Hancarville supplied them with a design that Flaxman modelled in 1778.[33] Certainly Townley himself was impressed by d'Hancarville's scholarship as well as his gift for conversation, and his success with *AEGR* (known to be complete at last, though still unavailable in England), as well as favourable notices of his scholarship by Winckelmann and others, must have helped to counteract his reputation as a swindler. In the surviving correspondence, Townley and Knight seem reasonably content to be subsidizing the *Recherches*—another theoretical discourse

[31] Constantine, 'Winckelmann' 65–6. D'Hancarville cites a letter Winckelmann wrote just before his death that resurfaced as he was writing his explication of the vase in question (Plate 71 of Volume I). If Constantine is right, then he is only pretending to be candid by disclosing how widely Winckelmann's explication (Hercules purchased as a slave by Omphale) of the same vase differs from his own (Cassandra foretelling the fate of Troy). The Winckelmann letter quoted above is one of several that confirm the German scholar's intention of writing the explications. In the absence of any surviving manuscript, it is impossible to confirm how much of this he accomplished: a reading of the explications suggests to me that they are d'Hancarville's work, but surely with more influence from Winckelmann than he acknowledged.

[32] Winckelmann to Volkmann, 27 March 1761 (*Briefe* II.131). Though Winckelmann had misgivings about d'Hancarville's moral character, he expressed his admiration from the first time they had contact. There are a substantial number of favourable remarks in his letters of 1767 (see especially III.242, 267, 317) and these as well as his published praise for *AEGR* in the second edition of his *History of the Art of Antiquity* (Constantine, 'Winckelmann' 71) must have helped d'Hancarville's career. To some extent the benefit was mutual: one of Winckelmann's letters shows how flattered he felt by a respectful reference to his latest publication in *AEGR* I (52–3)—although he assumes that Hamilton must have instructed d'Hancarville to insert the reference (*Briefe* III.278).

[33] *V & V* 184. For the Townley quotation, see Haskell, 'The Baron' 41.

about ancient art and religion composed 'under the pretext of writing a catalogue', this time of Townley's collection. But if Townley found his discourse 'more rational and more Satisfactory than all I had found in the absurd books of Antiquaries', Horace Walpole was quite caustic in a letter written shortly after d'Hancarville visited Strawberry Hill to draw one of his bronzes—for eventual publication, as it turned out, in Knight's *Discourse on the Worship of Priapus*. To judge from Townley's remark, Walpole may have felt that d'Hancarville was becoming too successful at a game that Walpole himself had mastered: gaining credit as an antiquary while pretending not to be one.[34]

D'Hancarville's employment as a draftsman did not earn him any respect from Walpole, but it confirms his talent for this crucial form of knowledge work. It seems very likely that he contributed many drawings himself to the *Recherches*, a three-volume quarto publication produced much more modestly than *AEGR*. D'Hancarville was in Paris from 1785–91 and then returned to Italy. Greville's anecdote about the chamber pot is only one of several clues indicating that he lived more comfortably, as a rule, after completing *AEGR* than before. If his production of this work did not exactly make an honest man of him, it launched a comparatively solid intellectual career. It made him an 'art historian', in Francis Haskell's words.

Griener's collection of d'Hancarville's letters provides a superb (and still under-used) body of evidence concerning the production of *AEGR*, a vital material context for d'Hancarville's often idealizing notions concerning the deep past. It is also a one-sided body of evidence, which makes it necessary to review the limited testimony available from Hamilton himself and from other observers, especially Winckelmann. There is nothing in this testimony to cast doubt on the general accuracy of d'Hancarville's account of the production process. It is easy to see why Hamilton was not persuaded by d'Hancarville's justifications of his budgeting difficulties and his numerous long delays. But it is also clear from these letters that d'Hancarville at times worked hard on the project, that he put all his own resources into it (such as they were), and that he became highly skilled in many aspects of book production. The volumes themselves confirm all this, and his detailed accounting of the cost of particular materials and processes enables a rich historical view of the work.

Early reactions to the work also indicate that he was at least somewhat justified in his own sense of its importance and magnificence. It would be unreasonable to expect confirmation of d'Hancarville's own sense of himself

[34] For the Townley/Knight correspondence, see Haskell, 'The Baron' 40–2. Haskell also reproduces Johann Zoffany's 1781 painting of Townley and d'Hancarville at work in the collection. Walpole reports on d'Hancarville's visit and reviews his new book in a letter to Lady Ossory of 9 July 1785 (*Walpole's Correspondence* 33.477–80). Rousseau discusses the political appeal of d'Hancarville's ideas for Knight, Townley, and Hamilton and also gives a useful account of Paul Henry Maty's sharply negative review of d'Hancarville's *Recherches* ('Sorrows of Priapus', esp. 115–17).

as a neglected or persecuted genius. Opinion remained divided on whether he was a brilliant historian of antiquity (as the letters of Townley, Knight, and Flaxman suggest) or merely a bombastic swindler (as indicated by the letters of Walpole and Greville, among others). In 1791, Hamilton still referred respectfully to d'Hancarville's theory as offering, together with Winckelmann's works, 'the most rational account' of Greek vases (*Collection* I.14). D'Hancarville, always a prolific writer, did his best to ensure that he had the last word.

The autobiographical topos of the persecuted genius infuses some of the most memorable passages in *AEGR* as well as the letters collected by Griener. The letters give us images of the beautiful printed folio sheets being devoured by rats (*Antichità* 133); of jealous machinations; of a bitter exile in Florence, where, he writes, 'all my effects have been seized, my wife is dying of hunger', and, without Hamilton's support, 'my work is ruined and I perish in prison' (137). The sense of persecution in the letters sometimes mounts to paranoia, as we have seen in the letter concerning the 'anthropophagic' conspiracy of his enemies in Naples to ruin his credit with slander. Both of the principals there are motivated by jealousy, and later on he learns that one Mr Dosten, a 'cruel man' and a jealous one, instigated the conspiracy that led to his banishment from Naples—a conspiracy, he adds mysteriously, that was governed from Spain (*Antichità* 142). In Florence, too, he is beset by malicious rumours that he will not finish the work, and that he has lost Hamilton's support—rumours that were surely true to some extent, and in any case had the same material effects whether they were true or not. Elsewhere in the letters he promises to triumph over these detractors, who thoughtlessly impede the publication of his work 'to the detriment of the republic of letters' (139). When d'Hancarville announces his own death in the preface to Volume III and speaks of 'the places where I have been burry'd, and the People who have instigated my burryal' (3), the paranoia and the triumph are more muted. But the complex conceit of his own death (Figure 3.5) allows him to target his detractors both in Naples (where he died, according to the engraved monument) and Florence (where his body lies).

D'Hancarville's mastery of the modesty topos allows him to make the case for his own genius more subtly in the published text as well. He is less decorous in the letters, particularly those written after the text of Volume III had been completed. Most extravagantly, d'Hancarville boasts of writing 'a very good book under the pretext of writing a catalogue' and freely uses words such as 'best', 'brilliant', and 'singular', to describe his new work (139–41). When he regrets, in his own epitaph, that he followed existing authorities too closely, he stakes his claim to originality in more familiar scholarly terms (*AEGR* III.2–3). Autobiographical passages throughout the four volumes present the author's genius and trials in a variety of more palatable literary inflections, often allegorical and thickly layered with modesty, as in the conclusion of his first preface. There he celebrates the fact that 100,000 characters have been cast 'on purpose' to compose his narrative but also entertains the possibility that his ideas may not warrant this lavish treatment (I.xxiv).

The 'preliminary discourse' on painting in Volume II offers another justly famous autobiographical excursus: 'O Painting! thou truly divine Art, it is to thee, I owe my comfort, that only comfort, which I enjoy'd in those gloomy days, when, by the sport of capricious fortune, the low intrigues of Courtiers, and the ambition of Princes, I saw, like Damocles, the fatal sword continually hanging over my head' (II.8). After developing this scene a little further d'Hancarville launches into a long ekphrastic digression on two paintings of Raphael (8–20) designed to illustrate the highest ambitions of painting through the extant canvases that most fully embody the principles of the ancients (a connection also made by Winckelmann). At the same time, he demonstrates his erudition—attested by one of his coyest footnotes (8–9n.)—and his sensibility. He devotes the chapter primarily to a critique of the contemporary art scene (hence 'preliminary discourse'), but the implied analogy to vase painting provides a hint that d'Hancarville found a similar retreat for his persecuted sensibility in his reveries on Hamilton's vases. As in the previous passage, then, he alludes to the production of his own work, and he often finds occasion to defend the slow production of his work as well (II.2–3, II.50, III.4). A more heavily allegorical passage ostensibly describes the state of the arts at the time of Daedalus but at the same time unmistakably promotes d'Hancarville's tragic view of himself. 'On rare occasions', he declares, 'men gifted with a Genius capable of returning to the sources of the Sciences or Arts and fixing their principles, find themselves alone among their species' (III.101). Daedalus, unlike d'Hancarville himself, had the even rarer privilege of belonging to one of the 'small number of fortunate centuries' (104) that enjoyed a happy synergistic concourse of genius.

My next chapter maps the contours of d'Hancarville's antiquity, a domain on the edge of prehistory in which semi-legendary figures such as Daedalus play a major role. Throughout the four volumes of *AEGR*, d'Hancarville's experimental anthropology of artistic practice and (especially) of religious art occasionally clashes with his much more conventional idea of ancient (especially Greek) politics and social structure. In both cases, however, we can read his account of 'most remote antiquity' as a metanarrative concerning the laborious production of this remarkable book. In this context d'Hancarville's fairly conventional civic–humanist account of Greek painters as public men allegorically suggests how the experience of its production informed the argument of his work:

> Born free they scarcely worked for any but their Equals, and had not the mortification of employing, in order to gain Protectors, a time which they must have taken from their Study's; so that, Genius was not confined by dependance,

> debased by want, or destroyed by the humiliating suffrages of a Proud Maecenas little capable of judging, and upon whom depends but too often with us, the fortune and reputation of an able man . . . Being to work for the Public only, who is seldom in a hurry to enjoy, because it is sure to exist for ever, he was at liberty to employ as much time as he pleased in his compositions; he could see, learn, examine and did not lose his reputation and his time in executing slowly, but in executing ill: hence painting fewer pieces, the Artist did not exhaust his spirit upon a multitude of different works, but employed his whole strength upon a small number: he did less, but he did better. (II.48–50)

D'Hancarville makes a show of distinguishing an elite or 'first class' among vase painters who might plausibly have achieved this public status (150), although more recent scholarship does not bear out this idea. But in this instance he is less concerned with ancient practice than with harnessing the power of historical empathy to defend a project that already in 1769 seemed to many readers to be dragging on and unlikely to be completed. D'Hancarville's experimental and ambitious approach to book production—and not just his lack of moral fibre, as scholars too often seem to assume—accounted for this and for many more delays, but did in the end produce monumental results. The technical challenges arising from his methods of production shaped the argument of the work decisively over a long period. My approach to that argument in the next chapter thus depends as much on the history and setting of the book's production as on the artefacts and the historiography with which it is explicitly concerned.

4

The Natural History of Art: Customs and Manners in *The Collection of Etruscan, Greek, and Roman Antiquities*

We have so many Etruscan vases in books, and the subjects represented on them shed so little light on the science of antiquity, that I am obliged to confess my indifference toward the overly expensive work that Mr. Hamilton is preparing. (P.-J. Mariette, 1767)

4.1 THE SHIFTING TERRAIN OF ANTIQUITY

The last surviving letter from d'Hancarville to Hamilton, dated 29 November 1774, announces that he has finally received permission to print Hamilton's copies of Volume III of the *Collection of Etruscan, Greek, and Roman Antiquities* (*AEGR*). The final volume followed early in 1776, confirming that d'Hancarville's writer's block was finally resolved, and Hamilton must have been hugely relieved to see the volumes in print (though their sale in England would not be authorized for another four years).[1] Hamilton lost no time in moving on to his next folio publication project, *Campi Phlegraei*, first mentioned in a letter to Greville of 6 June 1775 (*H & N* I.38). Although Hamilton complained of delays and escalating costs in this case too, these two volumes appeared at roughly the same time as *AEGR* III–IV in 1776. *Campi Phlegraei* was produced ten times faster, but the first of the incorporated letters dates back to 1766, the same year that Hamilton began working with d'Hancarville on the vase

[1] Jenkins and Sloan quote manuscript correspondence concerning this delay and Hamilton's consequent negotiations with the Florentine creditors (*V & V* 49). Paris booksellers had the new volumes by 1776, but Hamilton's bookseller, Thomas Cadell, had to wait until a new settlement was reached in 1780 before he received his complement of 100 copies. My epigraph is taken from a letter of the antiquary Pierre-Jean Mariette to Padre Paciaudi on 26 February 1767 (quoted in *Antichità* 9).

project. D'Hancarville's writing bears many traces of his engagement with Hamilton's evolving ideas about Vesuvius, with the volcanic phenomena themselves, and with natural history in general.

Notwithstanding obvious differences in form and content, the two book projects share certain key features in common. Both are topographical works engaged with the deep past and thematizing the conditions of their own production. Just as d'Hancarville writes of visiting Paestum with Hamilton and (probably) Bracci, so too Hamilton describes his fieldwork with Fabris, who in turn incorporated representations of their work into his plates (Plates 1, 2, 5). Taking a cue from d'Hancarville, Hamilton details the nature of his working relationship and his financial arrangements with the artist in an elaborate preface (*CP* 5–6). Just as Hamilton incorporated Fabris's expertise concerning regional customs and dress, d'Hancarville too immersed himself in the celebration of picturesque local landscapes and figures that provided a livelihood for many of his artists and initially—in the form of continuing sales of his 'golfe', his work on the Bay of Naples—helped to fund the production of the vase book itself.

In his final letters to Hamilton d'Hancarville also announces a radically new approach to his subject matter. Unlike the ostentatious published retraction of his own earlier arguments (III.3–4), these letters suggest a revision of the origins of art rather than a total departure from this central theme of the first half of the work, completed five years earlier. D'Hancarville's metaphysical interest in first principles remains tied, paradoxically, to the ethnography of local customs and the comparative study of ancient and modern cultures. He produces a prehistory of art by means of syncretism, triangulating surviving artefacts with ancient accounts of customs and manners that he compares to one another and to modern survivals. In the new volumes he drops his insistence on Etruscan priority, though ancient and modern South Italian peoples continue to figure in his examples. And the letters strongly reiterate his commitment to the prehistory of art: though now placed at the service of d'Hancarville's newfound Philhellenism, biogeography and environmental causes continue to play a major role in this prehistory, which despite its philosophical rhetoric continues to resemble a natural history as much as a natural philosophy (or a metaphysics) of art.[2]

D'Hancarville's letters provide a vantage point for looking back over all four volumes of *AEGR* and noting the persistence of motifs that link the vase work

[2] In this chapter I resist the temptation to read the four volumes of *AEGR* teleologically, as gradually building up the comparative approach that forms the core of d'Hancarville's last major work, the wide-ranging *Recherches sur l'origine, l'esprit et le progrès des arts de la Grèce* (1785). *AEGR* does present evidence that d'Hancarville gradually becomes more and more committed to a comparative approach to ancient religion and mythology, but the remarkable unity and consistency of the overall argument are too easily overshadowed by a focus on the increasing 'sophistication' of the individual volumes (Haskell, 'The Baron' 38–9, cp. *V & V* 99).

to natural history, to the common paradigm that also informs Hamilton's and Fabris's monumental natural history of the region where so many of the vases were found. Both projects rely on local knowledge. The new volumes tell a story of stylistic development in Greek art, correlated with developments in Greek religion and society. The present chapter focuses on the history of religion, while my next chapter takes up d'Hancarville's stylistic development hypothesis as well as points of intersection between these two narrative lines. D'Hancarville's observational raw material on religious and other customs, both ancient and modern, provides the local foundation for his natural history of art.

In his initial excited report to Hamilton, d'Hancarville emphasizes the philosophical dimensions of his pursuit of the deep past. In his letters that follow, extolling the virtues of his 'catalogue' (a rough draft of *AEGR* III–IV), d'Hancarville revisits the plan of the work as a whole. In these letters he is evidently responding to reminders from Hamilton that a proper catalogue of the collection, by now in the British Museum, is still lacking. Although he begins from a draft by Hamilton, apparently a sort of hand list (*Antichità* 141), and includes more descriptions of objects than before in the very extensive footnotes to the new volumes, there would be no proper catalogue before 1778, when d'Hancarville worked in the British Museum to compile the manuscript catalogue that is still preserved there (*V & V* 198–99). Instead, the plan of *AEGR* in its new incarnation is more comprehensive and more systematic than the plan of Volumes I–II, promising explanations for the 'profundity of the art of the ancients, for the delicacy of their expressions, for the subtlety of their intention, and for the correspondences [*accord*] among their poetry, sculpture, and painting' (*Antichità* 140). Because the new work is both a catalogue and a history of art, it will use Hamilton's collection to teach readers how to see other collections without any guide at all. For this reason, d'Hancarville suggests, the text can be printed in two octavo volumes, 'like Winckelmann's art history', and 'be of infinite service to travellers in Italy'.[3]

Volumes III–IV are more chronological in their organization and more systematic in their documentation than the first two volumes, which leads inevitably to a more exclusive focus on Greek works and ancient accounts of them. Yet in these valedictory letters d'Hancarville singles out for particular praise the early sections of his new work, the discussion of statues (139) and above all the discussion of types (*les types*), his best effort 'which alone explains everything' (141). In directing his first reader's attention to the parts of the book that deal with the shadowy prehistory of art—which eventually occupies

[3] *Antichità* 139. A cheaper edition of the whole work, in five octavo volumes, was in fact published by F.-A. David in Paris in 1785. This edition merits further study because of d'Hancarville's extensive revisions to the text and because of the innovative four-colour technique used by David, who re-engraved all the illustrations.

more than half the new text—he signals his continuing commitment to the deep past, to the farthest interior reaches of the 'vast country' of antiquity (*AEGR* III.3) that he began to map in the earlier volumes. The deep past of art—now relocated from Etruria to Greece—still anchors the quasi-evolutionary theory of representation that he calls his 'metaphysique de l'art' (*Antichità* 141). In these volumes d'Hancarville concentrates less on the 'improvement' of contemporary taste and more on demonstrating, by way of the vase paintings, how the ideas of the ancients were originally applied in practice.

In fact d'Hancarville's system is less a 'metaphysics' than a conjectural history, but his conjectural history is more grounded in direct observation of artefacts and customs than most others are. Rousseau, Adam Ferguson, and others in this tradition rely on travel narrative when tracing civil society to a putative state of nature. Like Rousseau's contemporary Marc-Antoine Laugier, d'Hancarville looks to the state of nature for the origins of architecture, but he goes farther in insisting that true architectural form must commemorate its natural origins, as the column still evokes the cut log.[4] This embedded history—seemingly an aspect of function as much as form—constitutes one of the five 'first principles' outlined in the first volume (*AEGR* I.74, 78). In this sense artistic form originates in nature and records that history within itself. In addition to the artefacts and ancient records that confirm this prehistory, authentic modern customs—such as Neapolitan gesture, dance, and costume—confirm it as well. Perhaps the most important point that d'Hancarville makes about his new work in this series of letters is that he will continue to look to popular customs and manners (rather than the decadent institutional culture of modern art) for evidence of remote antiquity: 'I speak objectively like a man writing in England, where there is neither Pope nor Inquisition, and I refer (*rappelle*) our modern customs to the ancient ones that are their source' (*Antichità* 140). D'Hancarville appeals explicitly to Enlightenment secularism to establish this point. What I described above as ethnographic actualism, the practice of deriving ancient customs from observable 'primitive' ones, remains crucial to his 'new science' of art history (*AEGR* III.3). This method would become a building block of modern anthropology as well.[5]

In his published prefaces to the new volumes, d'Hancarville promises to pay attention above all to the artefacts themselves (III.2–3; IV.vii), in much the same way that a naturalist might infer geological causes from observable

[4] Among other examples, see Rousseau, *Discourse on the Origin of Inequality* 93; Ferguson, *Essay on the History of Civil Society* 80; Laugier, *Essay on Architecture* 11, 14. See also Meek, *Social Science* 37–67.

[5] See, for example, Paul MacKendrick's 1960 book on the history of archaeology in Italy, *The Mute Stones Speak* (32, 57, 257, and *passim*). In what follows I focus primarily on the ethnographic aspects of this method, but as George S. Rousseau points out, d'Hancarville's comparative approach also relativizes the truth of Christianity (which explains the reference to the Pope and the Inquisition in this passage). See Rousseau, 'Sorrows of Priapus' 115–17.

processes—a technique that Hamilton pioneered with the intuitive example of volcanism. In practice, of course, d'Hancarville's observations were inevitably contaminated by his knowledge of both ancient and modern literature. Moreover, his access to objects and even books was strictly limited in Florence (*Antichità* 140–41). Unlike the naturalist working in the field, d'Hancarville was strictly confined by his creditors, and literally in prison at least some of the time. There are times, too, when his method is governed much more by neoclassicism, by his pursuit of the *beau idéal* and his engagement with the tropes—such as the 'concord of ancient poetry, sculpture, and painting'—of an idealist aesthetic theory, familiar today from Winckelmann, Lessing, and Reynolds, among others. But quite often he appeals to a field of evidence shared with natural history, ranging from physical geography to ethnography and material culture (like Parkinson's *Voyage to the South Seas*). Both the argument of his work and the history of its production vividly demonstrate the senses in which antiquarianism and natural history were cognate and contiguous sciences of antiquity.

D'Hancarville's volumes beautifully illustrate the proliferation of antiquities that is the heart of my concern in this book: it promotes two important non-classical antiquities in the form of volcanic phenomena and modern primitive customs, but also develops new subspecies of classical antiquity including the vases themselves—still regarded somewhat ambivalently as curiosities, rather than masterworks, by Winckelmann, who compared them to 'the smallest, most negligible insects' in a cabinet of natural wonders—and the new history framed to account for them.[6]

The special importance of d'Hancarville's volumes lies partly in the fact that they do lean more strongly toward anthropology than other early histories of art, so it is important to assess the extent of his debt to Octavien de Guasco and to situate both writers in the larger domain of Enlightenment discourse about 'primitive' peoples. D'Hancarville took several key concepts from Guasco without acknowledgement, but he also elaborated these concepts much more fully and inserted them into a different kind of art history. Guasco's *De l'usage des Statues chez les Anciens* (1768), published halfway through d'Hancarville's project, influenced him so much partly because it reinforced his own earlier claims. According to Guasco, 'Nothing could be more appropriate for a correct idea of the state of the arts, customs, and usages of the ancients, as yet little civilized, than to point out the practice of peoples who in our own times still live in a state of barbarism. . . . the same situation, the same needs, the same

[6] *History of the Art of Antiquity* 178, quoted in Constantine, 'Winckelmann and Sir William Hamilton' 72. Guasco too evokes the established relationship between natural specimens and natural history to bring home the importance of antiquities for the human sciences: 'the monuments of Natural History are, as it were, the medals of nature . . . [and] statues are the medals of ancient manners' (*De l'usage des Statues* x). This work is cited parenthetically in the text (as *De l'usage*) hereafter.

deficiencies in the arts produce the same ideas, the same practices, according to the climate.'[7] The last part of this quotation comes very close to d'Hancarville's assertion that 'Industry' before the Greeks 'was obliged to employ the very same means' as in Greece, which buttressed his initial claims for the antiquity of 'Tuscan' architecture (*AEGR* I.52).

Griener's analysis suggests that he came to prefer Guasco's version of these ideas. Griener also points out an alibi that must have seemed irresistible: by a curious coincidence, Guasco's title page is adorned with an image obviously adapted from a plate in d'Hancarville's first volume (my Figure 5.3).[8] The visual echo must have caught his eye, and Guasco's strong appeal to ethnography clearly resonated with d'Hancarville, who may not have been as well versed in the relevant literature—including Guasco's friend Montesquieu, Charles de Brosses, and other *philosophes*, travel narrative, and comparative mythography of the kind popularized by Jacob Bryant a few years later. Griener sums up Guasco's influence by saying that Volume III no longer 'accentuates the analysis of the primitive nature of art', becoming more concerned with the moral and symbolic content of images (*Antichità* 72–73). The titular emphasis on sculpture over other art forms in d'Hancarville's revised 'History of Art' (*AEGR* IV.vii) also provides a clue to Guasco's influence. However, as I hope to show, d'Hancarville remains committed to 'the primitive nature of art', and ultimately uses Guasco to buttress his own original contribution.

The key difference is that, despite d'Hancarville's global ambitions, his story is in many ways a local one. The constant presence of Vesuvius, its ancient buried cities, and the common people of modern Naples during the first years of the project; the scholarly interest of Hamilton and his other associates in those three entities; the vases themselves, which become more prominent in the text (or notes) even as d'Hancarville continues to evade the responsibility of dealing with them directly: all these continue to inform and unify his evolving argument. In the new volumes Etruscan art is said to be of Greek

[7] 'This should be a rule', Guasco insists, 'for antiquaries always to keep before their eyes, in order to judge intelligently the customs, usages, and the monuments of earliest antiquity' (*De l'usage* xiii–xiv, quoted in *V&V* 99). Griener was the first to discover d'Hancarville's plagiarism, but did not specify instances, focusing largely on d'Hancarville's first two volumes instead. Jenkins and Sloan, corroborating Griener's finding, locate particular passages in the later volumes that follow Guasco closely (*V & V* 99–100), and I have adopted their translation of the quoted passage. As Griener suggests, the sense of discovery in d'Hancarville's letters of June 1773 may in fact reflect his discovery of Guasco.

[8] Griener, who reproduces the title page (*Antichità* 82), supposes that it was chosen by Guasco's publisher—rightly so, I think, since Guasco refers to his failing eyesight near the end of his narrative (*De l'usage* 491). Guasco's volume probably appeared within a few weeks to a few months after the imprimatur was granted for his book on 3 March, 1768. Since finished copies of *AEGR* I were in circulation by January 1768, Guasco's publisher (de Boubers) might have seen the image and decided to re-engrave it as a frontispiece.

origin, but this revision helps to support the use of Greece as a case study in autochthonous art, and need not be seen as an exclusive claim for Greek priority. D'Hancarville's next work, the more broadly comparative *Recherches*, shows some affinity to Johann Gottfried Herder and other syncretists who rejected Winckelmann's 'obsession with Greek originality' and insisted on cross-cultural transmission as the key to ancient art.[9] This more wide-ranging account of ancient art and mythology, which takes in material from India and central Asia among other places, shows a continuing commitment to multiple points of origin first suggested in his conjectural defence of Etruscan antiquity.

AEGR as a whole promotes the idea that the ancient past is instantly accessible through archaeological sites, among which Naples stands out very clearly in this historical moment. Though influenced by Guasco's antiquarian injunction to attend to the 'savages of modern times', d'Hancarville remains less interested in generic 'savages' than in specific local peoples, especially the Neapolitans. Anecdotes of Neapolitan customs play a role in the explanation of objects unearthed in ancient Campania, and they echo the interest in popular superstitions expressed in Hamilton's *Campi Phlegraei* and much other contemporary writing about Naples. Just as Hamilton's analyses of lava flows suggest an ongoing process stretching back indefinitely into 'most remote antiquity', d'Hancarville's multi-layered history of art hews to an ethnographic idea of art as custom or ritual that contradicts any notion of a singular or 'catastrophic' origin of art, Greek or otherwise. At the same time, the complexity of this attempted synthesis helped to make the range of ancient and modern cultures and their practices more visible as potentially separate areas of study.

4.2 CUSTOMS AND MANNERS

The two new volumes bear out d'Hancarville's promise of June 1773 to 'refer [*rappeller*] our modern customs to the ancient ones that are their source' with many examples. A few are controversial enough to explain his reference in the same sentence to the Pope and the Inquisition (*Antichità* 140). Many of the ancient customs that d'Hancarville depicts also betray a romantic nostalgia for modern Naples and for the idealized classical past suggested by the vase paintings. Later ethnographic literature supports this comparative approach

9 Elisabeth Décultot, 'Comment l'art est-il venu aux Grecs?' 56. Herder's critical term for this ideology is *Originalsucht*. Guasco, apparently writing mainly in the mid-1750s, also sharply rejects the prejudice in favour of Greek originality (e.g. *De l'usage* 38)—but he favours a Mesopotamian point of origin (cp. *V & V* 100). Greece is nominally the protagonist of d'Hancarville's new prose (e.g. *AEGR* III.7–8), but he does not frame systematic arguments for Greek priority as Winckelmann does.

as well as the art-historical utility of modern customs as an aid in interpreting the gesture, ritual, and costume abundantly displayed in the vase paintings. At times d'Hancarville's romantic classicism obscures the empirical strength of his observations. Put another way, his knowledge work is grounded in the lived experience of analysing and classifying the vases in their modern Neapolitan setting, a recurring point of reference in his explications of the vase paintings.

Although by the 1770s he has shifted his allegiance wholly to Greek antiquity, and focuses on what he takes to be illustrations of Bacchic or Eleusinian rites, d'Hancarville continues to explain ancient images in light of Neapolitan customs. In several vase paintings, he argues, 'the Bacchantes carry a type of castanets . . . still in use in the Kingdom of Naples, where one rediscovers the authentic dances of the ancient Greeks' (*AEGR* IV.51n.).[10] One image (Plate 6a) prompts a longer excursus that makes Naples the scene of living antiquity and ideal beauty. In this vase painting, re-engraved from an earlier antiquarian work, Eros plays the castanets while dancing with two of the Graces as the third Grace plays the tune:

> This charming dance, represented by three pretty figures painted on the wall in ancient Herculaneum, still survives [*se conserve encore*] in the same country where these vases and these murals were discovered; I have often seen it performed in Naples and its environs, in the most exact measure. All its movements are inspired by pleasure, it knows no attitudes but those to which art is a stranger, or other laws but those of most simple nature . . . it seems inspired by the Graces themselves.[11]

This romantic pastoral engages simultaneously with local lore and with the global claims of Enlightened neoclassicism. D'Hancarville's long description of the dance (never named) calls on several neoclassical commonplaces. Since it is dictated by the laws of 'simple nature', no dancing master is required to

[10] In the original, 'les Bacchantes tiennent une sorte de Castagnettes formée de deux pieces de bois, dont le battement marquoit la mesure. Cet instrument est encore en usage dans le Royaume de Naples, où l'on retrouve les véritables Danses des anciens Grecs'. The painting in question here is Vol. III, Plate 90. A few pages later, d'Hancarville explicitly states that he is writing in October 1775 (66n.), a reference that suggests a probable date for these explications. Many readers have noted that d'Hancarville only chose three vases to explicate from the 130 plates in his first volume. In Volume II (154–68) he chose forty more vases from this volume using loosely ethnographic principles of selection (three of these [I.32, 45, 74] feature wedding scenes, for example, and several others are explained as scenes of theater or dance). In Volume IV he tackled the problem of providing the remaining explanations in a spectacularly long footnote (28–59n.), discussed in my next chapter.

[11] *AEGR* IV.57n. The vase is now lost, though the painting had previously appeared in antiquarian works by F. Gori and by G. B. Passeri. See Gisler-Huwiler et al., *Collection of Antiquities* 492. Here the castanets resemble those in the painting cited previously (*AEGR* III, Pl. 90), a case in which modern scholarship concurs with d'Hancarville's judgment (*Collection* 368). Elsewhere, however, he seems to have mistaken the jumping weights used by ancient athletes for castanets in his enthusiasm (*AEGR* II.166, cp. *Collection* 147).

learn it; its rules are grasped intuitively by 'the people' and even by children, though the upper classes have become too refined and too greedy to be receptive to the Graces who inspire it (IV.57–58n.). Alexander Pope's declaration that 'Nature and Homer were... the same'—a belief still echoed with conviction by Friedrich Schiller in 1795—puts this familiar idea in literary terms, albeit with less political emphasis. In Shaftesbury's more radical formulation, Greek culture itself was 'wrought out of nature', and Winckelmann gave the idea an influential art historical form in his *Reflections on the Imitation of Greek Works.*[12]

D'Hancarville's initial project was to show that not only Greek art but *all* art or culture is 'wrought out of nature'. In the later volumes, ancient Greece proved to be a more viable case study for this process than Etruria, partly of course because of the prestige associated with it by Enlightenment neoclassicism—by the rise of the 'imaginary Greek', as Michael Vickers has termed it. Winckelmann's careful engagement with the *philosophes*, as documented by Elisabeth Décultot, must have prompted his occasional analogies between Greeks and modern non-European peoples, such as his reference to 'swift Indian' in the *Reflections*. Although strongly influenced by Winckelmann in his Hellenic turn, d'Hancarville still turns to contemporary Naples for his analogies. In so doing, he foregrounds the problematic relationship between the imaginary Greek and the noble savage, between neoclassical aesthetics and Enlightenment ethnography.[13]

D'Hancarville's ethnographic bent often gives his neoclassical commonplaces a distinctive critical inflection. He certainly succeeded in giving Greek vases the status of high art by associating them with these commonplaces, as Vickers and Viccy Coltman, among others, have argued. But the vases were also documents for him, and he pioneered their use as evidence for an archaeological 'anthropology of everyday life', to adapt Alain Schnapp's term. Just as he revised the neoclassical view of the ancients as noble savages by insisting on a Mediterranean field of evidence, he also challenged the conventional representation of Neapolitan 'superstition' familiar from eighteenth-century travel narrative. Drawing directly on the vase paintings, he argued that *commedia dell'arte* was a direct descendant of Athenian satyr plays (II.162) and that the pattern of what he took to be Helen's bridal

[12] Pope, *An Essay on Criticism* (line 135), in *Poems* 148; Schiller, *On Naïve and Sentimental Poetry* 194–9; Anthony Ashley Cooper, Earl of Shaftesbury, *Treatise VI*, quoted in Décultot, 'Comment l'art est-il venu' 53; Winckelmann, *Reflections* 20 and *passim*. See also Reynolds, *Discourses*, esp. Discourse VI.

[13] Michael J. Vickers sets up the Enlightenment's 'imaginary Greek' as a foil to his imaginary Etruscans in 'Imaginary Etruscans' 155. For the 'swift Indian', see Winckelmann, *Reflections* 7. Much of the scholarship on Enlightenment race theory addresses its connections to neoclassicism, as exemplified in Petrus Camper's use of classical sculptures for anatomical norming. See, for example, Engelstein, *Anxious Anatomy* 5–6.

costume on one vase was 'still in use in the environs of Naples' (II.161). He compared ancient and modern horse branding among other everyday practices (II.168). Juxtaposed against these frequently occurring Neapolitan references, the exotic cases of Lapons and 'the negroes of Guinea' are just passing references in imitation of Guasco.[14]

Guasco, by contrast, relies exclusively on travel narrative. Although his 'savages' are almost always non-European, in one case he turns to a recent 'voyage d'Italie' for the image of a Fury who makes an ideal counterpart to d'Hancarville's dance of the Graces. Guasco paraphrases the unnamed traveller's account of an old Neapolitan woman berating a statue of the Virgin for failing to deliver the grace for which she had appealed; if the statue had not been protected by a grate, in Guasco's paraphrase, 'the old Fury would have dashed it to pieces'.[15] D'Hancarville deserves all the more credit for avoiding this stereotype because it was reproduced by Hamilton himself and others in his circle including Emma Hart (Lady Hamilton) and Wilhelm Tischbein.

While the views of d'Hancarville and Guasco seem to line up (respectively) with the 'noble' and 'ignoble savage' paradigms defined by Ronald Meek, the other characterizations are more ambivalent. Tischbein's 'fanatical' Neapolitans approximated the noble savage at least in their anatomy, as he found ample evidence of Greek nature among the Neapolitan working class. Emma Hamilton, as I argued above, appropriated the 'modern primitive' subject position herself in her Attitudes, in which she staged tableaux drawn from and accompanied by the vases from Hamilton's collection. D'Hancarville's probable influence on these productions—published in one case in a set of engravings 'faithfully copied from nature at Naples'—has been largely if not wholly overlooked.[16] Though his sketch of Neapolitan Graces may well have influenced the Attitudes, d'Hancarville did not meet the second Lady Hamilton, and he surely would have disagreed with her contention that the Neapolitans were 'great bigots' (*H & N* I.133). He is not free of condescension himself,

[14] *AEGR* IV.26–7. This is one of the comparatively very few references to non-European peoples in *AEGR*, and both d'Hancarville and Guasco (*De l'usage* 22) are mistaken about the 'Manitous' in question here (see Chapter 2, n. 12 above). D'Hancarville is also at odds with modern scholarship concerning Helen's bridal costume: according to the modern reprint of *AEGR*, the painting on this Attic hydria (Vol. I, Plate 32; now British Museum Vases E 225) may represent 'Apollo, Eros and the Muses' (Gisler-Huwiler et al., *Collection* 52). On Coltman and Schnapp, see further below.

[15] 'Elle étoit vraiment en fureur, & si l'image de la Vierge n'eut pas été dans une niche grillée, la vieille furibonde l'eut mise en piéces' (*De l'usage* 202n.). For Guasco the 'old fury' is truly a modern savage, a shocking relic in a Christian country of a practice more characteristic (in his view) of 'idol worship' in modern-day China and Siberia.

[16] On 'savages', see Meek, *Social Science and the Ignoble Savage*. On Tischbein and Emma Hamilton, see Chapter 2 (Section 2.7). Emma Hamilton's Attitudes are depicted in Friedrich Rehberg, *Drawings Faithfully Copied from Nature at Naples* (Rome, 1794). They were described by contemporaries including Goethe in terms reminiscent of d'Hancarville, who established the paradigm for conflating scenes on the vases with contemporary Nature.

but the grossly pejorative tenor of Guasco's narrative of 'idol worship' is largely absent from d'Hancarville's account of Greek religion and entirely so from his comparative references to Neapolitan customs.

Anticipating more objective studies, d'Hancarville's comparative ethnography largely avoids the trade in savages (noble or otherwise) so characteristic of much Enlightenment discourse. Schnapp argues that most of the credit for this new objectivism should go to Andrea de Jorio, a younger Italian clergyman and scholar influenced by d'Hancarville and Guasco, among others. Drawing on the research of Giancarlo Carabelli, Schnapp shows clearly how Italian scholarship differed from that of the 'French and English *curiosi*'. For foreigners as well as Italians interested in 'the comparative study of antiquity and ethnography', Schnapp observes, 'reliance on ancient sources and observation of social reality were understood as a joint means of understanding and verifying propositions'. But De Jorio in particular, in *Mimicry among the Ancients* (1832), studied everyday Campanian life 'from the inside rather than with condescending detachment': 'Neapolitans are no longer to be regarded as fossil remnants of an absurd and sometimes obscene ancient heritage; they are the only people to have kept alive a certain type of practical knowledge, a physical ability which was one of the achievements of the ancient world.'[17]

Schnapp is right to credit de Jorio and not d'Hancarville for inventing the 'anthropology of everyday life', but d'Hancarville too used his local knowledge of 'Neapolitan gesture' as a guide, and even an 'indispensable tool', for understanding vase paintings. De Jorio must have relied to some extent on d'Hancarville, if only because most of the vases he describes had left Italy. De Jorio refers explicitly to one of d'Hancarville's plates as illustrating an insulting gesture still in use in Naples, the 'fig hand' or *mano in fica*.[18] Significantly, d'Hancarville himself does not explicate this particular vase, but evinces the same type of interest in his description of the Neapolitan gesture of snapping one's fingers, offered as a gloss on one of the finds from Herculaneum (*AEGR* IV.233n.)—another frequent source of illustrations for him. The object in question, a bronze statue of a drunken faun presumed to be the copy of a Greek work, is presented to illustrate d'Hancarville's thesis that Greek art, in its penultimate stage of development, succeeded in making every anatomical detail unite in the naturalistic 'expression' of *sentiment actuel*—in this case

[17] Schnapp, 'Antiquarian Studies in Naples' 163–4, 166. For Schnapp, De Jorio is neither an 'ideologue' like d'Hancarville nor a 'polemicist' like Richard Payne Knight, but a genuine fieldworker who 'attempts . . . to prove that the language of Neapolitan gesture is a direct legacy of antiquity', that is needed to understand ancient art (164). 'By reversing the direction of the antiquarian's gaze towards the present', Schnapp continues, 'de Jorio's austere work brought into being a new discipline' (165–6). D'Hancarville's 'comparative' discipline (158), in effect, anticipates de Jorio's 'ethnological' one.

[18] De Jorio, *Gesture in Naples* 218. The plate in question is *AEGR* III, plate 68.

drunkenness.[19] As in his gloss on the Graces, d'Hancarville assumes direct continuity from classical Greece to first-century Herculaneum and Pompeii to modern Italy. In its general aspect, this assumption is conventional enough and can be traced from Winckelmann's discussion of the first bronzes from Herculaneum to Nathaniel Hawthorne's *The Marble Faun* (particularly apposite here). The local feeling for Naples makes d'Hancarville's version distinctive. This feeling energized his relationship with Hamilton and he encouraged it vigorously in Winckelmann, who despite his earlier visits and ongoing critical engagement with the excavations seems to have felt the greatest enthusiasm during his final visit under d'Hancarville's care.

D'Hancarville was well poised to put his stamp on the unique form of empathy, made possible by the discovery of Herculaneum and Pompeii, that revolutionized the European experience of classical antiquity. His feeling for Naples (where he spent six years) clearly persisted during his long and painful exile in Florence and accounts for the persistence of his Neapolitan glosses on the vase paintings, even as he ostensibly rejects his earlier repertoire of local and geological devices for amplifying antiquity and turns instead to ancient texts and the (largely non-material) Greek legacy. As Griener points out, the buried cities gave spectators unprecedented access to everyday life in the ancient world (*Antichità* 61–62). In d'Hancarville's case, immersion in the Neapolitan setting further inflected this historical empathy. The comparison to de Jorio helps to specify further what distinguishes d'Hancarville's archaeological sensibility: his entire history of antiquity remains tethered to the Bay of Naples, linked to ancient Greece by colonization and preserved by dead volcanic ash as well as living custom. Therefore his method of ethnographic actualism extends well beyond the domain of everyday life as I have indicated it so far: he compares ancient and modern Campanian vase-making; he devotes particular scrutiny to the ancient history and geography of the region, engaging also with the telluric nationalism of the Italian Renaissance; he situates modern ideological uses of art vis-à-vis the ancient world, and vice versa; and above all he compares modern with ancient religious ritual, as suggested by the vase paintings.

Though d'Hancarville continues to find room for analysis of the vase paintings, his observations on ancient and modern vase-making remain confined to the two earlier volumes. Coltman has shown how useful the neoclassical principles of imitation and ideal beauty were for England's ceramic and other design industries that used *AEGR* as a 'pattern book'. Her case studies demonstrate that d'Hancarville's formulation of these principles, in his prefatory discussion of Hamilton's goal of 'advancing the arts', empowered two generations of designers to adapt the designs from his vases in a wide range of

[19] *Drunken Silenus on a Lion Skin*, Villa of the Pisoni, Herculaneum, Inv. 5628, Museo Archaeologico Nazionale, Naples.

objects and built works. Coltman's argument brings out another set of implications in the licence that d'Hancarville grants to his artist-readers to work 'as if [they] had the originals themselves in [their] possession' (*AEGR* I.vi; cp. 152)—a claim echoed by Hamilton in his discussion of a Fabris plate depicting natural history specimens (*CP* XLIX). Since d'Hancarville and his artists themselves altered the original designs by transferring them to a two-dimensional print medium (as Coltman and others have emphasized), designers from Wedgwood to Henry Clay to Robert Adam did not hesitate to make further changes to the published designs, as dictated by technical or market conditions as well as the taste of their patrons. Coltman refers to this technique as 'faithful quotation . . . or imaginative interpretation', linking the latter to d'Hancarville's mandate to artists to 'invent in the same style'.[20] It is probably for the benefit of this same artistic or 'practical' audience that d'Hancarville alludes to the living tradition of Campanian vase-making in the same preface (*AEGR* I.xx), and he may have this audience in mind when he finally discusses ancient vase-making in detail in the second volume.

D'Hancarville first mentions what he takes to be the main centres of ancient vase-making in the preface and later builds his substantial history of vase-making on this essentially local foundation. Understandably, he focuses on Campania; the idea of vases as trade goods came later. The 'principal manufactures' of vases were probably in Nola, Capua, and Campana (near Cumae), he reasons, because 'the greatest quantity of antique vases are found' at these sites, and 'what ought to confirm [this opinion] still more, is that the inhabitants of this country, have still preserved the best forms in their vases' (I.xx).[21] D'Hancarville may have offered another hint to 'Manufacturers of earthenware' by alluding to 'the diversity and elegance of the most ordinary vases' in modern Naples; but if so, they did not take it up, and contented themselves with the engraved antique designs. The potters of modern Naples—untutored like its graceful dancers—imbibe perfection from their proximity to the 'purest' ancient source, and hence nature. The affinity between modern and ancient potters is much stronger in the early volumes; in the later volumes the

[20] Coltman, 'Sir William Hamilton's Vase Publications' 11. Coltman is right to suggest (through her choice of quotations) that artists and designers probably read no further than the preface, unless it was to take note of the example of a painterly imitation provided specifically for them in Volume II (168). Coltman also quotes a passage from Reynolds's *Discourses* (VI) elaborating the license granted to inspired imitation, a version of this principle that is especially apposite for the English context.

[21] D'Hancarville conflates imported Greek vases found in earlier Campanian tombs with those actually manufactured in South Italy after 430 BCE. Hamilton made the first concerted effort to prove that the vases found in South Italy were manufactured in Greece in his *Collection of Engravings from Ancient Vases* I.156–8, and this view held for nearly a century. Modern archaeology, however, has established that major colonial pottery centers, including Capua and Cumae, emerged in the late fifth century and supplanted Athenian exports by about 400 BCE. See Trendall, *Red Figure Vases* 17.

beauty of the vases is attributed abstractly to 'the progress of the arts' and concretely to Greek colonialism and educated patrons rather than to demotic wisdom and the Campanian genius loci. But the anecdote of the Graces shows that d'Hancarville never quite let go of the romantic notion that classical form is perpetuated through folk memory rather than high culture. This is one of the contradictions that makes his classicism fruitfully ambivalent or even anti-classical.

Ultimately the skill of the Neapolitan potters is of interest to d'Hancarville less as a practical example than as living history, one piece of evidence in an amateur's enthusiastic promotion of the region's antiquity. He also makes use of classical philology, drawing on everything from Homeric epic to local inscriptions and place names, and the archaeological evidence served him well. Even now, after 200 years of excavation in Greece—which had not yet begun in d'Hancarville's time—it remains the case that 'the bulk of the surviving Greek pottery has come to light in Etruscan and Campanian tombs'.[22] This archaeological fact leads him initially to explain away the 'Greekness' of the vases, or to argue at least that Greek colonists in the region came very early and assimilated harmoniously with what he takes to have been the high Etruscan civilization already existing there (II.136). D'Hancarville explains the presence of Greek motifs and inscriptions by postulating an export market, and since there is no available evidence of Campanian pottery exported to Greece he supports his claim by making analogies to Egypt and to China—two known cases that reverse the trajectory both of ancient history and of the modern porcelain trade: 'If some vases are found on which Egyptian ceremonies are painted . . . these Vases have been made on purpose to be exported to Egypt, in the same manner that they execute in China the designs that we send them from Europe, and which the Chinese return to us painted upon their China Ware' (134).

One vase in particular brings into focus both the continuity and the change in d'Hancarville's arguments over time. This is a large black-figure vessel showing a boar hunt, now in the British Museum and identified by Ian Jenkins as a column krater (mixing bowl) made in Corinth between 575 and 550 BC (Plate 6b).[23] Jenkins and Sloan make this vase the centrepiece of their account

[22] Vickers, 'Imaginary Etruscans' 155. Vickers's account (157) also sheds light on d'Hancarville's conflation of Etruscans and Phoenicians in the first volume, a piece of cultural nationalism related to many other strategic deployments of ancient peoples in early modern histories. D'Hancarville owes this particular notion to Jacob Martorelli, the Neapolitan professor of Greek who also (as it happened) provided the testimony needed for d'Hancarville to obtain an imprimatur for his book (*AEGR* I.178). The title cited by d'Hancarville, *Dell'antiche Colonie venute in Napoli*, was published by Martorelli (1764, 1773) under the pseudonym Michele Vargas Macciucca.

[23] Amyx confirms the find spot of this vase (height: 27.9 cm) as Capua and attributes it to the Andromeda Group. See his *Corinthian Vase-Painting* I.268 (NC 1478).

of d'Hancarville's changing arguments (*V & V* 149–55), and rightly so, since it is the only vase to be discussed in detail in all four volumes. This Vase of the Hunt features in the first four folio plates of Volume I (plates 22–25), and is the first of two vases chosen for the specimen explications provided at the end of the text (152–64). Of his numerous references to this vessel in the later volumes, the most remarkable is a long passage, written after a lapse of six or seven years, in which he revises and corrects the initial explication, in keeping with his pose of rejecting all of his own earlier arguments (III.204–11n.). The fundamental reason for his emphasis on this vase remains unchanged: it is 'one of the most ancient monuments' (I.164), coeval with the birth of painting itself. Despite his ostentatious revisions, d'Hancarville retained his original paradigm of the 'infancy...of the arts' (I.168) when he outlined his programme for the two new volumes (III.4). But the date of this infancy and hence of the Hunt Krater—perhaps not surprisingly for a subject located on the cusp of history—turns out to be much earlier in his second analysis, about 1250 (III.206) rather than 658 BCE (I.162). Jenkins and Sloan give d'Hancarville some credit for being an 'accomplished chronographer' according to the standards of his time (*V & V* 152), but ultimately they take a cynical view: d'Hancarville's 'real motive' in backdating the vase by 600 years was 'to glamorise his claims to have discovered the prehistory of Classical art' (155). I would point instead to d'Hancarville's engagement with natural history as another way of understanding his efforts to extend the anthropology of art into the domain of prehistory.[24]

He changed his mind as well about the motif of the Hunt. Although he soon found a way to explain the presence of Greek themes on vases made (as he thought) in Campania, in his initial reading of this vase he rules out any possible association with the boar hunts recorded in Greek myth, for simple geographical reasons. D'Hancarville summarizes the Calydonian boar hunt led by Anceus, recorded by Pausanias with other legends from the era of the Trojan War, to illustrate the historical importance of big-game hunting: there

[24] Although generous in many respects, Jenkins and Sloan's treatment of d'Hancarville is still tinged by a kind of art historical resentment concerning his cavalier approach to facts. (D'Hancarville's character as a 'rogue' has exacerbated this resentment and has certainly informed his reception over time.) His reversal on the dating of this vase seems less sudden and capricious when we consider that he not only allows for a much earlier date from the start (I.164), but also begins the process of revising the date when he recurs in Volume II to 'the Vase of the Hunt, which we shall soon demonstrate to have been anterior by some Century's to the foundation of Rome' (II.108). Thus the backdating is more gradual and deliberate than Jenkins and Sloan suggest. In this passage d'Hancarville does eventually assign a date of 300 years before the founding of Rome, though he is far from straightforward about it (II.122–4). What interests me about a passage like this is its geological aspect: d'Hancarville is less concerned with the real date than with establishing archaeological deep time. So too modern geology evolved by expanding, often dramatically, the periods held to be required for processes such as glaciation and mountain-building.

must have been many such legends, and such vase paintings might have served—like eighteenth-century history paintings—as 'monuments' of these events (*AEGR* I.154). He reasons that 'this [vase] having been found in the neighbourhood of Capua, and the manufacture distinguishable by the nature of the Clay of which it is composed, we imagine that the Hunt it represents might have been a celebrated one in Campania' (154). D'Hancarville identifies the figure labelled 'Antiphates' with a Lestrygonian king mentioned by Homer to locate this conjectural hunt near modern Mola, 'towards the extremity of Campania' (156). The history of vase painting in Volume II uses the same vase to define the first 'Epoch of the History of the Ceramick' and already assigns a considerably earlier date (II.108; cp. 122, 124). The revised explication in the next volume revisits the design and reverses the earlier decision not to link this vase with the famous Calydonian boar hunt. Here d'Hancarville's opportunism becomes apparent, as he reasons very arbitrarily that two of the hunters who (conveniently) remain unlabelled could only be the two most famous principals of the Calydonian hunt, Ancéus and Meleager (the Homeric associations of Antiphates are suddenly forgotten). The many other revisions vividly illustrate his new semiotic theory of art, but the historical and iconographic details are largely suspect.[25]

His first engagement with the vase is more satisfying: it is openly conjectural and ekphrastic, more attuned to the aesthetics of the image, and experiments in a pioneering way with a whole repertoire of archaeological techniques, including ethnography, epigraphy, and stratigraphy.[26] The history of vase painting in Volume II, building on these elements, suggests that d'Hancarville might have done a little fieldwork among the local potteries. He reports that ancient potters prepared the clay by sifting pebbles and silica out of the 'uncommonly fine' sand deposited around Capua by the Vulturnus river. It

[25] In the revision, d'Hancarville invokes the ancient authority of Isacius (though no passage is cited) to specify two distinct Calydonian boar hunts: the first hunt, depicted here, in which Anceus is killed, and the second, 'more famous' one, in which the boar is finally defeated by a veritable parade of worthies including Theseus and Nestor (III.206n.). The flowers signifying that the hunt began in the (Campanian) plains (I.156) here become vines to signify that Anceus is defending his vineyards in Epirus. There the vulture flying in a contrary direction directly above the boar means the boar will be defeated; here the same bird (now an owl) 'passing over the head of the boar shows that this animal *was not killed* in this hunt' (III.205n., emphasis added). For Pausanias's version, see *Description of Greece* viii.47.2.

[26] Both explications of this vase engage in detail with the forms of the Greek letters inscribed on it, though here as elsewhere d'Hancarville draws two different sets of conclusions. These 'dubious epigraphic arguments', as Jenkins and Sloan call them (*V & V* 155), are beyond the scope of my present discussion. But it is worth noting that d'Hancarville himself addresses these inconsistencies: he suggests that the resemblance of Etruscan letters to Pelasgian letters (the originals), and the complex subsequent history involving Cadmean letters and various dialects, have made it easy for many Greek inscriptions to be read as Etruscan, and consequently for all clay vessels to be attributed indiscriminately to Etruria (III.208n.). He now repudiates this attribution clearly as a mistake [*méprise*] (209n.).

was then turned on the wheel and coated with iron ochre while still wet (II.136–38). This account of vase painting is far from accurate, but it helps to explain his enthusiastic response to the Hunt Krater and other vases, while also developing the idealized account of art as inspired improvisation: since the clay had to be painted when wet, he argues, vase painters 'must have been directed by the eye alone, and we may be assured that all the figures which are on the Vases of the Ancients were began and finish'd just as we see them, without any other preparation than practice and intelligence' (140). It is also of interest that d'Hancarville reads the two sides of this vase as a temporal sequence (I.156), a view that is out of step with the exclusive association between temporality and verbal art made famous by Lessing's *Laocoön* (1766), but probably more faithful to the design. This observation is shelved in the revision, along with most of his earlier, more impartial observations and even his enthusiastic conclusion that 'the action of the figures is full of fire... neither the Men nor Animals are wanting either in expression or character' (I.158). In the theoretical vocabulary of the new volumes, 'expression' designates a stage of development that this vase has not yet achieved (III.208), and the theory (very loosely based on Guasco) dictates a greater emphasis on the crudeness [*grossièreté*] of its 'signification' (202). Despite all these revisions, a curious vestige of local natural history remains in the hunting dogs, unspecified at first but identified in the revision as an ancient race of Epirus, still flourishing in modern Calabria (205–06n.). Calabria was part of the Kingdom of Naples, so the dogs may be another local reference at odds with d'Hancarville's strenuous effort to re-Hellenize this vase.

This ambivalence about the origin of the vases attests to the ways in which natural history and antiquarianism gradually disrupted the literary and high-cultural mode of identification with classical antiquity that developed from the humanism of the Renaissance. Although he valued the pursuit of local knowledge, d'Hancarville was strongly drawn as well to the neoclassical veneration for Homer and literary sources and to the high-minded classicism of Winckelmann in particular. Volume II of *AEGR* is more notable than the others for its reflections on historiography, including several iterations of the orthodox view that the cultures of classical antiquity continue to matter because their achievements are timeless. This was also the attitude taken by academic history painting, so it is not surprising when d'Hancarville reports that the Greeks 'had the good sense to represent nothing but their Histories, or Fables, which were a sort of History' in painting (II.50). But the Greek history painters enjoyed many advantages, including 'the History of their Liberty' as material and the stable intelligibility of their subjects across time: 'they were dressed, they spoke, they conducted themselves nearly by the same principles, in the time of Pericles, as in that of Harmodius and Aristogiton' (50–52). Surprisingly, he cites figures who lived only two generations apart, though this choice becomes more plausible in light of the political discussion that follows. The changing fashions of the moderns, it turns out, are due to the evils of feudalism: 'our History's ... are in effect but the lives of our Masters', and thus a medieval Frenchman today 'would appear as strange to us as an inhabitant of the Indies or China' (52).

In this sense the ideal of Greek democracy justifies the demand for timeless art as well as the method I have been calling 'actualism': the present is, or ought to be, the key to the past. The explosion of antiquarian scholarship in d'Hancarville's time made the past newly and broadly accessible in ways that were increasingly bewildering even as they appeared to advance the neoclassical ideal. As in his declaration that 'antiquity is a vast country', d'Hancarville draws repeatedly on navigational metaphors to shadow this temporal expansion. The cultures of ancient Greece and Egypt are now known in the land of the Incas, he reflects: 'Every mind is in some measure become contemporary; and times, stopped as it were in their progression, seem to be reunited' (II.6). Here he credits navigation, writing, and painting for this achievement, but painting most of all, because of its power of inducing empathy. This argument reflects the archaeological turn initiated by Herculaneum and also prefigures d'Hancarville's later theory of painting as *écriture figurée*, a 'sort of writing' or 'historical discourse' (III.204n.). Alluding to his hoped-for future readership, he reflects that he may one day be cited as one of the 'ancient authors of 1769' (II.2). In this light the present is merely a future antiquity. The new material evidence of classical antiquity both augments and diminishes historical distance at the same time, not unlike a navigational aid that brings distant cultures more rapidly and more forcefully into contact.

There was a danger that the effort to 'refer [*rappeller*] our modern customs to the ancient ones that are their source' might result in presentism, a historical fallacy against which Guasco, citing Montesquieu, warns explicitly (*De l'usage* xi). But in natural history, it was proving to be the case that a preoccupation with present processes could be a reliable guide to the past, in the domain of lava flows if not of canine breeds. And in the domain of culture the comparative method sometimes made the present strange even as it made the past familiar, as in the case of religious ritual. D'Hancarville's ongoing fascination with the early 'Pelasgian' forms of Greek letters captures some of these nuances. His claim that in this early form Greek letters most closely resembled Roman characters, and hence 'our own', effectively naturalizing the current alphabet, seems grossly presentist (*AEGR* IV.69n.). But when he reflects that he is writing his own text in characters that have been in use for 3,472 years (74n.), the same claim appears as a kind of archaeological anthropology.

Most of the examples given so far do little to explain why the Pope and the Inquisition might object—as d'Hancarville intimated in his letter to Hamilton—to his linkage of ancient and modern customs (*Antichità* 140). The one exception is his claim that 'our History's', unlike those of the Greeks, 'are only worthy of being read by Enthusiasts or Slaves' (*AEGR* II.52), which alerts readers to the egalitarian politics that inform some aspects of the argument. The 'equality nature has placed amongst Men', which he originally cited as the motive for his natural history of art (I.76), appears more political in this light. Both d'Hancarville and Winckelmann, in Griener's view, promoted 'a certain vision of society' by representing antiquity as a 'time in which the state occupied itself with men like them, without abandoning them to the caprices of private men' (*Antichità* 73). Griener's intuition that d'Hancarville came to Naples 'disguised as a gentleman engaged in the Grand Tour' (36) underscores the problematic nature of his entanglement with the class structure that this faux baron set out to subvert—an entanglement complicated by Hamilton's patronage. Most of d'Hancarville's political references are more muted, and further complicated in Volume III by his courtship of a new patron, Pietro Leopoldo I of Tuscany.[27] The Sublime Porte of Constantinople, a safely Orientalized despot, becomes the analogue for the first ancient rulers who took on divine attributes (*AEGR* III.14). Galileo named Jupiter's moons after the Medici, just as the Olympian pantheon was established by naming heavenly bodies after 'the princes descended from Uranus' (III.28). In the same comparative ethnographic mode d'Hancarville also makes a statement that may indicate the elitist drift of his third and fourth volumes. 'The people', he writes, 'who are the same in all times and all places, like children', demand to be fed on fable; but since the people always 'govern their masters', they get it (III.12).

More than this political strand, however, d'Hancarville's growing interest in comparative religion justifies his reference to the Pope and the Inquisition. The history of vase painting in Volume II already proves the archaeological usefulness of comparing ancient and modern Mediterranean religion. Of four major 'classes' of vases outlined there, the largest comprises 'votive vases' for both public and domestic worship (II.66). The smaller domestic ones were consecrated to household gods and used like 'the images of Saints, which we meet with in almost every house in Italy, and before which, it is the custom to light lamps' (II.70). According to the second round of explications written in 1774–75, one vase (Vol. I, plate 45) depicts such a scene of domestic devotion in great detail (IV.38n., 33n.). If popular Catholicism is rife with pagan practices, this is not, for d'Hancarville—as it is for Enlightened clergymen like Guasco—grounds for condemning ancient or modern superstition.[28]

[27] The volume is dedicated to Pietro's father, the recently deceased Grand Duke Francesco II (d. 1765). For additional instances of Florentine flattery, see *AEGR* III.29 and d'Hancarville's letter to Hamilton of March 1770 (*Antichità* 137); see also *Antichità* 51. D'Hancarville perhaps had mellowed since his first publication, the openly anti-aristocratic *Essai de politique et de morale calculée* (1759); Griener suggests that he found a more socially acceptable form for these sentiments in the veneration of Athenian democracy (73).

[28] D'Hancarville charges a patristic writer, Clement of Alexandria, with being 'very much mistaken' in his prejudice against ancient ritual (IV.31–32n.), thereby implicating Guasco as well. Elsewhere d'Hancarville warns against any prejudiced rejection of ancient

He is more apt to condemn the role played by superstition in the public arena of taste and patronage, as revealed in a passage on engraved gems that probably represents his most critical engagement with modern religion. D'Hancarville argues:

> As at all times the number of superstitious people infinitely exceeds the number of people of taste, we are indebted to superstition, much more than to the ancients' taste for the Arts, for the prodigious number of gems that survive; in the same way, the last three centuries of painting show us that the piety of the moderns, rather than their love for painting, gave employment to the great artists who flourished then. (IV.60)

He gives the criticism a more political inflection by likening efflorescences of ancient gem engraving to the flood of 'portraits and medals' produced whenever a new saint is canonized by the Church. D'Hancarville arguably became more anti-clerical after he advised Richard Payne Knight on Knight's *Discourse on the Worship of Priapus* (1786) and published his own *Recherches* (1785); Schnapp and G. S. Rousseau have situated this work against the larger backdrop of Enlightenment anti-clericalism. Griener notes that Hamilton's discovery of another set of votive objects—the wax phalloi of Isernia—reinforced d'Hancarville's 'utopian' interest in fertility cults, an interest that becomes apparent in *AEGR* IV and is central to the *Recherches* (*Antichità* 78). In *AEGR* the comparative religion is less polemical and vividly brings ancient religion to life. As I shall argue in Chapter 5, d'Hancarville dwells on religious origins to establish the earliest possible date, a quasi-evolutionary epoch, for the origins of art.

D'Hancarville also uses comparative religion to historicize current practice. He finds himself reaching for *nos usages modernes* most often when he is trying to explicate, or re-explicate, the vase paintings. The comparisons are often more vivid than persuasive, though they do attest to close observation both of Italian customs and of the vase paintings. D'Hancarville's sense of how the vases came to be grave goods remains relatively consistent across the four volumes. Thus when he delivers the second set of explications to support his new (and ambitious) archaeological thesis concerning the ritual meaning of the vase paintings—that most of them illustrate the Eleusinian Mysteries or Bacchic rites—he retains his sense that each vessel is consecrated to the tutelary spirit of the person with whom it is buried. According to his first history, the deceased are buried with the votive vases used in their 'domestick Oratorys' (*AEGR* II.70) and 'consecrated to their Household Gods', although the vases are then used for a different purpose—to hold 'provisions' for the afterlife (II.74). In the later explications, the deity represented or evoked by a

religion—'polytheism', he insists, and not 'idolatry'—as an all too likely anticipation of our own fate in the eyes of a more enlightened posterity: 'do we not have enough absurd practices to fear, with reason, that posterity will treat us as we treat those who preceded us?' (III.12n.).

given vase painting is associated with the profession of the person buried with that vase. If a given painting represents a scene from the rites conducted in the cult of Demeter at Eleusis, this means that the person buried with it was an initiate of the Eleusinian Mysteries (and d'Hancarville explains too how those rites nonetheless remained secret).

This is where modern custom comes in: the ritual use of parasols and certain kinds of vessels, as depicted in the vase paintings, is confirmed by the use of similar objects in modern funeral rites (see Plate 7): 'The *ombrelli* or parasols used today in Italy to minister to the dying... allow us to recognize that the parasols used by the ancients in their funerals served to cover the capsules or cases in which they placed the *indications* of particular guardian spirits.'[29] D'Hancarville privileges this example by restating at some length his comparative method of 'returning [*remontant*] by way of present customs to those of earlier times' (IV.34n.). Similar modern customs used to explicate the vase paintings include the festive use of large round votive shields, made of paper flowers and suspended by garlands over the portals of Roman churches (35n.), and the Spanish practice of having oneself interred in the garb of St Francis (36n.), which he likens to the (ostensibly) ancient practice of burying initiates with vases that displayed scenes of their cults.

4.3 DOMAINS OF PREHISTORY

Schnapp credits de Jorio with having created the 'anthropology of everyday life' in 1832 because the Italian scholar outlined a domain of inquiry that belonged partly to antiquarianism and partly to natural history, but managed to separate it firmly from both. D'Hancarville's earlier and more amorphous disciplinary innovation—termed 'comparative archaeology' by Schnapp, and 'art history' by Francis Haskell—draws from the same sources. Natural history remains present here not only in the form of attention to local customs and geology, but also as the main subject of his most important ancient source, Pliny's *Natural History*. Pliny's unique account of many ancient artists and lost works was necessarily central for Winckelmann, Lessing, Guasco, and all early art historians. D'Hancarville draws critical attention to art history's dependence on this source. In the early stages of his work he uses Pliny most heavily for the history of vase painting (*AEGR* I.160–62; II.84–92, 110–22).

[29] D'Hancarville's usage of 'les mourans' is somewhat ambiguous: 'Les *Ombrelli* ou *Parasols* dont on se sert maintenant en Italie, pour aller administrer les mourans', along with the Ciborium, which inherits its name and form from Greece, and the material (*étoffe*) with which this funerary urn is covered—'tout nous laisse entrevoir, que les Parasols employés par les anciens dans leurs cérémonies funebres, servoient a couvrir ces Capsules ou étuis, dans lesquels on plaçoit les *indications* de quelques uns des Dieux *Gardiens*' (IV.34n.).

Though later he resolves to use even ancient sources more critically (III.2), he relies on the same stadial paradigm from Pliny, in which the discovery of outline is followed by monochromatic and finally by polychromatic painting, for his revised narrative of the 'birth of painting', which occupies the latter half of Volume III and much of Volume IV as well (1–59).

D'Hancarville turns to Pliny again in reviving his quest for the 'maxims' of the ancients, begun in his first volume, noting that several lost treatises on art survive only in Pliny's quotations and allusions (IV.vii, cf. II.48–54). We can 'easily' distinguish the original theories, he claims, because Pliny is 'very superficial, when he speaks on the arts in his own voice, but very profound, when he compiles what Apelles and Pamphile wrote on painting' and what Polykleitos and others wrote on sculpture (IV.13–14n.). D'Hancarville faults Pliny for tracing the history of art 'only to the time of Cyrus; viz: thirteen hundred years after its beginning', and for lapsing into uncertainty and error even so: 'It is then this vacancy of near fifteen hundred years, and this uncertainty that we have endeavoured to fill and rectify here' (v). D'Hancarville draws on other ancient sources, such as Dionysius of Halicarnassus, to fill this gap, but mainly he proposes to draw on 'the monuments of Art itself' (vii). Among other problematic passages, he specifies Pliny's two contradictory dates for the birth of painting as 'the cause of so much error in the History of Art' (2n.). His most fundamental criticism, in disciplinary terms, is that Pliny's art history is accidental: he 'regarded the history of Art as an accessory to his natural history' and introduced the arts merely to illustrate the properties of 'metals, stones and colors' (119n.).[30]

D'Hancarville thus sets his 'History of the Arts' apart both from natural history and from the 'romance' of art history as perpetuated by Pliny and his successors (iii). This firm distinction between history and romance is less an empirical reality than a rhetorical gesture at this point, though a crucial one for discipline formation. I have argued that Neapolitan geography and ethnography are central to his account of the vases, and there are other ways in which natural history proves hard to escape. Guasco owes a great part of his art history to the ethnographic observations of contemporary voyage accounts, and Friedrich Schlegel among other contemporaries spoke casually of the 'natural history of art' as an area of study, some vestiges of which survive today in the anthropology of art.[31] Travel narrative proves hard to escape for

[30] For Pliny's narrative of the origin of painting, see *Natural History* XXXV.5. Pliny cites Apelles later in this book (XXXV.36); he also mentions a Pamphile (XI.26), but d'Hancarville may be conflating her with Pamphile of Epidaurus, a historian. See also Dionysius of Halicarnassus, *Roman Antiquities* I.33.

[31] Schlegel observed in a letter to his brother August Wilhelm on 5 April 1794 that 'the history of Greek poetry is a complete natural history of the beautiful and of art, and for that reason my work is—aesthetics' (*Kritische und theoretische Schriften* 229). For a recent example, see Gell, *Art and Agency*.

d'Hancarville as well, primarily because Pausanias's *Description of Greece* provides the only remaining record of many ancient monuments.

The traveller's field notes have a special resonance for the anthropology of art in the era of the scientific voyage narrative. Guasco makes this resonance especially clear by juxtaposing Pausanias regularly with Engelbert Kaempfer, Jakob Roggeveen, and dozens of other contemporary voyagers represented in the fifty volume [octavo] *Histoire Générale des Voyages* that Guasco owned and duly cited in his notes in 1768. Within five years the publication of Bougainville's and Cook's voyages made even more material available. Although he relies as heavily on Pliny and Pausanias, d'Hancarville uses modern travel narrative much more sparingly, and his skepticism informs some compelling disciplinary questions that Guasco and others do not address: does art belong to natural history? Does ethnography? Is the history of art a history of writing? d'Hancarville sets himself apart as well from antiquarianism, or at least from its prevalent negative stereotype, by insisting on a critical use of both ancient and modern sources. Even so, he remains deeply indebted to both natural history and antiquarian scholarship, and the intensely situated nature of his archaeological work—an aspect of archaeology recovered and embraced by some recent practitioners—suggests that Pliny's fusion of art and natural history may have been more than accidental after all.[32]

D'Hancarville used Pliny and Pausanias throughout his ten years of writing *AEGR*, but his Hellenic turn of 1773 confirms an increasing reliance on these sources in the later volumes. His escalating claims to originality notwithstanding, this emphasis on ancient sources brings him in some respects closer to the mainstream of art history (such as it was) at that time, and commentators have rightly pointed out that the later volumes are indebted as well to his contemporaries, Winckelmann and Guasco. Haskell especially emphasizes the sophistication of d'Hancarville's 'theory of the sign', which forms the overarching narrative of Volumes III–IV, while Jenkins and Griener suggest that this theory owes much of its sophistication to Guasco. Griener also notes that many scholars followed Winckelmann in crediting the Greeks with a 'rational theory of art', suggesting that d'Hancarville stands apart only for the confidence he felt in his 'unique' ability 'to recover the lost text of this theory'.[33] This lost-text motif is one of several neoclassical commonplaces that become increasingly prevalent in Volumes III–IV and make it harder to sustain my position that d'Hancarville's prehistory of art, at its most extreme, is anticlassical.

In spite of his Hellenic turn, however, it would be a mistake to say that d'Hancarville makes two distinct arguments about ancient art. The new emphasis on Greek materials is in the end more pragmatic than ideological. The example of religion shows how d'Hancarville's revision of his own earlier

[32] On situated archaeology, see Tilley, *Materiality of Stone*.

[33] See, respectively, Haskell, 'The Baron' 39; Jenkins, *V & V* 99; and Griener, *Antichità* 72–3.

work actually constitutes the new arguments. Several other features indicate how fully the two later volumes evolve out of the earlier ones. The essential point is that d'Hancarville's 'theory of the sign'—his history of the transition from 'sign' to 'figure'—provides a new way of specifying his original thesis that art records its own history. Unlike his broad earlier claim that architectural and sculptural forms continue to refer to the natural materials from which they originated, the new version lends itself to much more detailed illustration from the monuments at hand, not only vase paintings but also engraved gems and sculpture. In the new version, primitive 'signification' gives way to figurative 'expression' only by way of a complex transitional process in which several stages can be specified (IV.12n.). But d'Hancarville continues to cite the basic principle as articulated in his first volume, that classical form 'recalled the history of the Art' (I.76), and the continued quest for such 'first principles' or ancient 'maxims' itself underscores the continuity of his project throughout the four volumes. So too does his repertoire of ancient sources, even as these are given increasing prominence: in addition to Pliny and Pausanias, Dionysius of Halicarnassus plays an important role throughout and even Sanchoniathon—who gains a new importance for d'Hancarville after his reading of Guasco (*V & V* 99)—is already cited in the first volume (*AEGR* I.62). In his final volume he explains how the chronology of Dionysius gives us historical access to epochs that are relegated to prehistory by Pliny and Winckelmann alike (IV.69–70n.).

Artefacts themselves provide the main avenue of access to 'remotest antiquity'. D'Hancarville does a great deal more with the vases in the later volumes, but he also returns dramatically to one key artefact that indicates the impact of his archaeological fieldwork—limited as it was in scope—on his theoretical imagination. This is the fragment of an ancient inscription found inside a more recent (but still ancient) wall in Paestum, which had prompted his apostrophe to Etruscan architecture in the first volume (see Figure 5.1 below). When he returns to this artefact in the last volume his enthusiasm, and his conviction that it represents 'the first ideas of art', have not diminished. The only change is that he now holds the characters of the inscription to be Pelasgian rather than Etruscan (76n.).

In the new volumes d'Hancarville insists just as staunchly on an autochthonous art of the Greeks as he insisted on autochthonous Etruscan art in the old ones. He shifts his attention to the Pelasgian Greeks, roughly corresponding to Mycenaean civilization, because there is more evidence concerning them available in ancient texts and because they can be causally linked with the classical Greek civilization that his contemporaries valued.[34] Yet d'Hancarville remains focused on the pre-Homeric 'infancy of the arts'. By relocating

[34] My understanding of d'Hancarville's chronology and his ideas concerning the Pelasgians is very much indebted to Jenkins and Sloan's learned explanation in 'D'Hancarville's Exhibition'

them in origin nearer to classical Greece, he can incorporate the acknowledged masterpieces of classical Greek sculpture. Even so only the tail end of his book is devoted to this era of 'expression' and ideal beauty. The bulk of the new material remains focused on early developments more or less arbitrarily relocated to Greece (to Arcadia and the colonies emanating from it). Though now organically connected to the familiar history of Greek sculpture, d'Hancarville's narrative of origins still depends on implicit contrasts between pre-classical and classical art, which echo the explicit and deliberate contrasts of the first volume—between 'Tuscan' and Doric architecture, for example, or the supposedly stronger influence of Etruscan (rather than Greek) precedent on Roman mythology (I.62). Initially, then, he rejects what might be called a catastrophic origin, in which culture begins with Greek civilization, in favour of a model of gradual accretion—not unlike Hamilton's insistence on the uniform action of volcanic processes as opposed to the catastrophism promoted by theories of the earth. His assertion that ancient Sicily, too, had a 'genius, which equalled it to Greece' (xxii) reinforces this sense of cultural polygenesis.

By making the cultural bedrock of South Italy Pelasgian rather than Phoenician, d'Hancarville appears to reverse himself, especially in light of his earlier contempt for Greek pretensions (I.94). But he backdates the Pelasgian migration, and with it the origin of the arts, to 1700 BCE, using historical rather than cultural distance to perform the same kinds of prehistoric manoeuvers. This is a past so remote that it retains the same conceptual shape even as its affiliation changes. In the case of the Paestum inscription, he refunctions his former 'prejudice' in favour of Etruscan antiquity to support an even more profound Pelasgian antiquity (IV.75–76n.). This kind of argument is far from the mainstream of Enlightenment views of ancient Greece, and d'Hancarville at times seems ambivalent about his newfound Hellenism, as when he reasons that since signs are natural, the arts can only follow one evolutionary path (III.7–8)—the Greeks, though not the most ancient of peoples, were simply the first to find this path (IV.113–14). By embracing the thesis of a Greek discovery of art, d'Hancarville puts himself at odds with Guasco, whose attacks on what he calls the 'prejudice' for Greek originality stand out as especially forceful for this era (*De l'usage* 38). Although d'Hancarville certainly took inspiration from Guasco's distinction between the sign and the figure, he almost seems deliberately to oppose Guasco's political thesis that Greek religion and art originated as a series of impostures perpetrated by Phoenician (and later Egyptian) colonizers (39, 53). In Guasco's view, the Greeks were slow to absorb the lessons of their Eastern 'masters' (57–64) and truly Greek art began only 200 years after Homer (397).

(*V & V* 149–55). Jenkins also illustrates these ideas specifically in relation to the history of gem engraving (96–100).

D'Hancarville defends Greek originality, yet he also downplays the importance of national culture in the new volumes. His new 'History of Arts' began, he writes, with the recognition that 'there was but little ground gain'd by having ascribed to some Nations a style, that ought not perhaps, but rather to such or such a period of the People's Art who transmitts us these Monuments' (*AEGR* III.2). Rather than placing artefacts by national style, we should place them in time, relative to the history of civilization. So Caylus misidentified a 'crude' statuette of a golden bull as Egyptian because he inferred that the artist lacked skill, rather than thinking of the primitive state of metalworking when it was produced (96). Even more important, the bull has unnaturally long horns, a vestige of primitive signification confirming that 'this little *morceau* is one of the most ancient monuments of art' (97–98).

D'Hancarville focuses almost entirely on the two earliest stages of art in his new volumes, adducing many similar examples and thus accelerating the proliferation of antiquities that he launched with his first volume. The prestige associated with classical Greek sculpture serves this project well insofar as it justifies a long prehistory, but the small amount of space he leaves for the period after 600 BC (IV.203–52), and his cursory treatment of it, indicate where his real interests lie. These volumes carry a new running head, *The History of Greek Sculpture,* but again the prestigious idiom of sculpture is displaced by a massive digression on painting (and within that digression, another long digression on gem engraving) that makes up the core of the work. D'Hancarville's real interests in prehistory and ethnography drive his strategic deployment both of neoclassical pieties and of a nascent archaeological empiricism that sometimes supports and sometimes works against neoclassical aesthetics.

In the absence of an archaeological discourse, d'Hancarville seems to oscillate between the history of civil society, as practiced by Montesquieu or Ferguson, and a localism influenced by Hamilton and more akin to natural history. His focus on customs and manners owes much to natural history, but the conceptual framework of the 'progress of mind' seems more strongly affiliated with the history of civil society, or conjectural history. This philosophical discourse offers itself above all as a dignified medium in which to merge the literary practice of classicism and the empirical investigations that together make up d'Hancarville's eclectic practise of classical antiquarianism. In his first preface he declares his intention to 'follow the steps of the human mind' (I.x), an intention reaffirmed as 'the object of this book' in his last (IV. v). Toward the end he specifies that the 'progress' chronicled by art takes the form of a movement from religion to art to laws to letters, 'the course followed by the human mind among almost all peoples' (74n.). By framing art as a 'natural writing' and mythology as an 'aesthetic religion', d'Hancarville created 'a narrative of the cultural origins of art preceding historical times', in Griener's words, a narrative powerfully supported by images of excavation

(*Antichità* 64; see Figure 5.4 below). In Griener's view, d'Hancarville's secular outlook and his critical use of religion for art history make him a representative Enlightenment thinker (*Antichità* 59). Jenkins and Sloan seem to agree, suggesting that d'Hancarville's 'primitive' artist, 'like Rousseau's noble savage, stands intelligent but culturally naked before us' (*V & V* 151).

There is a limit to these affinities between d'Hancarville and the conjectural history of the *philosophes*. His writing shows their influence less strongly than Winckelmann's, whose art history tends to become 'a general history of ancient peoples', as Décultot has argued.[35] Quite possibly with Winckelmann in mind, d'Hancarville makes a point of deriving the history of nations, conversely, from the history of the arts (*AEGR* IV.5): he returns one last time, in a very long excursus note, to the succession of southern Italian peoples, revised according to the new insights he has gained into their monuments (IV.73–96n.). Guasco, like d'Hancarville, takes 'the steps of the human mind' for his subject and likewise insists that antiquarianism becomes philosophical when joined to the history of manners (*De l'usage* ii, v). But for Guasco manners are progressive, and the history of pagan art is merely the history of superstition. Guasco premises an originary monotheism, of which all forms of polytheism or 'fetishism' are the decadent descendants (9–11). He constructs a broad class of ignoble savages who place total reliance on their 'fetishes', including antiquities as well as the colossal heads of Easter Island (193) and figures of the virgin in Naples. Guasco's strictly literary use of travel narratives for support, clearly inspired by the *philosophes*, seems to be largely rejected by d'Hancarville as armchair ethnography.[36] But the greatest difference between the two appears from the architecture of Guasco's book: one half deals with mythological figures or idols, 'which make philosophy blush', and the other half turns to honorific statues of great men, 'which reason may avow' (229). Guasco's Catholicism makes him perhaps atypical, but the horror of superstition is a very general sentiment of the Enlightenment, one that d'Hancarville does not share.

D'Hancarville practises his own brand of armchair ethnography throughout his work, but it is counterbalanced by a strong empirical interest, communicated by the plates as well as by narrative strategies. When at times he insists on monuments over texts as the key to the past, he participates consciously in the scientific spirit of influential antiquaries such as Bernard de Montfaucon,

[35] 'Comment l'art est-il venu' 45.

[36] For example, Charles de Brosses' *Du culte des Dieux fétiches* (1760) is cited many times by Guasco, but only once by d'Hancarville (*AEGR* IV.51n.). In fact d'Hancarville, though admittedly disingenuous about his influences, cites very few contemporary writers of any kind apart from antiquaries such as Winckelmann, Caylus, and Gori. Modern 'primitive' peoples who have not yet found their way 'back' to monotheism do not fare well in Guasco's comparative ethnography, being dismissed as 'nations... sauvages' and 'peuplades naissantes' (193), uniformly 'stupide' (16, 30), and everywhere the same (12, 22).

who also inspired Gough's *Sepulchral Monuments of Great Britain*. Far from observing a strict distinction between texts and material culture, a distinction so important to innovators like Gough and to later historians of archaeology, d'Hancarville actually conflates the two when prompted by his love for classical philology or by some strategic purpose. By treating Pliny's or Pausanias's descriptions of lost monuments as though he were analysing the monuments themselves, d'Hancarville is only following the practice of neoclassical theorists such as Lessing.[37] It is still a form of armchair ethnography when he declares, judging merely from the plates in a book about a recent archaeological discovery in Germany, that these Obotrite artefacts allow us to 'rediscover the earliest vestiges of Sculpture from the most ancient epoch of Greece' (III.113). His engagement with the scholarly literature on this discovery, however, seems prompted by the conscious emphasis on monuments in the later volumes. Like Neapolitan customs, these engraved medieval artefacts provide a kind of proxy empirical evidence for the state of Greek culture 'at the time of Daedalus'.

D'Hancarville's expanding repertoire of tactics for amplifying antiquity might be described as an archaeological drift, or at least as a consequence of backdating monuments and peoples in the effort to create an archaeological deep time. Customs and manners, artefacts, and texts are all necessary to this project.

[37] All in all d'Hancarville relies less on eighteenth-century antiquarian sources and more on ancient writings about art in Volumes III–IV, often treating descriptions of artworks that no longer survive as equivalent to the works themselves. Much like Lessing in the *Laocoön* (1766), he forms elaborate aesthetic judgments from ancient descriptions (e.g. *AEGR* III.35–36n.) that are hard to distinguish at first from ekphrastic responses to existing works. For analogues in Lessing, see *Laocoön* 15–16; 162–4; 200–2. Momigliano distinguishes the 'armchair historian' and the 'true traveller' in *Classical Foundations of Modern Historiography* 51.

5

'Their History, Written By Themselves': Ancient Religion, Deep Time, and Embedded History

5.1 VIRTUAL ANTIQUITY

The different portions of d'Hancarville's title—*Collection of Etruscan, Greek, and Roman Antiquities*—are warranted in different ways. Etruscan antiquities appear primarily in narrative form, while the objects depicted in the colour plates are understood as more or less Greek from the start. Roman antiquities appear occasionally in the ornamental in-text engravings. Inscriptions, including the Eugubine Tables and the supposed fragment from Paestum, are almost the only Etruscan antiquities depicted in the work. In the narrative, however, d'Hancarville conjures up virtual antiquities to support his forays into prehistory. He accounts for Etruscan architecture, religion, and more shadowy areas of social organization largely from early modern scholarship and by inferences concerning absent originals. Initially, though, he seeks to give these virtual antiquities an importance equivalent to that of the physical Greek artefacts: the assertion of Etruscan priority takes place in the domain of theory and especially architectural theory. William Blake's visions of 'stupendous originals now lost' make a more explicitly anticlassical appeal to virtual antiquities, and Jonah Siegel has shown how effective this strategy was for romantic artists in general.[1]

Etruscan antiquity was a special case because it remained largely 'imaginary', as Michael Vickers has put it, before the advent of Etruscan archaeology, and because of its association with a specific cultural nationalism. Hamilton

[1] Blake, *Complete Poetry and Prose* 530; Siegel, *Desire and Excess* 80–9. The only authentic Etruscan antiquities depicted in *AEGR* are the Eugubine Tables, some of which are in Etruscan character (I.48–51); some of the gems engraved for the later volumes (III.194, IV.24; see also *V & V* 96–100); and a *bucchero* vase that is part of a larger composition (I.112; my Figure 5.2, bottom right). The fragments from Paestum (Figure 5.1) are probably neither Etruscan nor authentic, but as Ceserani has noted, 'Paestum's architecture was quite alien' to 'the canon of the classical' ('The Antiquary Alessio Simmacho Mazzochi' 257).

himself later felt compelled to report new archaeological evidence from Greece that disproved the Etruscan origin of painted vases once and for all.[2] But in his Etruscological phase d'Hancarville insists that Etruscan art was 'not borrow'd' from the Greeks (*AEGR* I.52) or the Egyptians (110), claiming that the Greeks were 'barbarous . . . before the Siege of Troy' (86). Drawing on the speculations of early modern scholarship, he declares that the original 'Tyrrhenians or Etruscans' came from Phoenicia (66). By dating the 'Pelasgian' or first Greek migration afterwards, he makes it historical. Later he abandons this Etruscan substrate and makes the Pelasgians original (IV.74n.), but in either case d'Hancarville pushes the frontier farther back into prehistory, anticipating the deep human time of archaeology.

As Greek mythology and the vase paintings themselves became more central in the later volumes, d'Hancarville found a new kind of virtual antiquity in myth. Mythological scenes depicted on the vases could be explained in light of modern customs, as we saw in the last chapter. More important, vase paintings document the customs of a pre-literate, ancestral Greek culture, as d'Hancarville argued especially in his new explications of over 100 individual vase paintings (IV.28–59n.). Myth itself provides a record of political history and the mythological scenes further document both archaic religious practice and earlier forms of visual representation, according to the final version of this argument. The present chapter shows how d'Hancarville's new emphasis on Greece allows him to situate the linked origins of art and religion more firmly in the expanded timescale he had posited.

Etruria's legendary qualities made it a powerful signifier of high antiquity and in 1766 it was an obvious choice of subject for a work nominally about ancient vases; Winckelmann only pioneered the thesis of their 'Greekness' in 1764.[3] But d'Hancarville's Etruscology is partly inspired as well by his visits to Paestum: though he deploys it as evidence of Etruscan priority, the intensity of his Paestum narrative also demonstrates the power of the Etruscan framework to promote an aesthetic experience of antiquity. The antiquity here is not all virtual, and he makes some attempt to assimilate the real temple complex to his Etruscan thesis. Although he links them to the vases only indirectly, through the architectural principle of proportion, he is able to cite physical artefacts that he and Hamilton discovered during four visits to the site,

[2] Vickers, 'Imaginary Etruscans'; Hamilton, *Collection of Engravings* I.16–22, 156–8; II.4–10, 96.

[3] Winckelmann, *History of the Art of Antiquity* 175–6. Guasco sums up a common perception of the time when he observes that 'ancient Etruria comprised a great part of Italy' (*De l'usage* 63). For the complicated historiography of Etruria, see Lyons, 'Neapolitan Context of Hamilton's Antiquities'. This essay also provides an invaluable survey of techniques and arguments that d'Hancarville inherited, directly and indirectly, from A. F. Gori, F. Mastrilli, and other Italian intellectuals. From this point of view, some of d'Hancarville's ideas, and some of Winckelmann's, are 'pioneering' only insofar as they transmitted Italian connoisseurship to a wider European public.

including two inscriptions and a 'fragment of a Tuscan Entablature' that appears to match them in scale and style (I.96). Since the three temples are 'of a sort of Architecture whose Members are Dorick, altho' its proportions are not so', he infers that these structures, and the whole Greek colony of Posidonia, must have been built on Etruscan foundations (96–100).[4] These imaginary Etruscan works are virtual antiquities in the same sense as the ancient Gaelic manuscripts (never authenticated) from which James Macpherson supposedly translated the poetry of Ossian in the same decade.

Unlike Macpherson, d'Hancarville abandoned virtual antiquity for a real early Greek alternative that provided much more corroborative evidence for his conjectures in the form of both physical artefacts and texts. The temples and inscriptions from Paestum (Figure 5.1), some of which were engraved for the volume (27), are in fact more accurately described as Greek, but d'Hancarville, who remained an ambivalent empiricist at best, continued to rely on a range of rhetorical strategies to amplify or exaggerate the antiquity of the objects he discussed. It is worth noting, too, that classical archaeology depends on virtual antiquity in every case of an absent original, as in the modern view that most Hellenistic and Roman sculptures are based on lost Greek originals.

Beginning in 1773 d'Hancarville rhetorically marks his conversion from Etruscan to Pelasgian proto-antiquity by confessing his former errors and indulging in pointed sarcasm at the expense of the authorities who misled him. This claim to have found the true origins of Greek and South Italian civilization, on more solid evidentiary grounds, is itself an assertion of the high antiquity of his scholarly domain. If it is merely a matter of Greek history beginning earlier than we thought, then the most ancient artefacts can be incorporated into a more coherent classicizing narrative without sacrificing their antiquity. In fact certain synecdoches of high antiquity—including Paestum, the walls of Tyrins, and sacred standing stones—keep on surfacing in his text, and d'Hancarville cross-references his earlier account even as he gradually replaces Etruria with Arcadia, the home of his Pelasgian Greeks. On the one hand, this substitution invokes the methods of Enlightenment science by replacing the legendary with the real (though this Arcadia, to the modern reader, may seem hardly more real). On the other, it underwrites the claim of high antiquity more solidly by referring regularly to Pliny and other ancient sources rather than appealing vaguely to the Tuscan order. The first step in this process is the (first) history of vase painting in Volume II: here he

[4] D'Hancarville cites Vitruvius on the 'antiquity of the Tuscan order' (I.76), but a more careful reading would have suggested that original temples or prototypes were wooden structures (*De architectura* IV.2), meaning that the 'fragment of Tuscan entablature' he found could only have been a Roman imitation. It is theoretically possible, however, that since-vanished wooden prototypes were still extant when the Greek colonists arrived (in d'Hancarville's conjectural scenario) and modelled the surviving structures on them.

concedes that there is no evidence of Etruscan painting prior to colonization (136) and represents the vase workshops as institutions of a hybrid colonial culture (128–134). Nonetheless, d'Hancarville continues to refer in this chapter (100–02), and even in later volumes (IV.74n.), to his first narrative of the Etruscan origins of art, and particularly his favourite example of the walls of Tyrins, built in the most primitive architectural style (I.108).

He turns more and more explicitly against Etruria, however, with an elaborate identification of a Virgilian character, Dardanus, as Pelasgian (III.69–70; cp. *Aeneid* III.167; 227–38), and with re-interpretations of examples formerly held to be Etruscan, such as the inscriptions on the Hunt Krater (*AEGR* III.209n.). These characters turn out to be clearly Pelasgian, and d'Hancarville confesses that he was labouring under a general mistake (*méprise*) in his first essays on this and other vases (cp. IV.38n.). In the final volume he openly ridicules Etruscology, sarcastically referring to Etruscans from the sky:

> Some Antiquaries, wanting the whole world to have been created in Etruria [and] not content with making the Etruscans descend from the heavens, make them go forth again from Noah's Ark; they seem to have written only for the majority of the creatures that found refuge in this Ark, and Philosophy sends them back there again to seek admirers for their opinions and their systems. (IV.18n.).

The passage concerns a painting (pl. 12) of the Choice of Hercules myth from a black-figure vase (now lost), in which Hercules and other figures are labelled in characters supposedly representative of very early Greek letter forms. D'Hancarville takes advantage of the opportunity offered by this vase to repudiate Etruscology and flatter his new patrons, the Florentines, by 'recogniz[ing] their Etruscan, Pelasgian, and consequently Greek Origin' (IV.18n.).[5] Much of *The History of Greek Sculpture* (*AEGR* III–IV) revolves around this central revisionist thesis of a Pelasgian antiquity even older than Etruria. This argument is no less interesting, in terms of disciplinary history, for being compromised occasionally by ulterior motives or manipulation of evidence.[6] The challenge for an art history seeking to map the 'most remote antiquity' is as much rhetorical as evidentiary.

[5] The form of the initial *eta* in 'Hercules', so d'Hancarville argues, denotes the aspirated 'h' passed down by the Pelasgians to the Etruscans and preserved in the Latin spelling of (H)Etruria and in modern Florentine pronunciation. The 'antiquaries' ridiculed here probably include Thomas Dempster as well as d'Hancarville's own earlier work and the Neapolitan academic Jacob Martorelli, cited heavily in the first volume. In that volume part of his concern was to placate the Neapolitan authorities, but this text shows him currying favour with the authorities in Florence, where he had found his new (if precarious) refuge.

[6] Jenkins and Sloan have argued that d'Hancarville may have altered letter forms to support his philological argument in a parallel case in which the vase—with evidence of alterations—survives (*V & V* 158). The 'Choice of Hercules' vase described here does not survive, but modern scholarship has adopted the provenance given by d'Hancarville, 'formerly Palermo, Jesuits'. See Gisler-Huwiler et al., *Collection of Antiquities* 420–1.

The key point for d'Hancarville is that ancient artefacts provide the means for tracing the origin of the arts back, by way of religion, almost as far as the point of human origins. Thus he continues the same argument that he launched in his Etruscan phase: art records its own history. What is significant is that he develops this argument so much more fully in the new volumes, not that he applies it to Pelasgian rather than Etruscan antiquity, or that he now uses terms derived from Guasco ('sign' and 'figure') to mark the stages of that history. The imitation of ancient architectural form by the Dorian architects of the second colony at Paestum, who built the extant temple complex, provides an early example of the 'conservation' of prehistoric forms in historical works. When he finally turns to the vase paintings in the last volume, d'Hancarville argues that they not only 'conserve' archaic formal features but also preserve the memory of primitive religious rites in pictorial form by incorporating them with contemporary ritual in scenes of worship. In this volume he also returns to Paestum for a sublime vista on remote antiquity, recast as Pelasgian and therefore Greek; this site becomes the focal point of a substantial interpolated essay subtitled 'Historical Remarks on the Origin of the Pelasgians' (IV.68–96n.).

The effort to dismiss antiquarian arguments for Etruscan origins leads d'Hancarville to a new body of evidence, engraved gems, incorporated in his usual digressive manner into the history of sculpture in the same way that the vases are incorporated into the history of painting.[7] He also uses inscriptions and cross-cultural comparisons to bolster his Pelasgian case. By the time he returns to the scene of Paestum, then, d'Hancarville has prepared his reader carefully to reject the Etruscan 'prejudice' and accept the monuments on this site as even older. He still maintains that the city was founded by Oenotrus, the grandson of Pelasgus (cp. I.104), but is now content to follow Vitruvius (*De architectura* II.2–5) in making the Pelasgians the inventors of the first primitive huts. Instead of Etruscan civilization (originally assumed by d'Hancarville to cover much of the Italian peninsula), these first Greek colonists now find the easily displaced 'barbarians' remembered only by the name 'Ausonian' (IV.75n.). His narrative otherwise retains the earlier structure, relegating Posidonia—built on the foundations of Oenotrus's Phistulis—to a much later, second wave of colonization following the Trojan War. Far from retracting his earlier views, d'Hancarville offers the same revisionist history of

[7] Another in-text plate (IV.24) shows six gems re-engraved from works by Winckelmann and Caylus (d'Hancarville abandoned his own collection of gems, noted by Winckelmann, when he fled Rome in 1764, and Hamilton's had gone to London in 1772). In an extensive discussion thoroughly analysed by Jenkins and Sloan (*V & V* 96–8), d'Hancarville reassigns to Pelasgian antiquity some examples held to be Etruscan even today (*AEGR* IV.25–71), correcting Winckelmann along the way (62–66n.). The discussion of gems in fact begins in Volume III (193) and may have occupied d'Hancarville for the better part of a year, as the end of the discussion bears an internal date of October 1775 (IV.66n.).

colonization extending to 'remotest antiquity' (I.98), a history pieced together out of Strabo, Dionysius of Halicarnassus, and Pliny. His empirical description of the plan and walls of Paestum, too, follows closely his initial description of 1766: the current walls are reared on more ancient foundations and enclose older, unhewn stones (IV.75n.), and the irregular distribution of the towers shows them to have been watchtowers predating classical military science (76n.).

For his most crucial piece of evidence, d'Hancarville rereads an engraving from his first volume, revisiting the site with us in imagination. This is the only cross-reference to his earlier work that is not derogatory: situated archaeology remains the *ne plus ultra* in d'Hancarville's quest for antiquity and his encounter with these ruins must have been transformative. It seems plausible enough today to suggest that the site has a peculiar power to inspire young tourists to pursue careers in archaeology. Although Paestum was becoming a tourist site in the 1760s, 'archaeology' was still synonymous with 'antiquarianism' and could be imagined as a professional path only, as I have argued, from the point of view of the knowledge worker, the scholar for hire. In this sense d'Hancarville anticipates a modern archaeological narrative. In reviewing the evidence for Pelasgian origins he returns, characteristically, to an inscription, this one from immured [*enclavées*] stone tablets, originally engraved as a title vignette (I.27), which he now calls 'possibly the most ancient monuments of our writing' (IV.76n.). More careful study has revealed the letters as Pelasgian, not Etruscan; he returns particularly to the coin in the lower left-hand corner of this vignette (Figure 5.1), a coin that he now interprets as a medal commemorating the Pelasgian origins of 'Phistulis' and conserving its original alphabet.

In fact, the large inscription is almost surely a pastiche. In the engraving we see Bracci's characteristic Piranesian tactic of monumentalizing the fragment, though in the original accompanying text d'Hancarville specified actual measurements of one palm for the height of the letters and six French feet for the width of the blocks (I.96, 98). In the new account he calls these characters a 'disorderly mixture', which indeed they are. Even if he forged the inscriptions and made up the story of finding them inside a wall that was being dismantled for building material—it is still an archaeological story. This vignette, like many others in the work, showcases d'Hancarville's ability to design antiquity; if the inscription does have some authentic foundation, he is nonetheless concerned not with its meaning but with persuading us that his expanded concept of antiquity has both a material and an aesthetic basis.[8] Paestum is the

[8] As the fragment does not appear in Rix, *Etruskische Texte*, it seems likely that d'Hancarville forged it, which would make this fragment at best an ironic marker of authenticity. Professor Rex Wallace of the Etruscan Texts Project has confirmed this suspicion, dismissing this inscription as 'rubbish' (pers. comm.) According to Professor Richard Foley (pers. comm.), the first and third lines read right to left and the second line reads left to right (*boustrephedon* style), and may be transliterated as 'duvsmokvo/??? khnus/popen', assuming the resemblance to Etruscan characters to be more than coincidental. On the smaller engraving, see Ramage, 'Initial letters' 448.

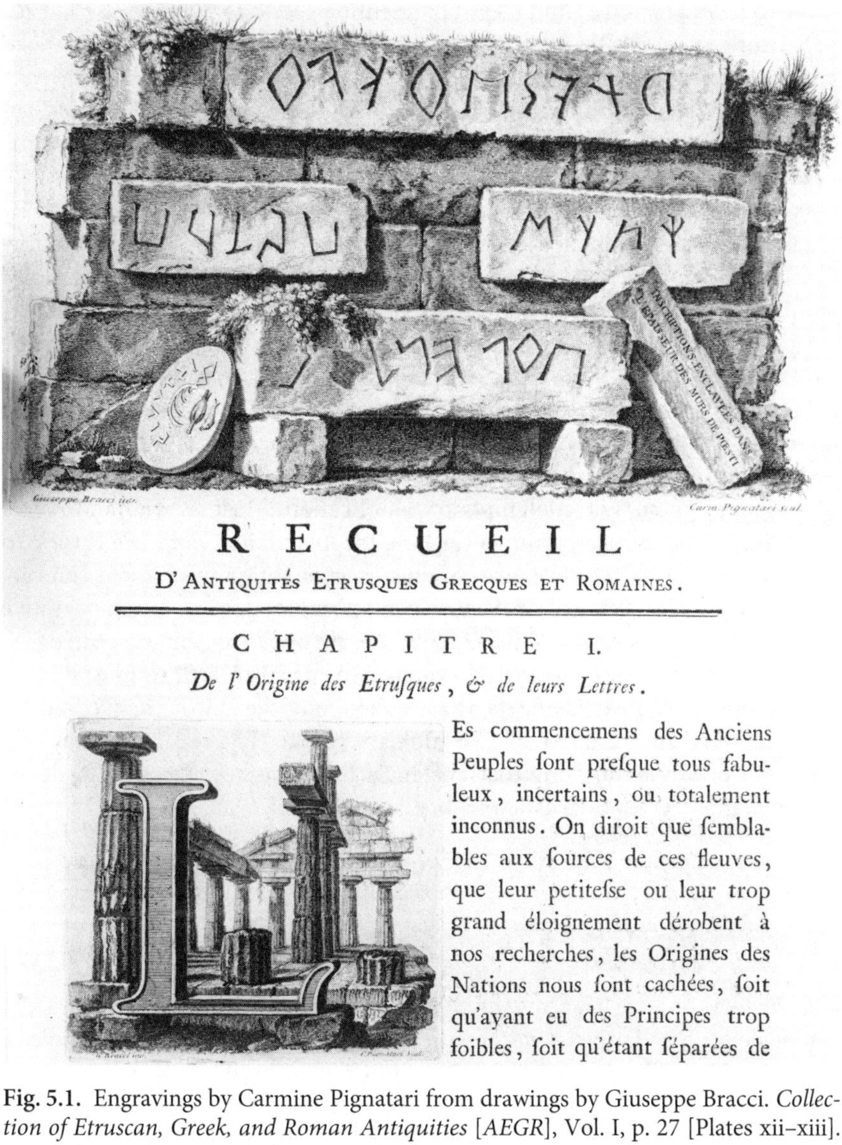

RECUEIL

D'ANTIQUITÉS ETRUSQUES GRECQUES ET ROMAINES.

CHAPITRE I.

De l'Origine des Etruſques, & de leurs Lettres.

Es commencemens des Anciens Peuples ſont preſque tous fabuleux, incertains, ou totalement inconnus. On diroit que ſemblables aux ſources de ces fleuves, que leur petiteſſe ou leur trop grand éloignement dérobent à nos recherches, les Origines des Nations nous ſont cachées, ſoit qu'ayant eu des Principes trop foibles, ſoit qu'étant ſéparées de

Fig. 5.1. Engravings by Carmine Pignatari from drawings by Giuseppe Bracci. *Collection of Etruscan, Greek, and Roman Antiquities* [*AEGR*], Vol. I, p. 27 [Plates xii–xiii]. Ostensibly an 'Etruscan inscription found in the wall of Paestum' (*AEGR* I.176); the capital L is superimposed over an image of the Temple of Neptune at Paestum, anticipating G. B. Piranesi's more famous engraving of this temple. Courtesy of Margaret M. Bridwell Art Library, University of Louisville.

privileged site that gives him a kind of scientific access to prehistory: his usual patchwork of ancient authorities is reinforced here by situated ekphrasis, empirical documentation, and local observations including natural history.[9]

This two-stage account of colonization in Paestum supports d'Hancarville's new derivation of all ancient art from the earliest Greeks. Etruscan antiquity becomes easier to celebrate once it turns out to be Greek, and 'Pelasgian' makes a better umbrella term for the culture(s) of 'remotest antiquity'. By excavating Paestum in this quasi-archaeological manner, d'Hancarville furthers his new project of making Pelasgia (or Arcadia) a fixed historical point within prehistory, and simultaneously develops his repertoire of techniques for backdating ancient artefacts—a central project of all four volumes. The two-stage migration thesis is one of many arguments designed to 'stretch' time and promote an ethno-archaeological deep time that parallels or shadows emerging geological time. 'Deep time' in its archaeological sense has some concrete affinities with the unfathomed antiquity of volcanoes that Hamilton was excavating in his contemporary publications. Just as Hamilton uses repetition, especially in commenting on the plates in *Campi Phlegraei*, to establish abundant local instances of unsuspected antiquity, so too d'Hancarville multiplies instances of artefacts that support a more remote and primitive origin for the arts than Winckelmann or anyone else had suspected. As Hamilton points to archaeological evidence for prehistoric eruptions, d'Hancarville points to the deliberately archaic version of the name engraved by the later colonist who crafted his Posidonian medal (Figure 5.1), tracing the antiquity of arts by arguing that ancient artists routinely recorded or 'conserved' earlier stages of artistic practice in their works (III.22).

5.2 OLD ARTEFACTS, NEW HISTORIES

The history of vase painting in the second volume begins to expand the timescale by correlating artefacts with ancient literary sources. The primitive iconography of the vase paintings that draw d'Hancarville's special attention, or rather the conjectural dates that he derives from their formal properties, may be corroborated with reference to Pliny. Although he never refers explicitly to painted black- or red-figure Greek vases, Pliny repeatedly alludes to a special subset of earthenware vessels much more precious than the common tableware produced by the eight major pottery manufacturing centres of his time (*Natural History* XXXV.46). D'Hancarville makes a number of inferences

[9] In one such observation, perhaps inspired by Hamilton's collection of fish specimens, he identifies the fish design on the medal with a species, *Sêche* (genus *Polypos*), still 'very abundant' in the Bay of Paestum (IV.77n.).

concerning the 'Earthen Vases, which according to Pliny were dearer than even the Murrhine Vases' (II.86), some of them being coveted luxury goods and others being used for pouring libations (88). At a loss to explain their special status, he reasons that here 'Pliny means to speak of Earthen Ware Vases, much more Ancient than those of his time' (86), and which could not be counterfeited. The chain of inferences links these more or less cultic objects to the most monumental of the vases engraved for the book, such as the Meidias Hydria, loosely grouped in this analysis as the fourth category of ancient vessels, 'their richest vases' (82). He concludes that 'they were Earthen Vases, like those we now publish, that he speaks of in his book' (90), vases already valued in Pliny's time for their high antiquity.

D'Hancarville doubles this temporal distance by arguing that the art of vase painting itself had been lost for centuries prior to Pliny's time (92), perhaps since the Roman conquest of Capua (106), which was already known as an archaeological site in Caesar's time (94) and was the source of numerous Hamilton vases. In this way he finds ancient precedent for the status of painted vases as prestige commodities or 'Master pieces, that in all times have really merited the attention of men of taste' (106). Vickers rightly notes that this conclusion is far too favourable to the book's promotional aim to be taken at face value. Vickers also shows vividly how influential d'Hancarville's argument has been across time, despite this apparent conflict of interest.[10]

Having underscored the antiquity of even the more recent painted vases, d'Hancarville moves to establish substantially earlier dates for the vases he considers most primitive (in a positive, more or less Rousseauvian sense). Though initially content with a date of 658 BCE for the Hunt Krater examined in Chapter 4, he hurtles backwards another 600 years in the third volume. In the second volume, he establishes the principle of historical exhibition and the canon of formal properties attaching to the earliest vase paintings, both of which are essential to the two later volumes and developed more fully there. Jenkins and Sloan observe that these discussions, with their emphasis on stylistic development, simulate a display of specimens arranged by date, which they proceed to replicate visually in their splendid catalogue: 'D'Hancarville's "exhibition" is here reconstructed with the vases arranged according to his chronology' (*V & V* 155). Building loosely on Pliny's account

[10] Vickers points out that in the crucial passage Pliny is satirizing special earthenware vessels manufactured by a recent emperor, a fact that seems unfavourable to d'Hancarville's thesis ('Hamilton, Geology, Stone Vases' 268). Vickers is right to caution us 'about imposing anything d'Hancarville said about antiquity on antiquity itself'. Vickers's own provocative view of the vases as 'kitsch' or 'mock silver' (270) does not address the archaeological problems that d'Hancarville was attempting to solve in this passage, however (Pliny's references and the absence of painted vases from the Roman archaeological record). Lyons shows that Italian connoisseurs had already made the argument that the vases were precious in antiquity ('Neapolitan Context' 233).

of stages in the development of painting, d'Hancarville moves from the Hunt Krater to a more recent vase fragment to a very early Roman painting described by Pliny, increasing the amount of time required for each stage: 'if we measure the great distance that there is between them and the progress the arts must have made to arrive from the one to the other, we shall be easily persuaded that it could not have been compass'd in less than two Centuries' (*AEGR* II.122). Foreshortening is one of several technical developments that had to intervene, he argues (124). In similar fashion Vitruvius traces the history of architecture from a surviving 'relic of antiquity'; the topos of antiquity as d'Hancarville inherits it from Vitruvius, Pliny, and other writers is itself a species of *differance*, a constantly aggregating deferral of origins.[11]

In Volume III d'Hancarville elaborates this narrative exhibition of the Hunt Krater and its two predecessors and creates parallel exhibitions of sculpture and engraved gems. At the same time he establishes ancient analogues for this principle of exhibition, and in each case traces the progress of representation from the 'sign' (iconic and allegorical images) to the 'figure' (naturalism and ultimately ideal beauty). In each case he dwells on the 'most ancient pieces that can exist' to foster the expansion of the timescale, or archaeological deep time (IV.vi). D'Hancarville presents this extraordinary formulation in the fourth preface, offering collectors of drawings the opportunity to make their historical 'series' complete with his 'faithful copies' of the vase paintings. Here antiquity is guaranteed by the durability of ceramic, as opposed to other media. His 'most ancient' specimens of sculpture and gem engraving share this durability and, in the latter case, the abundance that also characterizes the archive of vase painting. His stylistic argument—more properly a developmental aesthetic theory—unites these examples, which become mutually reinforcing. The golden bull supposedly misidentified by Caylus (as Egyptian) actually belongs to the second historical phase of Greek sculpture, making it 'one of the most ancient monuments of Art' (III.97–98). I return to this example here because the exaggerated length of the horns on this gold statuette provides the earliest surviving example of the sign (*signe*) as d'Hancarville understands it: the artist made the horns disproportionately long to signify that this was a particularly large and splendid bull, and orbicular to signify that he was sacrificed to Diana. Yet the sign here is 'allied to the figure' (97), since the rest of the bull has its due proportions, to the extent permitted by the rudimentary state of the art.

[11] Antiquity is constantly deferred and yet constantly present. Vitruvius writes, for example, that 'in Athens on the Areopagus there is to this day a relic of antiquity with a mud roof', comparable to the huts built in Vitruvius's own time at Marseilles and by 'foreign tribes' generally. *The Ten Books of Architecture*, trans. M. H. Morgan (New York: Dover, 1960), 40, 39 [De architectura II.i.5]. On *differance*, see Derrida, *Of Grammatology* 62–5.

The earliest specimens of gem engraving and vase painting, too, document this fundamental shift in representation. A cameo representing the Seven Against Thebes—considered by Winckelmann as not only 'the most perfect [monument] of Ancient Etruscan art', but also 'the most ancient monument of Art in general'—may be recognized as not actually earlier but later than a 'Pelasgian' gem considered late Etruscan by Winckelmann (IV.63n.). This second gem, an image of Tydeus, is evidently still 'regulated by the sign' (64n.); it is one of several 'early' cameos (23, Figures 1–4) that place their athletic subjects in an anatomically improbable or 'exaggerated attitude' to 'signify suppleness'—a 'vestige of the sign' that is no longer present in the Seven Against Thebes gem as d'Hancarville reads it (67–68). The issue, then, is not whether these gems are Etruscan, as Winckelmann and Caylus argued, but what is their stage of development: d'Hancarville's history substitutes the 'particular style' of a period in ancient art for the 'general style of a particular nation' as the category of analysis (65n., cp. vii). Yet his quotation reveals that Winckelmann inspired his quest for remote antiquity in the first place, and this quest constitutes a style of argument that is fundamental for the history of the discipline. The Tydeus gem, like the golden bull, records or 'conserves' an earlier form of representation—'indication' as distinct from 'expression'—while the temples and inscriptions at Paestum similarly conserve earlier forms of architectural order and lettering. The multiplication of these examples from the 'infancy' of their respective arts contributes rhetorically to the proliferation of antiquities. The birth of painting, which follows sculpture and carving, replicates these stages, thus conserving the collective history of arts.

The great importance of the vase paintings is thus not merely aesthetic—a supposition that Vickers rightly challenges as unhistorical—but archaeological. Their antiquity, abundance, and (often) degree of preservation allow them to document stages in the history of the other arts, and of religious ritual, that are otherwise lost. This archaeological premise informs the historical nature of d'Hancarville's 'exhibition'. The principle of conservation (III.22), perhaps the most fundamental through-line of *AEGR* as a whole, ensures that less ancient vases will replicate more ancient techniques of painting to commemorate the history of painting itself; and since each new art form manifests the same stages of development, these vestiges of the sign or 'indication' will also illustrate the early stages of sculpture and carving from which no actual specimens survive. D'Hancarville's long-promised explications of the vase paintings finally appear, ironically, in his longest excursus note on gem engraving, because he finds himself drawing his evidence from objects depicted in the vase paintings.

These correspondences among the arts account not only for the digression on gem engraving but also for the long interpolated narrative of the birth of painting that occupies the latter half of Volume III and the beginning of Volume IV. In analysing the gems, d'Hancarville reminds us that he has already identified a subset of the vase paintings, supposedly contemporary with Cimon of Cleonae, that display the same 'vestiges of the sign' and document the same transition from 'attitude' (form) to 'movement' (harmonizing form and content) (IV.67). The earliest of these paintings, the Choice of Hercules (Pl. 12), uses human figures themselves as signs: the unnaturally large hands and feet of Pleasure, for example, indicate to d'Hancarville both the irresistibility of her charms and the evils that follow in her wake (18). Two other paintings (I. Pl. 63 and III. Pl. 103) can be assigned a slightly later date because signification is confined to the 'forced' or 'exaggerated' poses of the figures (IV.23–24), as in the Tydeus gem. The absence of proportion in the first case, and of perspective and ideal beauty in all three, are marks of high antiquity, of an authentic naivity reflecting the progress of the arts rather than the primitive state of a particular nation or artist.

In this way d'Hancarville expands the domain of antiquity and legitimates archaic forms, continually honing and revising a particular canon of 'earliest' vases the 'Choice of Hercules' belongs with the sequence of 'Song and Meditation' (II.125), 'Oedipus at the Crossroads' (Vol. I, Plates 93–94), and the boar hunt, a sequence he sums up as follows: 'far from being repulsive [to readers], the very crudeness of these monuments will greatly interest them' in the light of his analysis (III.202). Despite his strenuous efforts to distance himself from his early work in these later volumes, their basic function is to elaborate a claim that he established in the second volume: the body of evidence offered by the vases is 'the only one that can shew…the progress of human industry in the finest of the Arts of its invention' (II.128). In this passage, d'Hancarville emphasizes the role of inscriptions, a common element of vase paintings that increases their archaeological significance.

To the inscriptions he later adds other landmarks of stylistic development that complete the homology between this development and the anthropological evolution of figurative out of non-figurative representational practices. Moreover, he finds an ancient analogue, a set of sculptures in the main square of Pharae, for his practise of narrative exhibition, a technique he introduced with his first set of explications (I.152–72). As reported by Pausanias, the statue of Mercury in the centre of the ancient marketplace here (near modern Kalamata) was surrounded by bronze lamps and thirty square standing stones. Incorporating a nearby fishpond sacred to the god and a sacrificial hearth (interpreted conjecturally as a non-figurative Vesta), d'Hancarville reads this collection as a comprehensive 'Suite de l'histoire de la Sculpture' (III.35–36): the pond 'shows the primitive fashion of *designating* the Gods by objects'; the standing stones or *cyppes* show the earliest manner of *indicating* a divinity; the lamps are *signs* allegorizing a divinity (Vesta in this case); the Hermes itself, lacking a pedestal, resembles the 'term', an early form of sculpture halfway between the *sign* and the *figure*; and the expressive classical statue or *figure* of the Vesta (not really present in Pausanias) completes the sequence.[12]

5.3 ART AS HISTORY, MYTH AS HISTORY

Although Haskell refers to d'Hancarville as an art historian, he is unlikely to displace Winckelmann as the founding figure of that discipline. D'Hancarville's historicism is less empirically sound than Winckelmann's, but is in some ways more radical and more systematic. Art is history for d'Hancarville, both because artefacts themselves encrypt the origin and progress of the arts and because they translate into intelligible myth the social and political events of a pre-literate past that are otherwise irretrievable. The freewheeling conjectures that he offers to support his correlations between myth and history suggest a genius more literary than historical. At the same time, his exploration of ancient religious ritual—still conjectural but more firmly tied to empirical evidence—gives an anthropological dimension to art history that is less developed in Winckelmann. The methods that d'Hancarville uses to establish the antiquity of artefacts, though of great intrinsic interest, are ultimately less important than the stories that the artefacts tell: the story of a stylistic development from sign to figure and a story of the simultaneous development of ancient Greek religion and society. Where the two stories intersect—where the progress of art is given a crucial spur by a political event, such as the Trojan War—I will attempt to draw out the anthropological implications, returning to the last chapter's theme of customs and manners. D'Hancarville's innovative antiquarian method shares at least two substantive traits with archaeology: the strong interest in dating artefacts and the willingness to take ancient religion seriously as religion.

The influence of Guasco is strongest at the beginning of Volume III, where d'Hancarville picked up again after his long hiatus. Guasco offered him a set of terms for specifying his original thesis in a new way, as well as a history of ancient religion that provided the spur to d'Hancarville's antithetical one, which replaces the pejorative rubric of 'superstition' with a secular aestheticism. He takes (without acknowledging) Guasco's narrative of sculptural signs that designate, leading to figurative statues that express, and makes it self-reflexive: like the log-shaped column, the statue that incorporates conventional allegorical signs also narrates its own origins. Having introduced this

[12] See Pausanias, *Description of Greece* 7.22-4. Anticipating early archaeology in another way, d'Hancarville next maps an evolutionary sequence of structures on a sacred site in Syria that began simply as a mountain consecrated to Zeus (36), though again he draws on ancient texts rather than field evidence.

reflexivity in the first volume as a maxim of ancient artistic practice—initially as an architectural convention of alluding to the natural origins of structural forms—d'Hancarville now uses Guasco's terms to illustrate how in sculpture, too, ontogeny recapitulates phylogeny. The new application manifests the same romantic organicism that is characteristic of d'Hancarville's work as a whole. Although his elaboration of Guasco's sketchy narrative owes much to its source, d'Hancarville ultimately subordinates that narrative not only to his own organic paradigm, his habit of correlating the progress of sculpture with that of painting and other arts—he also makes artistic form a causal factor in the development of religion, as opposed to Guasco's symptomatic narrative of decline from an original monotheism. In this way, d'Hancarville finds evidence in the vase paintings for the origins of both painting and sculpture and for the role of artefacts in an evolving repertoire of devotional practices.

In his last broad iteration of this organic progress from iconic to mimetic to expressive representation, he argues that the devotional logic of the early Greek cults itself dictated a conventional iconography ensuring the preservation of earlier forms. In statues of the Dioscuri as well as engraved gems from sundry eras of antiquity (IV.6–7), Pollux has his arm around Castor not only to mark his divine descent (unlike his brother's) from Jupiter, but also to recall the cross-piece (*tenon*) used to join two square beams (*poutres*) which 'signified' the Dioscuri at an earlier stage (5). The *signe* belongs to an earlier stage of representation, but also to an earlier stage of deification, in the sense that figurative statues even after the birth of sculpture were permitted only for full-fledged divinities, while heroes were designated in the old non-figurative way (as exemplified by the *poutres*). Since the birth of sculpture antedates the Trojan War, the creators of very early memorials of the Dioscuri must have made them non-figurative by choice (one such work is reported in Plutarch's *On Brotherly Love*). If their apotheosis post-dated the war by only thirty years, as d'Hancarville suggests based on the *Iliad* (4), it is reasonable to suppose that the first sacred, hence figurative, sculptures of the twins originated as direct imitations of the joined beams, turning the joint between them into Pollux's outstretched arm—which now 'expresses' the relation formerly 'signified'. D'Hancarville illustrates 'this manner of conserving the idea of primitive signs' from several further examples including the Three Graces, incorporating Lamberti engravings of a Herculanean wall painting and of a neo-Attic sculpture (the San Ildefonso Group) to document visually the *beau ideal* that he now supposes to be the culmination of the many vase paintings engraved for his book (6–9).[13] The 'maxim' of conserving original form that he first

[13] In this case—unlike the Etrucan gems—d'Hancarville does correct a mistake in Winckelmann, who identified the San Ildefonso Group as Orestes and Pylades in his *Monumenti Antichi Inediti*. See Haskell and Penny, *Taste and the Antique* 174n.21. Cp. Guasco's in some

presented in his defence of the 'Tuscan order' (I.80) here becomes a matter of religious and political protocol, a reflection of the fundamentally public and social role of art (III.22). The embedded history of form and technique that ancient artefacts offer us is thus at the same time an embedded history of ancient politics and religion.

Though the origin of painting follows, and formally recapitulates, the origin of sculpture in d'Hancarville's narrative, both are swept up in the rapid transformation of art and religion that follows the Trojan War. On their return the Greeks 'consecrated a great number of Temples and Statues [as] monuments of their victory and of their gratitude toward the gods' (IV.2). D'Hancarville strongly implies that vase paintings, too, began to proliferate during this ancient Renaissance, when Greek 'Religion was occupied with deifying the heroes [of the Trojan War], their poetry sang them, their architecture raised temples to them, and sculpture, painting, and engraving laboured to make them monuments of every kind' (11–12). Citing several crucial innovations in the early history of painting, he gives these developments the same early date and in so doing offers a rare acknowledgment that the hundreds of colour plates composing half his work might have some significant bearing on his narrative. From the vase paintings that 'fill this work', he writes, we can get an exact idea of the pioneering technique of a very early group of painters, who set their monochromatic designs off against a contrasting ground (1).[14] This and related innovations date 'either from before, or at the latest just after the ruin of Troy' (2), and the implication is that they became widespread when the arts began to flourish immediately afterward.

D'Hancarville does not claim that all the vase paintings date from this time, but because of the 'liaison' or correspondence among all the arts, including their reflexive character (III.150), the vase paintings can tell us much about the rapidly evolving relationship of religion and the arts in that era, when it became 'natural' for artists 'to seek to recall, by means of the figures that were substituted for the Signs, these ancient objects of public veneration, which seemed to render Art more respectable, by means of the very dignity that it conferred on them' (IV.7). This formal convention became a 'maxim… common to Sculpture, Painting and Architecture,' and 'preserved the memory of what they were even before these arts existed…: *their history written by themselves* is recorded in these monuments'. The grandest works of art are

respects similar treatment of these two examples (*De l'usage* 49–50). For Plutarch's account of the primitive monument of the Dioscuri, see *Moralia* 478a–b.

[14] Though he takes his account of these innovators from Pliny, he maintains that their true 'antiquity [was] unknown in Pliny's time' (IV.1). D'Hancarville notes elsewhere that 'of all the materials that have ever been painted, fired clays [*les terres cuites*] are incontestably the most durable' (III.151n.68). No cave paintings were known in Europe at this time.

'exalted by the very feebleness of their origins', a 'feebleness' recalled by formal allusion.[15]

From the beginning of his work, d'Hancarville relied on the conventional neoclassical idea of a correspondence among the arts. François Lissarrague and Marcia Reed, tracing this correspondence as a visual motif through the four volumes, argue that the many in-text engravings serve to illustrate 'the principle of analogies among the arts on which the book is based. To rephrase and extend Horace's dictum: as it is with ancient painting, so it is with vases. This is why the title words convey the idea that this is a collection of many kinds of antiquities, not just vases'.[16] D'Hancarville achieved this variety, they argue, partly by having images of gems, sculpture, and wall painting (such as the Graces) re-engraved from earlier antiquarian books. As a formal principle, this notion of analogy or correspondence allowed him to explain 'the art of making Vases' initially as a synthesis, 'having a dependency upon Architecture by the proportions it employs, upon Sculpture by the outlines, & on Painting by the Designs with which its productions are adorn'd' (I.70). In the course of writing his four volumes, d'Hancarville became more and more invested in a historical understanding of this formal principle: his effort to contextualize the analogy among the arts eventually led him to trace each one to a religious point of origin. His main concern was not, like Caylus, to find visual-verbal parallels or, like Lessing, to distinguish representational modes categorically. He gradually built up the paradigm of an organic process out of which all the arts evolved. Sculpture became central when d'Hancarville began to see not only the outline or shape of the vases, but also the representational content and ritual significance of the paintings, as deriving historically from sculpture.[17]

Initially, though speaking in terms of the 'maxims' and 'souvenirs' later echoed in the fourth volume, d'Hancarville constructs an unspecified deep time as the setting for his prehistory of art, a sort of golden age lacking any cultural specificity, like the early stages in conjectural histories of civil society. The bronze candelabra with their horizontal nubs, depicted in his first visual

[15] Here is the original passage: 'Cette attention passée en maxime, & devenue commune à la Sculpture, à la Peinture, & à l'Architecture, nous a conservé la mémoire de ce que les choses furent, même avant que ces Arts existassent: Ils semblent, par cette pratique, avoir voulu nous conserver le souvenir de ce qu'ils ont été dans leurs commencemens, celui du chemin qu'ils ont parcouru, & des obstacles qu'ils ont du franchir: *leur histoire écrite par eux mêmes* se lit dans ces monumens, & quand on considère de quels foibles principes ils partirent, pour arriver à ce point de grandeur où ils parvinrent dans la suite, on voit que ce n'est pas sans raison, qu'ils paroissent s'être glorifiés de la foiblesse même de leur origine' (*AEGR* IV.9, emphasis added).

[16] Lissarrague and Reed, 'Collector's Books' 283.

[17] In the preceding quotation d'Hancarville is also justifying the structure of his book. His turn to sculpture explains why the original history of sculpture (I.112–52) later expands to fill two volumes, in which the histories of architecture and painting are not so much expanded as incorporated into the history of sculpture. The original passage essentially establishes d'Hancarville's method along with his tripartite organization of the arts.

treatment of this evolutionary theme (Figure 5.2), commemorate the stripped canes from which the first candleholders were made in this early state of society (I.80). Mythology provides the only avenue for historicizing these origins more precisely. Accordingly, Volume III begins with myth and religion and moves next to sculpture, because of its affinity with the natural objects consecrated to the earliest divinities. Engraving and painting, which follow, reflect and mediate the naivety or original 'feebleness' by contrast to which the progress of the arts appears so formidable. The maxim or 'inviolable law' of 'conservation' (III.22) carries over into the history of sculpture, but the nature of the 'souvenir' conserving art's history in a given work changes from the mimetic evocation of material by form (the canes) to the formal evocation of spiritual history (Pollux's arm). 'Man owes his most ingenious inventions to his taste for society'—so this history of sculpture begins—and the arts began as a recognition of the 'advantage' to society that accrued from communicating sentiments, laws, and histories, particularly those of 'heroes, who eventually became the gods' (III.5–6).

The earliest Greeks, then, like other peoples at various stages—including contemporary Native Americans—began with 'indications' and 'signs' to serve this end (8): rocks, rivers, and trees were consecrated to gods and heroes, were called by their names, and eventually signified the divinities themselves. What began as a history of heroes, art converted to myth and religion; soon enough, standing stones were held to mark 'less the figure than the actual presence of the gods' (11), as in the case of a massive tree trunk that was placed in a Thespian temple as the embodiment of Juno (16; cp. I.120). To the extent that he departs from his earlier account, d'Hancarville is clearly influenced by Guasco, adopting his distinction between *designer* and *exprimer*, along with other terms, to mark stages in this history. He contradicts Guasco, however (again without naming him), by insisting that classical statues do not truly represent the gods (or 'idols') any more than these stones and trees did, except to the vulgar; but the 'sublime idea' of divine presence that they convey should teach us that the ancients had 'a bizarre Polytheism rather than an absurd Idolatry' (12n.).

Polytheism, however bizarre, is religion rather than idolatry, and d'Hancarville emphasizes the metaphysical content of Greek worship when dealing with the origin of sculpture as well as other topics such as (later) the Eleusinian mysteries. In Guasco's account, mythology *in toto* is imposed on the Greeks by wily Phoenician colonizers.[18] For d'Hancarville it is not a specific hegemonic project but rather a universal quality of the will to power that leads rulers (both ancient and modern) to arrogate to themselves superhuman qualities and god-like stature. The divine attributes of the massive *cyppes* or standing stones, for

[18] Guasco, *De l'usage des Statues* 38–42. Cited parenthetically (as *De l'usage*) in the text hereafter.

example, such as strength and durability, render these vaunted human qualities truly sublime. Thus sculpture as it became figurative 'rendered metaphysical ideas in material form' (*AEGR* III.15). In this way the 'principles of the Religion of the ancients' became indistinguishable from those of their art ['*ne laissoit pas d'écarter celles de l'art qui devoit les produire*'] (16).

This first order of signs (i.e. tree trunks, standing stones, and the like) is neither uniform nor exact (17), and so the *cyppe* evolves into the more conventional *terme*, a vertical standing stone with a rounded stone placed on top for a head (18). Sculpture is born in this moment when signs are united with forms, because the new method reveals, like nature in the germination of plants and animals, all the parts that will in due course be carried to perfection (19). As sculpture moves from this germinal phase into 'infancy', written characters are added for identification as depicted in d'Hancarville's synoptic exhibition in Volume I. These are finally sloughed off by the image—in sculpture as later in vase painting—striving for independence from language. Drawing independently on Sanchoniathon (an ancient Phoenician author now held by some to be a second-century CE invention), d'Hancarville's first volume already uses standing stones and *termes* to define the origin of sculpture (I.116–26), as Guasco does. But d'Hancarville's refinements, as reviewed here, show the influence of Guasco's inventive reading of this source (*De l'usage* 5–6, 14). Guasco envisions an even more systematic progress toward naturalistic representation (1–2, 28–32), which lends itself well to d'Hancarville's newly refined organicism, his notion that the ontogeny of individual steps in the progress of art (such as 'germination') recapitulates the phylogeny of the medium as a whole.

D'Hancarville's synoptic exhibition (two engravings reproduced as Figures 5.2–5.3) condenses visually almost all the thematic elements that will concern him throughout the work, including the Dioscuri, the *cyppe*, and the *terme* as the origins of sculpture; and natural objects including coconuts, horns, and ostrich eggs as the origins of pottery. The egg is particularly apt, since the second engraving includes the Horatian inscription *orditur ab ovo* (*Ars poetica* 147): perhaps the irony is intended, but clearly d'Hancarville is not following Horace's advice to start in the middle of things when he traces vase painting back to the first dawn of the arts (going back to the egg, as it were). In his gloss on these images d'Hancarville also points out the outline of a profile traced on the wall in the first engraving (left background), alluding to the origin of painting (*AEGR* I.177).

He points out the formal allusions to natural materials in many of these artefacts, allusions that he describes as 'conserving' the prehistory of their fabrication. One of the smaller jugs (Figure 5.3, left foreground) alludes particularly to the form of the ostrich egg from which such vessels were originally fashioned, and the cow horn (Figure 5.2, centre-right) recalls the natural history of the drinking horn. The Roman bronze candlesticks recall the

Fig. 5.2. Unsigned engraving. *Collection of Etruscan, Greek, and Roman Antiquities* [*AEGR*], Vol. I, p. 112 (Chapter III.ii headpiece [English text]). Courtesy of Margaret M. Bridwell Art Library, University of Louisville.

Fig. 5.3. Unsigned engraving. *Collection of Etruscan, Greek, and Roman Antiquities* [*AEGR*], Vol. I, p. 113 (Chapter III.ii headpiece [French text]). Courtesy of Margaret M. Bridwell Art Library, University of Louisville.

prehistoric moment when canes or young trees were stripped of their smaller branches for this purpose. The statue with Greek characters engraved on its left leg (just beyond this horn in Figure 5.2) shows the fourth stage of the emergence of sculpture, preceded by the more 'primitive' Egyptian statue without articulated limbs, the *terme* or column showing only the head and feet, and finally the standing stone or *cyppe* and the 'symbolic column'—both evoked by the horizontal log (all Figure 5.3). The log also alludes to the origin of architecture—a favourite example for d'Hancarville—which originally deployed cut logs before stone columns were devised in the same form, a transformation signalled by the far end of this column with its carved necking.

The outline on the wall (Figure 5.2, left background) alludes to the legend of the Maid of Corinth, which d'Hancarville revised to anchor his own foundation myth, in so doing, articulated the evidentiary importance of mythology for art history. Pliny first reported that the daughter of Dibutades (or Butades) invented painting by tracing her lover's shadow on the wall on the eve of his departure *(Natural History* XXXV.43). D'Hancarville returns more than once to this foundation myth, which is also celebrated in Joseph Wright of Derby's painting *The Corinthian Maid* (1782–84), among other neoclassical works. D'Hancarville, following Pliny, argues that the incident inspired innovation in sculpture before it gave rise to painting: Dibutades, being a potter by trade, modelled the profile in clay and thus created the first relief sculpture (*AEGR* II.116). Pliny also claims that Dibutades pioneered the use of 'red earth' to colour works in ceramic, from which d'Hancarville infers not only that this is the red ground familiar from black-figure vases but also that Dibutades must have begun applying reliefs of the human profile to his pots. When he traces the occurrence of ornamental profiles on the Greek vases (such as Vol. I, Pl. 38) to this supposed invention, arguing that the artists must have been observing the 'rule' that they must 'call to mind [the] Origin' of their medium (118), the reasoning is suspiciously circular.

When he develops the same origin myth later, however, as a basis for the history of engraving and other arts, he argues that it operates in the larger arena of mythology, an evidentiary domain in which ancient beliefs can only be extrapolated from artefacts. History, moralizes d'Hancarville, preserves no record of the potter's birthplace, Sicyon, of its rulers and their wars and vain ambitions. The period lives instead in 'fable', which celebrates the humble potter Dibutades who, unlike those rulers, is now immortal (III.73–74). Dibutades's pottery, on d'Hancarville's uncertain reckoning, pre-dates the potter's wheel, and it was his 'habit' of using molds (based on animal horns and large eggs) to shape pots that led him to the discovery of modelling (75). Having made a mold of the departed lover's profile, he began to attach not only profiles but other reliefs and three-dimensional ornaments to his pots. Again d'Hancarville points to his plates—the 'monuments displayed in the different volumes of this work'—as 'conserving' both the primordial forms of these early vases and Dibutades's pioneering ornaments, in the form of dog's-head handles among other examples.

On this reading, art transformed history into myth—not only its own history, but political and social history as well. D'Hancarville treats Dibutades (a shadowy figure comparable to Daedalus) as a semi-divine hero in his own right, tracing not only vase painting and decoration but also chiselling, gem engraving, metal engraving, coins, casting, and even (indirectly) printing to his original discovery (77). Although Dibutades's daughter's role is minimized in this account, in contrast to the version prevalent among academic painters, d'Hancarville notes Pliny's report that her original outline was preserved by the Corinthians as a sacred relic for almost 1,300 years. Just as Dibutades becomes a kind of secular saint to whom d'Hancarville offers up thanks for enabling (somehow) the discovery of printing, the medium in which he himself completes the artist's apotheosis (77–78), so too the drawing as relic anticipates the religion of art that d'Hancarville establishes with his engravings.

The art that generated Greek mythology, he argues, not only made gods of heroes such as Castor and Pollux. It also made divine intervention part of historical truth, 'altering the truth of history by the very means that were used to conserve it' (9). D'Hancarville's emphasis on myth as history strongly suggests the influence of Giambattista Vico's *New Science*, which claims that 'fables . . . were the first histories of the gentile nations'.[19] Possibly following Vico, d'Hancarville claims that poets and artists were the first theologians and the first historians (9–10). The remains of art therefore provide 'monuments . . . more certain than those of history' for the historical events encoded in them (26). The constellations themselves provide such a monument, the Sabism of the Orient having been adapted by early Greeks to serve a political project of naming heavenly bodies after their leaders (here he makes the analogy to Galileo's names for the moons of Jupiter) (28–30). The deified Olympian rulers thus appropriated the divinity of the stars, while poetry and sculpture collaborated in altering the record of their originally political, now mythological opponents, the defeated Titans (31). One key to this unorthodox approach is to take mythological dating literally: art's 'more certain' monuments underwrite an allegorical reading according to which the first of many rulers called Jupiter (i.e. Zeus), later amalgamated into the god, reigned until 1778 BC (26). If his contemporary Prometheus 'formed man out of clay' then that fixes a date for the origin of sculpture. A complex narrative unfolds here, but it is sufficient to observe that art and religion co-evolve on this account from the social instinct, and mythology provides both the religious and the historical content of art. As d'Hancarville sums it up, 'myth [*la fable*] itself here serves to determine the epoch of History which it envelops, and of which it could alter the circumstances without, however, changing the foundation' (24).

D'Hancarville's project of recovering history from art and myth commits him above all to a history of religion. His methodical translation of mythic time into human prehistory provides the first stage of this history, what might be called an evolutionary narrative concerning the co-adaptation of religion

[19] *New Science* 33[I.51], cp. 21–22.

and art.[20] When the Greeks said that the gods lived 'in the time of Inachus', they meant, or unconsciously recalled, that the historical Uranus lived in this period, two generations before King Apis, d'Hancarville's ur-patron of the arts (26). Uranus, according to Sanchoniathon, 'was the first to use *boetiles*', or god-stones—a foundational claim for both Guasco (*De l'usage* 22–23) and d'Hancarville. These stones were the first devotional objects, antedating the birth of sculpture (in the enlightened reign of Inachus's grandson Apis), and so signal the advent of religion understood as a physical mark (*marque*) of deification (*AEGR* III.27). The progress from these *boetiles* and other 'forms' to the more sophisticated 'sign' was quite rapid, but the earliest stage is crucial for d'Hancarville and he returns to it at length elsewhere (IV.26–28).[21] Here the mode of representation also marks the period as historical.

This paradigm elevates the importance of primitive art and of pioneering artists such as Dibutades and Daedalus. Daedalus, on d'Hancarville's account, invented proportion as a way of establishing naturalistic representation in sculpture about 1308 BCE (III.104). If the nineteenth century BCE was the time of the gods, then the fourteenth century was the time of the demi-gods, a sublime epoch that also witnessed Hercules, Tiresias, Jason, and others whose extraordinary merit led them to be deified. Guasco, so scornful concerning Greek religion, also debunks the myth of Daedalus in no uncertain terms (*De l'usage* 391–92), but d'Hancarville's literalism, though less than skeptical, should not appear entirely naive. His Daedalus is an artist in the eighteenth-century, mechanical sense: he invented the plumb line, the *tarriere*, and other instruments for incorporating the 'laws of nature' into art (*AEGR* III.106–07). By means of these techniques, 'the sign was totally banished from the arts it had disfigured'. D'Hancarville illustrates this point by contrasting the thirty-breasted Diana of Ephesus with a figurative sculpture (108–09), but his rejection is surprisingly strong considering that the earlier portion of the narrative celebrates the appearance of this same 'sign' as the authentic form

[20] Though pre-evolutionary, d'Hancarville's thesis that art and ritual are coordinated adaptive behaviours is in some ways quite close to contemporary evolutionary aesthetics. See, for example, Coe, *Ancestress Hypothesis*. In a remarkable gloss on the 'circle called mythic', d'Hancarville historicizes the period described by Proclus as extending from Uranus to Ulysses (III.206–07n.). The names of these characters, like early sculpture, allegorize their essential personal or biographical traits. By pairing this metonymic 'discourse' with the infant 'forms' of sculpture, he recodes mythical time as human prehistory, as an evolutionary stage in the history of art. On the general problem of dating in d'Hancarville, see *V & V* 149–55. Vico's *New Science*, again, provides a relevant model in many respects.

[21] Guasco actually considers the *boetiles* to be *signes* (*De l'usage* 51), while d'Hancarville does not—a difference worth setting against the imputation of simple plagiarism. The only truly prehistoric art object included in *AEGR*, an Italic *askos* (2800–2400 BC), is depicted twice in the collection, but without any explanatory text (*V & V* 234–5). The *askos* does feature in the constellation of proto-aesthetic or transitional objects that I have called his 'synoptic exhibition' (Figure 5.2, lower left).

of naive art.[22] Daedalus discovered 'expression', or figurative representation, in the correlation between sentiment and realistic action. In retrospect, the sign appears 'monstrous' both for conflating human and animal attributes, contrary to nature, and for using nature merely as an accessory to representation (100).

The 'infancy of sculpture' nominally ends with Daedalus, but elsewhere it becomes clear that the transition is gradual and not an instant result of the Daedalean revolution. Partly this transition is a matter of aesthetic necessity: expression is the telos of art for d'Hancarville, and as long as the sign dominates the figure, sculpture cannot express anything (99). The sign remains authentic in transitional efforts, preceding Daedalus or struggling to come to terms with his discovery, in which the sign and the figure are employed together; and as a 'souvenir' included in later, expressive works. Thus the very crudeness of the many small 'Pelasgian' bronzes, for example, proves that they are the naive but vigorous efforts of a developing art, and not the barbarian displays of decadent, mercantile art (100). By making these distinctions, d'Hancarville may intend implicitly to answer Guasco's charge that *all* sacred art from antiquity is decadent in its deviation from original monotheism (*De l'usage* 9–11). By adhering so rigidly to the teleology of expression and ideal beauty—which he had actually questioned in his first volume—d'Hancarville further aligns himself with a secular neoclassicism; but he does so with the instrumental view of creating a favourable reception for his arguments concerning the earlier stages of art, where his interest (in my view) genuinely lies. Although he claims a kind of scientific revolution on Daedalus's behalf, the remainder of his narrative lingers in the era of the Trojan war, a period of transition that seems to continue about 300 years.[23] Hence his particular attention to the subset of vase paintings including the boar hunt and the Choice of Hercules, a 'species of figurative writing [*écriture figurée*]' (III.203n.) in which 'the figures... [still] *speak* more

[22] Compared to figurative sculpture, the Diana of Ephesus with her thirty breasts is a monstrosity, conflating the animal with the human for allegorical emphasis (108–9). D'Hancarville acknowledges that the sign—now an impediment to art—was necessary to make it intelligible in the first place. But his pejorative tone (*défiguroit*) here sets the stage for his prejudicial treatment of the barbarians whose art never transcends the signifying mode. The Greeks, by contrast, eliminate monstrosity from sculpture—and from poetry—within two centuries.

[23] The second half of Volume III and all but the last fifty pages of Volume IV thus focus on the period from roughly 1300 to roughly 1000 BC, which also coincides with the birth of painting. According to d'Hancarville's history of painting, based very loosely on Pliny, Daedalus's contemporary Euchir was the first painter, whose style corresponds to sculpture in its infancy (III.150). Cimon of Cleonae was the first 'totally to banish the sign' and 'reform' pictorial representation in the same way that Daedalus revolutionized sculpture (IV.24), though the example of sculpture meant that the process did not take as long in painting.

than they *express*, they reveal the artist's *intention* more clearly than the *sentiment* of their subjects' (IV.12n.)

5.4 ART AND RITUAL

The art historical importance of primitive forms hinges on the recognition of their ritual significance. When d'Hancarville returns to the *boetiles* [or *baetyles*] in Volume IV he uses ritual to account for their continued presence in more advanced arts, including vase painting. In a long excursus on religious rites as depicted in vase painting—the most substantial and focused engagement with vases anywhere in the four-volume work—he argues that the entire corpus can be classed in two main groups, 'purely historical' paintings and those depicting ritual (33n.). Here he turns from the former class, which includes the boar hunt and a handful of other mythological scenes that have occupied him through much of the first three volumes, to the latter class, explicating (or re-explicating) over 100 vase paintings in this thirty-page double-column footnote (28–59n.). At the end of Chapter 4, I cited one example, the depiction of a ritual parasol that supposedly prefigures Italian funeral customs. This painting (Plate 7) turns out to feature a larger set of sacred objects or *indications*. D'Hancarville reads the three small *boetiles* or 'globules' emerging from the opened box as 'indications' of the Three Graces, a frequently recurring trinity in his work that has a particular significance for the cult of Agdistis, which he claims is represented here. These objects 'indicate' in both a representational and a devotional sense: as visual proto-signs they evoke the three perfect and identical bodies of the Graces and as sacred signs they have their own cult, being elaborately enclosed in the box [*cassette mystique*] or the leaf-shaped *étui* in the figure's left hand and covered with the parasol (IV.33–34n.). In documenting this ritual, the vase painting also documents an earlier art historical phase, one in which the objects of art and religion were more nearly identical.

For d'Hancarville, the archaeological significance of the vase paintings is twofold: they illustrate earlier art historical developments for which no original sources survive (e.g. Daedalean sculptures); and they provide a record of religious mysteries. The vase painters, he argues, recorded scenes from contemporary cults that continued to pay devotion to the objects originally taken as embodiments of the deity, alongside the abstract deity itself. Besides the *boetiles* and carved amulets originally associated with a range of divinities, the larger set of *indications* also includes the sesame seeds and pomegranate associated with Demeter, the vine leaf and pineapple associated with Dionysos, and many other examples derived primarily from Clement of Alexandria, a patristic scholar of ancient religion (32n.). As d'Hancarville urges elsewhere, 'these monuments cannot deceive us concerning the customs they represent' (III.127): a painting's embedded history can teach us about objects that no longer survive, including ritual objects and the primitive forms of carving and sculpture to which they gave rise. Just as painting gives a formal record of the earliest stages of art (150), so too religious rites in the classical period incorporated archaic devotional practices that can be distinguished, along with primitive signification, in the scenes recorded by vase painters. Understood in this way, vase painting lies on the cusp between prehistory and history.

Prior to this excursus prompted by the *boetiles*, d'Hancarville cites individual vase paintings as illustrations of an art historical thesis derived from textual sources or from pure conjecture. Within this essay, however, d'Hancarville draws on the entire corpus of vase paintings to support an argument about religious practice, and that argument takes precedence here. The turn to *boetiles* is an archaeological turn in this sense, because it prompts an engagement with the vase paintings as ethnographic source material, and not just as corroborating evidence to support his narrative about stylistic development. This turn is prompted by a digression in the main text that purports to integrate gem engraving into the history of sculpture, and hence into the larger narrative concerning the 'sign' and its replacement by the figure. Gem engraving, as a medial stage between sculpture and painting, incorporates *indications* of divinities with their figured images, a synthesis found in other kinds of jewellery as well. Though 'this practice' goes unmentioned by ancient authors, 'the monuments that remain suffice to assure us' that ancient artists commonly incorporated both modes of representation in one artefact (IV.29–30). The note on vase painting, however—so excursive that the main text on gems is confined to one line per page for over twenty pages—demonstrates that vase paintings rather than gems are the 'monuments' d'Hancarville has primarily in mind.

Engraved gems and other types of jewellery evolved, he argues, as increasingly portable forms of the *boetile*, as amulets that protected the wearer against evil spirits [*mauvais Génies*] (28). As with many other artefacts examined in the course of the work, these amulets are both 'assuredly of the highest antiquity' (27n.) and yet—by the logic of ethnographic actualism—the object of a 'custom still practiced today in my native country' (27n.), presumably the country around Nancy, where d'Hancarville was born. The vase paintings, however, show us the context in which *boetiles* and other *indications* continued to feature not just in private devotions but in public worship. The first *boetiles* were anointed with oil (here d'Hancarville borrows Guasco's illustration from Genesis 28:18) and 'wrapped in bands of wool', a practice that accounts for the myth of Kronos eating a stone wrapped in cloth and taking it for the infant Zeus (26–27n.). A whole reliquary system evolved from this ancient practice, and various cults enclosed their divine indications in 'artfully arranged *bandelettes* of wool', *étuis* or *capsules* that were themselves enclosed in *cistes mystiques* or willow baskets—eventually represented by any round container, a motif occurring in many vase paintings (28n.). These *cistes* sufficed for domestic devotions, a practice already outlined in the second volume, but in more formal settings they too were enclosed in *armoires* or *tabernacles* (29n.) or replaced by caskets like the one shown in the painting of Agdistis or Eros (Plate 7).[24]

Having shown how a number of vase paintings may be explicated in light of this reliquary hypothesis, d'Hancarville turns to the mysteries of Eleusis for a unifying logic to confirm the importance of sacred containers in vase painting. He alludes to substantial ancient testimony from Cicero and others for the importance of this 'most sacred institution' (31n.)—a cult that also drew the attention of other Enlightenment free thinkers and secret societies such as the Freemasons (cp. 49n.). The mysteries of the cult remained famously mysterious, except that Clement of Alexandria, in a 'misguided' effort to discredit the beliefs of the cult, produced what d'Hancarville takes to be an accurate guide to the material props used in its worship. He draws substantially on Clement for the list of *indications* mentioned above (32n.). The emphasis on immortality in the Eleusinian cult, perennially fascinating to intellectuals in the Christian era, seemed to corroborate both his general notion of the vases as grave goods and the thesis that many of the vase paintings were created for initiates of this cult and alluded to its rites. This hypothesis proved surprisingly durable. Hamilton himself made the stronger claim, twenty years later, that painted vases 'were placed only in the sepulchres of those who had been initiated in the religious ceremonies of Bacchus or Eleusis', and the art dealer James Christie devoted most of a treatise to this view in 1806.[25]

Many of the *indications* or sacred objects listed by Clement are organic and d'Hancarville dwells on their perishability, though this provides at best circumstantial evidence for the documentary function that he ascribes to the paintings. Nonetheless, he argues that tombs in which the vases were found originally contained the sesame seeds, pomegranates, ivy, and other objects depicted on the vases—including the willow basket reliquaries which, he suggests, originated at Eleusis. These have decayed, of course, and the *boetiles*, too, become unrecognizable once their elaborate cloth wrappings (*bandelettes*, which d'Hancarville sees everywhere in the vase paintings) have decomposed: 'the forms of these *étuis* and these envelopes could not have been preserved except in these vase paintings; most of these *indications*, being nothing but

[24] Trendall attributes this Apulian *kylix* (British Museum 1772,0320.695) to the 'Stoke-on-Trent Painter' (*Red-Figured Vases* 888–97). By citing vase paintings as evidence for his claim that containers like this box were sometimes carved with the images of the 'indications' that they enclosed, d'Hancarville constructs vase painting as an archive of images originally produced in other media, no longer surviving in their original form. Guasco's very different account takes the myth of Kronos tricked into eating stones as an allegory of Phoenician colonialism (*De l'usage* 39).

[25] Hamilton, *Collection of Engravings* I.30; Christie, *Disquisition upon Etruscan Vases*. The ancient text that d'Hancarville refers to is Cicero, *De legibus* 2.36.

Fig. 5.4. Engraving by Carlo Nolli from a drawing by Giuseppe Bracci. *Collection of Etruscan, Greek, and Roman Antiquities* [*AEGR*], Vol. II, p. 57 (Chapter II headpiece [French text]). Tomb at Trebbia. Courtesy of Margaret M. Bridwell Art Library, University of Louisville.

ordinary stones, must have mingled with the stones of the tombs themselves, and could not have been recognized by those who opened the tombs' (32n.). The scene of excavation recalls the interior of a tomb at Trebbia depicted in Volume II in the famous Nolli engraving (Figure 5.4), which according to Griener 'signals the beginning of a new era' in archaeological representation (*Antichità* 59). D'Hancarville describes the tomb at length in this note (*AEGR* IV.42–43n.), examining the organization of its contents for evidence concerning ancient belief in the afterlife.[26]

The resulting division of vase painting into ritual and 'purely historic' classes is not factually accurate, but it is certainly less speculative than the classification offered in Volume II and closer to archaeology in its concern with excavation and decay. This two-part division is a blunt instrument and d'Hancarville simplifies matters further when he conflates the rites of Demeter

[26] There is no accompanying text in Volume II, but his systematic survey of the plates in Volume IV brings him to a symposium scene engraved from the kylix-krater found in this tomb (Vol. II, Pl. 74). The painting itself does not support his Eleusinian thesis, but he draws attention to the aureole of iron spikes around the skull as a possible sign of beatification (IV.43n.).

and Dionysos, apparently assuming that the latter was an equally important dedicatee of the Eleusinian rites. However blunt the instrument, it energizes his engagement with the paintings to a degree that is especially remarkable when we consider that he only had access to one plate at a time while writing this volume (the set being held by his creditors). To some extent, the new religious classification—the large subset of vases 'represent[ing] the cult of [certain] Deities, those of their *indications*, and at the same time the different symbols enclosed in the *cistes* that are consecrated to them' (32n.)—must have been prompted by a consciousness of his long-deferred promise to explicate the vases. To judge from the internal date of the next excursus note (October 1775), the very long notes especially characteristic of Volume IV must have been written last, and the charge that he failed to explicate the vases, dating back to 1768, must have driven him at last to find a venue for accommodating the explications to his high argument. At the same time, the anthropological tendency of this argument points increasingly to a religious context for the origins of art. Thus the internal logic of the argument, as it evolves over all four volumes, also justifies the return to vase painting as a religious practice.

Modern classification of the vases based on their use—*hydria*, *krater*, *lekythos*, and so on—more or less invalidates this notion of using the images painted on the vessels as a guide to their original functions. D'Hancarville's achievement—a dubious one in some scholars' eyes—was to gain acceptance for vase painting as high art. In the wake of his book, Greek vases quickly became prominent in European scholarship, museums, and commercial design. Formally, he achieves this transformation by pictorializing the vases, as Caecilie Weissert and others have impressively demonstrated. He translates them into the two-dimensional medium of engraving.[27]

D'Hancarville initially created a more or less academic, neoclassical art-historical context for the vase paintings in his ekphrastic digression on Raphael in Volume II. In Volume IV, however, he explains the new classification more broadly with reference to spaces of worship, both public and domestic. Many vases were adorned with ritual scenes just as areas of ancient houses were devoted to sacred images and libations, and just as 'paintings in our temples' today 'represent the histories of the saints to which they are dedicated' (32n.). The first example probably alludes to the excavations at Herculaneum and Pompeii, where such spaces were in the process of being identified in ruined villas. D'Hancarville's familiarity with these sites might also account for his statement that Ceres/Demeter (his conflation) was most commonly depicted and worshipped with the household gods. There is no evident basis for the analogy (between wall painting and vase painting), but it

[27] See especially Weissert, 'Kunsttheorie versus Wissenschaft' 9–10. The issue is taken up more briefly by Griener (*Antichità* 75); Coltman, 'Sir William Hamilton's Vase Publications' 11; and Brylowe, 'Two Kinds of Collections' 28.

provides a second argument for the presence of images related to her cult. In addition to those of Ceres—goddess both of Eleusis and of the hearth—he also lists the *fêtes* and *histoires* of Bacchus, Mercury, Apollo, and Hercules as 'frequently recalled in these vase paintings'.

Perhaps anticipating a jaded readership, d'Hancarville forcefully defends his new set of explications as the logical result and culminating illustration of his argument concerning the formal evolution of early sacred art. As modern readers, we can more readily accept that this excursus note is both an expedient way of inserting the delayed explications and a legitimate illustration of the larger argument. His authorial anxiety prompts a digression that warns uninitiated readers away from the explications to follow, insisting that his readings of some 103 images can only be understood in the context of the three stages of signification he has outlined:

> The preceding observations, shedding a new light on the monuments engraved in this book, put me in a position to correct some of the explications already given and to add some new ones . . . : but I am obliged to warn my readers that . . . if they have not examined with attention what has been said in Volume III of the times when Gods were represented by *Baetyles*, or *indications*, and of the times when they were designated by *symbolic columns*, they will not be able to understand what follows: . . . only the succession and correspondence [*la suite et l'accord*] of the monuments . . . can render intelligible what I have been compelled to write on these matters. (37n.)

More simply put, the excursus note containing these explications is part of the drift from style to content, from the history of forms that supposedly links gem engraving and vase painting to the religious significance of the images that both media share.

He builds the initial case concerning the iconography of *indications* by correlating new examples with engravings from the first volume, specifically scenes he had already explicated as wedding ceremonies. The image of Agdistis with his three *boetiles* and parasol, for example, echoes the parasol and other motifs already noticed in the first set of engravings (Vol. I, Pl. 45), motifs that now take on a broader meaning. Revisiting these examples, d'Hancarville renews the pattern of revising his own earlier readings that typifies his approach in the later volumes. In his visual quest for *indications* and their containers, he looks to wedding scenes first because he had already recognized their ritual character in part. His full recognition of the ritual aspects of these paintings, he now declares, prompted the whole suite of explications (IV.31n.), including the long digression on Clement and the Eleusinian Mysteries.[28]

[28] What had at first appeared to be a box for wedding gifts in a painting showing the nuptials of Paris and Helen (Vol. I, Pl. 32; II. 160), for example, now appears as a sacred container

The first analysis to be informed fully by the discussion of Clement again returns to an apparent marriage ceremony (II.162–63), which d'Hancarville now shows to be replete with ritual objects and acts (Plate 8). The initial reading, in this case, turns out to be wholly invalidated because of a 'ridiculous mistake' (IV.38n.) concerning one of the *indications* featured in this image, the Ionic capital in the centre. In the new extended gloss, the balls held by both female figures become 'spherical envelopes', made of rolled *bandelettes* or strips of wool, to hold *boetiles* (here indicating the Fates); the large ring or trochus above the central figure's head is an *indication* of Bacchus (not a bridal coronet), and 'sesame seeds' appear beneath the feet of all the figures (IV.33n.).[29] D'Hancarville makes these examples correspond with Clement's account. Once again, the rectangular cases appear as the sacred containers from which these *indications* have been removed. The action of the male figure, evidently pouring libations, introduces a new element very likely borrowed from Guasco: he is pouring oil from a dedicated pitcher (*Gutturnum*) onto the column that serves as a seat for the central figure.

According to d'Hancarville's history of forms, this column must be a *colonne symbolique*, a sacred stone several stages removed from the *boetile*, but nonetheless still subject to the archaic practice of anointing stones with oil. He documents this practice with a reference to the story of Jacob anointing the stone of Bethel (Genesis 28:18–19). The example plays a central role for Guasco (*De l'usage* 4–5), and since he too associates the stone of Bethel with Sanchoniathon's *boetiles*, the charge of plagiarism seems justified here. There is an important difference, though: Guasco insists that Jacob's essentially secular, commemorative act was only later corrupted to idolatry (24), whereas for d'Hancarville this act provides the impetus for the history of sculpture because it is a sacred act. That history is arguably the central concern of the new explication. The woollen balls in the two women's hands indicate the Fates, and Clotho specifically, while the figure at left can be taken to represent a cognate, younger divinity, such as Nemesis or one of the Eumenides: 'though we find the latter *figure* painted here, it is natural to find *indications* of the former here [as well], for although this practice predated the use of figures,

(IV.31n., cp. 37–38n. and Pl. 38). Another example from the first volume (Pl. 74) turns out not to represent a wedding at all, but rather a scene of worship in which a series of *indications* have been removed from a *cassette mystique*, most obviously the 'sesame seeds' scattered on the ground (IV.31n., cp. II.165). A round floret in the centre of this image (corresponding to the shoulder of the vase on the front side) also provided his first example of an affinity between engraved gems and certain motifs in vase painting, and he credits this vase for prompting the whole sequence of new explications.

[29] These rows of white dots are viewed today as groundlines. This vessel (1772,0320.393) is the subject of an unusually lengthy gloss on the British Museum website.

they are nonetheless preserved together [*se conserverent . . . en même temps*], as shown by a thousand examples in this work' (*AEGR* IV.33n.). Likewise, the *boetiles* and the *colonne symbolique* preserve two separate representational stages in one and the same image. D'Hancarville's own *méprise ridicule*—taking this column for a doorjamb—leads to the unravelling of the whole notion of a marriage scene: 'this painting represents nothing but a domestic religious ceremony, in which a Greek woman worships her household gods in the customary fashion' (38n.).

In returning to the vase paintings, d'Hancarville not only delivers the promised explications but also makes the vases more central to his natural history of art. This discussion also comes closest to fulfilling his more recent promise to base his history on 'the monuments of art itself' (IV.vii). By featuring vase paintings as a material source in his *History Of Greek Sculpture*—even if under the guise of an explanatory footnote—he elevates the vases to a monumental status that they did not have before. At the same time, they become privileged aesthetic objects as well as archaeological monuments that document religious practice. They also become monuments of a prehistoric phase in the development of art, expanding the historical domain by providing visual and formal cues for conjectures about the 'infancy' of art. By giving us 'their history, written by themselves', d'Hancarville continues to build on the original thesis laid out in his first volume. His use of textual evidence to gloss the vases, for example, refines the comparative method outlined above: 'as in the Physical Sciences, experiments often repeated can alone account for certain Phenomena, which at first sight, did not seem to require any, so, it is, that by force of comparisons only, we can arrive at clearing up certain points of Antiquity' (II.58). D'Hancarville means his characteristic method of comparing monuments and texts, as well as comparisons across media. In his first history of vase painting, he makes inferences concerning earlier clay vessels from Pliny's descriptions of gold and silver vases (64)—inferences challenged by Vickers, who argues forcefully that ceramic vessels must have been modelled on those of precious metals, and not the reverse.[30]

In explicating the vase paintings, d'Hancarville offers us this 'comparative' method in a more sophisticated form, mapping art history onto an anthropological timescale. He locates exact iconographic parallels between the vase paintings and descriptions by Clement and by Aristides the Athenian (IV.48–49). Although he evidently suppresses Guasco's name, *De l'usage des statues* become an important modern source of textual support. D'Hancarville borrows Guasco's connection between the stone of Bethel and Sanchoniathon's *boetiles*, using it to gloss the 'domestic religious ceremony' (Plate 8) and another vase painting (III. pl. 90) that supposedly shows a stone being consulted

[30] Vickers, 'Hamilton, Geology, Stone Vases' 270. See further n. 10 above. D'Hancarville draws a similar kind of evidence from Suetonius (*AEGR* II.102).

as an oracle (40n.).[31] D'Hancarville's inferences from this connection—that the stones were sacred to begin with, that their worship is documented in vase paintings, that they evolved into sculpture—are original contributions inspired by 'the monuments themselves' and by his ethnographic perspective on the Neapolitan setting. From Guasco's scriptural-literalist perspective, developments post-dating Jacob cannot be prehistoric, even if portions of the interval are lost to the historical record. D'Hancarville's timescale, expanded by his geological interests and by his secularism, allows him to backdate this episode and expand tremendously the period during which *boetiles* were primary ritual objects, a period perhaps corresponding very loosely to what we now think of as the Neolithic era. He finds additional ancient authors such as Democritus (as cited in Pliny) to support the antiquity of this practice as he understands it (40n.).

The rich and unfamiliar content of the vase paintings, little explored at that time except by Italian connoisseurs, allowed d'Hancarville to synthesize many such sources on ancient art and religion into one coherent narrative supported by images.[32] Although his highly conjectural history of representational practices—from a primitive *indication*, barely discernible in the verbal record, to an ethnographically verifiable, non-naturalistic *signification*, to the *expression* perfected in figurative classical sculpture—was inspired by the vases that filled his Neapolitan workshop when the plates were produced, he seems to leave the vase paintings far behind in the course of his long argument until he suddenly returns to them in Volume IV. Having elaborated at such great length on the implications that these vases, with their arresting synthesis of form and function, seemed to hold for the anthropology of art, he returns to find an allegory of these origins in the paintings themselves. Some examples, such as the 'domestic religious ceremony' (Plate 8), bring the several stages of

[31] Guasco uses an anecdote from Arnobius (fourth century CE), which d'Hancarville also appropriates, to support his contention that *boetiles* became incorporated into pagan religious practice—a claim that influenced d'Hancarville greatly (*De l'usage* 23, cp. *AEGR* IV.37n.). Other moments of indebtedness to Guasco in these explications include an uncharacteristic valorization of monotheism (*AEGR* IV.39n.) and an extended reference to Charles de Brosses, *Du culte des Dieux fétiches* (1760) (51n.). More fundamentally, Guasco originated the idea that 'signs' come first, as symbols that 'designate' rather than 'express' a divinity (*De l'usage* 1). 'Real images' of divinities evolve out of these primitive signs to become 'idols'. D'Hancarville does not adopt these latter terms, however, nor does he adopt Guasco's intriguing if rather mechanistic developmental hypothesis (29–32).

[32] One of his few references to existing vase literature occurs when d'Hancarville makes the intriguing argument that the difficulty of achieving three-dimensional depth in vase paintings has led representations of sacred containers to be 'taken for mirrors or *paterae*' by other commentators (*AEGR* IV.34n.). The influence of Italian connoisseurs was much more substantial than this passing reference would suggest; see Lyons 'Neapolitan Context', esp. 233. Lissarrague and Reed have shown just how many images d'Hancarville and his artists re-engraved from other antiquarian books, which suggests that he probably drew on their texts more often than he cited them.

representation together in one image. At least one vase even incorporates all three forms of the same divinity, represented in various parts of the composition as both a sphere and a Doric column, and personified as a laurel-crowned youth with a bow: 'thus we rediscover in this single painting the three manners of representing Apollo, by *indications*, by *symbolic columns* [colonnes signifiantes], and as a *figure*' (34n.).[33]

These paintings literally incorporate art history, then, but they also resemble history painting as conventionally understood, as d'Hancarville argues in the second volume. The composition of history paintings owed something to theatrical stage sets, and he evokes this connection by making some of the vase paintings document the Eleusinian mysteries as they were celebrated, with great theatrical pomp, in monumental shrines such as the semi-legendary Temple of Demeter. By correlating the images with descriptions of these ceremonies by Aristides and other authors, d'Hancarville also reinforces his thesis that ancient sacred art transformed historical events into fable. He argues that relevant motifs on the vases, such as the lamentations of Persephone and Demeter, are drawn from 'Mystic spectacles' staged in these temples, and by pictorializing these scenes his engravings accentuate their putative theatricality (48n.). The alternating scenes of history and fable 'painted on the vases in this collection', therefore, 'can clarify what [ancient] authors tell us and can reveal many things to us that they [the authors] could not communicate'. D'Hancarville demonstrates his synthesis of ancient texts and vase paintings in a visionary passage, purporting to paraphrase Aristides and other ancient authors collected in 'Meursius Vol. IX', that describes the sublimity of ancient sacred spectacle. He claims that one especially remarkable painting (Volume III, Plate 47, taken from an Apulian *pelike* now in Catania) mimics such a spectacle in its depiction of 'the arrival of Ceres at Eleusis, which in the course of time gave rise to the institution of her Mysteries'.[34] The fourteen figures on this vase are arranged in a particularly theatrical manner, though d'Hancarville is probably responding to neoclassical painterly convention after the fact rather than realizing intentions explicitly stated when the engraving was done eight years earlier.

[33] This is Plate 109 of Volume I, and not Volume II as the text states. In his first attempt at explication, d'Hancarville was totally nonplused by this vase (II.166). The vase itself is now lost (Gisler-Huwiler et al., *Collection* 134); this image is one of about 85—about 20 per cent of the total—identified in the modern reprint as coming from a lost or unidentified vase. (Less than half of these are colour plates; since most vases are depicted in more than one engraving, the actual number of lost or unidentified vessels will be somewhat lower.) Some other vexing typographical errors in Volume IV, partly explainable from d'Hancarville's poor working conditions in Florence, include the substitution of the plate number 45 for 43 in the explications (IV.38n.) and the absence of note 61 (see pp. 96–7).

[34] The *pelike* in question is held by the Museo Civico del Castello Ursino (MB 4404 [L 770]).

5.5 PRE-CLASSICAL ANTIQUITY

Griener aptly proposes 'the rose and the thorns' as a fitting emblem for the 'long production' of d'Hancarville's book. In a historical view, his work certainly represents a flowering of connoisseurship and the fine art book, but d'Hancarville's authorial path was a thorny one during the roughly ten years occupied by its production. (Hamilton's own purgatory, as noted by Jenkins and Sloan, continued several more years until his copies of Vol. III–IV were finally released by the creditors.) D'Hancarville knew how to comfort himself in extravagant prose, if not always in other ways. From the beginning, his text alludes frequently to his authorial difficulties and his elaborate production techniques, described in Chapter 3. His authorial persona transgresses more and more egregiously against all canons of decorum, ancient or modern. His theatrical retraction of his own earlier arguments gives way to increasingly exalted claims on behalf of the new work. Suitably for a crypto-prehistorian, d'Hancarville only reaches the time of Homer about three-quarters of the way into his *History of Greek Sculpture*. His notion that Homer blazed a trail for the visual arts, which were impeded by their materiality (IV.114), is much more conventional than his ethnographic thinking, and of a piece with Winckelmann and Lessing. He presents the Shield of Achilles (*Iliad* 18.478–608) as the poet's prophetic vision of the progress of the arts over the next 500 years—but then, suddenly comparing Homer's project with his own, d'Hancarville departs radically from any sort of neoclassical piety:

> This Shield is perhaps the most singular monument of the intelligence of the ancients, and just as . . . I make the effort to discern, through the shadows of antiquity, the steps they followed in these very distant times, so too Homer, guided by the sublimity of his genius . . . gazed into the obscurity of the future and prophesied . . . the great things accomplished by the Arts, several centuries after his time, when they would have acquired all the strength of which he judged them to be capable. (117–18)

Another metanarrative digression, perhaps less outrageously self-aggrandizing, helps to account for the particular fascination that the Eleusinian mysteries held for him near the end of his long and heroic (if not Homeric) labours. For his explication of a frieze of animals from a Corinthian *oinochoe*, d'Hancarville draws on the dialogues of Plutarch for an ancient scene of excavation. When the Spartan king Agesilaus opened the tomb of Alcmene, he found a variety of remains including vases and 'hieroglyphic' inscriptions in bronze (44n.).[35] D'Hancarville uses the episode to authenticate the antiquity of the tombs being opened in Naples in modern times and to bolster his argument

[35] The *oinochoe* is now held by the British Museum (GR 1772.3-20.498 [OC 397]). D'Hancarville cites Plutarch's early dialogue *On the Sign of Socrates* (*De genio Socratis*).

that it was common practice for tombs to be equipped with references to religious mysteries in which the occupants had been initiated. He urges readers to examine the hieroglyphic character of these animal friezes in the British Museum, where by now this *oinochoe* was installed with the rest of the Hamilton collection (45n.), and thus to recover for themselves part of the historical content that ancient authors 'could not communicate' in writing. At the same time he explains the painter's ability to record mysteries in which the painter himself was not initiated. He restates the point elsewhere by analogy to the painters hired in modern times by the Freemasons (49n.). Here he uses an analogy with his own authorial practice, again asserting a spiritual continuity between ancient art and his own project:

> It was possible to paint all these things, nonetheless, without betraying the secret, and without communicating the theological, moral, or religious sense to those whose hands executed the painting, just as the Italian compositor, employed to assemble the characters on this page, though ignorant of the languages in which it is written, can communicate my reflections to those who understand these languages, without understanding anything himself. (46n.)

By the end of *AEGR*, diligent readers have been through something like initiation into an art-historical cult. (D'Hancarville's interest in cults and comparative ancient religion became central in his next book, the *Recherches*.) The celebrated colour plates offer a rich visual experience of a form of ancient art that was still very unfamiliar to northern European readers in the 1770s. The text circumscribes more than it describes the vases and their splendid paintings, but this indirect strategy serves to establish them as a hidden key for unlocking a vivid life-world of religious ritual and artistic production, just as the excavations of Herculaneum and Pompeii had opened a vista on the unimagined daily life of the ancients. Drawing on natural history for his geological and ethnographic methods, d'Hancarville puts forward what amounts to an archaeological thesis: the history of art is 'written' by the artefacts themselves. His religious account of the vase paintings is the last in a logical sequence of strategies for accessing the deep past, inspired by the natural seriation so visible in the landscape of Vesuvius. Guided by the Vitruvian axiom that form must follow function, he investigates first the 'natural necessities' that must have given rise to material culture. Inspired by the excavations of tombs that yielded these vases, as well as larger archaeological discoveries from Pompeii to Paestum, he turns next (in Volume II) to the funerary customs that, for him, gave vase-painting its religious inflection. *The History of Greek Sculpture* (III–IV) then delivers the aesthetic implications of this prehistoric or archaeological turn.

D'Hancarville saw clearly that the *beau idéal* venerated in unhistorical fashion by his neoclassical contemporaries must have its origin in an epoch far beyond the written record. He used the means available to him to access this prehistory. Not content with conjectural histories of society or even archaeological sites, he finally returned to the monuments at hand to interpret—however extravagantly—their contribution to the story of origins that, from an anthropological perspective, art history must tell.

Although his explanations of the vase paintings often appear far-fetched, even to the non-specialist, d'Hancarville deserves credit for creating a kind of archaeological deep time. In his volumes, the ancients become knowable in their difference. By locating the distant epochs in which the vases were produced, he locates customs and manners very different from our own. Even as these become intelligible by analogy to modern customs, according to the method I have called 'ethnographic actualism', they provide glimpses of a new and unfamiliar culture, and not—as in the humanist tradition—of shared values or a shared humanity. In this sense d'Hancarville's mysterious vases carry some of the strangeness of the erotic and obscene images that were being excavated in Pompeii and Herculaneum. As Jonah Siegel has observed, these discoveries caused the gradual unravelling of neoclassical aesthetics as antiquity became defamiliarized and romantic artists began to appropriate its legacies in new ways.[36] In this sense d'Hancarville's project may be understood as anticlassical. Hamilton's vases, too—as promoted by d'Hancarville, and by Wedgwood and other designers—contributed to the proliferation of antiquities that was emanating from Naples, from Athens, from the newly discovered South Pacific islands, from the sedimentary rocks of northern Europe, and even from its Gothic cathedrals. In the next chapter we turn to the funerary customs and the iconography—less ancient, but not much less strange—that were being recovered from England's Gothic cathedrals by a newly self-conscious science of national antiquities.

[36] *Desire and Excess*, ch. 2, esp. 64–72.

Part III

England's Ruins

Interlude: Classical to Gothic

The final section of this book turns from more cosmopolitan scenes of knowledge production to British antiquities and natural history as they developed in tandem in the late eighteenth and early nineteenth centuries. In a declaration quoted at the outset, Thomas Chatterton 'defies' the ancient epic poets to equal his own heroic version of the British fifteenth century, a declaration that underscores the nationalist and populist associations of British antiquity. The Cook voyages and Hamilton's publications, among other European and global influences in the last quarter of the eighteenth century, confronted the strident anticlassicism of Chatterton and John Carter (among others) with many new possibilities. British antiquity was only one alternative to neoclassicism among the range of non-classical antiquities opened by scientific voyaging and early archaeology in the Pacific and the Mediterranean. The British plate books under investigation in my last two chapters—Gough's *Sepulchral Monuments in Great Britain* and Englefield and Webster's *Description of the Isle of Wight*—reflect these cosmopolitan influences in their references to Tahitian burials and island geology, respectively. The collaborations initiated by Banks and Hamilton fostered intellectual exchange between Britain and Europe, cross-cultural investigation of human and natural phenomena, and comparative analysis of classical and Gothic or other 'primitive' artifacts. This brief interlude returns to the institutional setting of the Society of Antiquaries to describe the British synthesis, as well as the antithesis, of classical and Gothic.

If Greek vases became antiquities in the process of being recognized as art—as I have argued in the last two chapters—so too medieval churches attained the status of antiquities when they were recognized as architecture. Gothic antiquarianism, though a relative newcomer, was no less contentious than the science of classical antiquities. The letters of Horace Walpole, whose vast antiquarian correspondence included Hamilton and Gough (1735–1809), situate these major figures within the national network while participating avidly in antiquarian contention. When Walpole condemned the Society of Antiquaries' journal *Archaeologia* as 'old women's logic', he particularly had in mind the essays on Northern and Eastern antiquities and on medieval

Europe—newer interests cultivated alongside, and sometimes in place of, classical antiquity.[1] These criticisms, together with the insistence on Grecian purity that seems to motivate them, are surprising in light of Walpole's lasting association with the Gothic through his neo-Gothic mansion, Strawberry Hill, and his novel *The Castle of Otranto* (1764). Gough, as Director of the Society of Antiquaries, came in for his fair share of abuse in Walpole's private letters, in which he repeatedly insists that antiquarian pursuits must be regulated by essentially neoclassical canons of good taste. Yet when Gough sent Walpole presentation copies of the two volumes of his *Sepulchral Monuments in Great Britain* (1786; 1796)—which praises Walpole's *Anecdotes of Painting* as a source of inspiration—Walpole called it 'the most splendid work I ever saw' and refrained from casting doubt on the tastefulness of such deeply archaeological attention to the Gothic.[2] Walpole's ambivalence helps to explain the evolving relationship between classical and Gothic antiquity and sheds light on the social tensions associated with the complex division of labour that was required to make the Gothic visible as antiquity.

Walpole attacks the study of non-classical antiquity on two fronts. As in his letters to Gough, he may have been more courteous in his direct communication with d'Hancarville, but in writing to others he impugns d'Hancarville's intellect, partly on the basis that he is a draftsman and a client as well as an author—in short, a knowledge worker. Although Walpole does not mention it, Gough, too, relied on the knowledge work of an antiquarian draftsman and budding author-for-hire, Jacob Schnebbelie. In an important antiquarian letter to Lady Ossory, Walpole simultaneously develops his critique of the Society of Antiquaries and attacks d'Hancarville for his promiscuous comparisons between Greek works and those of other cultures. Indeed, d'Hancarville's renegade neoclassicism incorporates some potentially anticlassical primitivism, but British antiquarianism at the time offers an even more decisive anticlassical turn in its democratizing pursuit of national antiquities. In this 1785 letter Walpole lumps the empirical research programme of the Society of Antiquaries, pioneered by Gough, together with d'Hancarville's *Recherches sur l'origine, l'esprit et le progrès des arts de la Grèce* (1785) and similarly syncretic works on antiquity by Jacob Bryant and others. Walpole condemns

[1] Walpole to Cole, 28 July 1772 (*Horace Walpole's Correspondence* I.270). These non-classical antiquities, in Walpole's view, fostered the superstition and credulity that he meant to implicate by calling these antiquaries old women.

[2] For the texts of these letters, see *Walpole's Correspondence*, XLII.168–70 and 444–5 (quoting 444). As early as 1812 John Nichols noted that 'it is remarkable that Mr Walpole should treat Mr Gough with such profound respect in his answers to him; whereas, in his correspondence with his friend Cole . . . he is often very unceremonious in his notices of our learned Antiquary' (*LA* VI.290n.). See also the chapter on 'taste' vs. 'learning' in Evans, *History of the Society of Antiquaries* (148–69), and the opening paragraphs of Martin Myrone, 'Society of Antiquaries and the Graphic Arts', on Walpole's attitude.

d'Hancarville and Bryant along with numerous members of the Society of Antiquaries as a large tribe of primitivists who 'are busied in inquiring how long the world has blundered without discovering what was worth knowing'.[3]

The complex affinities among 'primitive' cultures, both ancient and modern, heightened the interest of comparative research for a broad range of inquirers influenced by some combination of neoclassicism and the history of civil society, including Banks, Hamilton, and other figures examined in this book. Walpole, nursing a long-held grievance against the Society of Antiquaries, here takes the dissenting view that these affinities merely heighten the contrast between primitive cultures and 'the one nation worth studying', ancient Greece. He begins this letter with a pretence of recommending the new volume of *Archaeologia* (vol. 7), but it soon becomes clear that only a few articles compare favourably to d'Hancarville's 'impertinence' and only one of them qualifies as 'a very sensible discourse'. Walpole singles out for praise an essay by Edward Ledwich 'on the religion of the Druids, in which the writer, unlike his companions, demolishes fantastic reverence for barbarians, instead of discovering arts and sciences among rude nations'. He reminds Lady Ossory of his own earlier 'translation' of *Archaeologia* as 'old women's logic', quoting a letter written thirteen years earlier at the height of his resentment, which led him to suspend his involvement with the Society. Criticism of his own work on medieval history had led to the rupture, and his diatribe in this letter passes over ongoing research on the English Middle Ages—both his own work and at least nine medieval articles in the 1785 volume—in favour of a polemical antithesis between the Greeks and 'the Phoenicians, Pelasgians, Vics, Egyptians, Edomites, Scythians, and Gentoos'. Scholarship on these peoples, Walpole declares, is no better than the romances and fictionalized memoirs indicative of current popular taste; moreover, 'such books' pretend that 'the Grecians borrowed from the Egyptians, Tartars, Indians, etc. That is, they stole the genuine principles of all beauty and all taste from every idea of deformity and absurdity!'[4]

[3] Walpole to Lady Ossory, 9 July 1785 (*Horace Walpole's Correspondence* XXXIII.477–80, here 479). D'Hancarville would seem to be in agreement with Walpole's sentiment here when he condemns the 'Gothick stile' as 'barbarous' (*AEGR* I.92, cp. xxii). The antithesis was a neoclassical commonplace, though in fact many antiquaries were interested in both classical and non-classical antiquities. Walpole himself was a serious student of the Gothic, while d'Hancarville preferred a nebulous preclassical antiquity, but Walpole's polemical point is that Greek primacy should remain unquestioned.

[4] Walpole alludes specifically to 'Madame de Genlis', whose *Tales of the Castle* appeared in English the same year, and to 'Mrs Bellamy', whose five-volume *Apology for the Life of George Anne Bellamy, Late of Covent Garden Theatre*, also appeared in 1785. Walpole's dismissal of 'Indians' and 'Gentoos' is especially salient in light of Francis Haskell's argument for the revolutionary impact of 'first-hand knowledge of the religions and literature of India and the Far East' on d'Hancarville's *Recherches*. See Haskell, 'The Baron' 42. On Walpole's secession from the Society of Antiquaries, see Evans, *History* 166–9.

Walpole finds the comparative and primitivist varieties of antiquarianism incompatible with the ideals of neoclassicism, including social distinction. His argument situates the global antiquarian pursuits of d'Hancarville, Banks, Hamilton, and others relative to the British institutional context. On Walpole's admittedly biased reckoning, British antiquarianism as a whole was dangerously infected with the anticlassical tendencies manifested in these comparative projects. These projects also reflect the collaborative methods in favour with the Society of Antiquaries. The collaborative, not to say conspiratorial, nature of much antiquarian research provides a second, less obvious source of discomfort in Walpole's letter. D'Hancarville is on his mind partly because he visited Strawberry Hill with the object of drawing an 'ugly morsel' in Walpole's collection: 'I have lately dipped into D'Ancarville's two volumes, in which he ascribes universal knowledge and invention to the Scythians, as Bryant did to the Lord knows whom; but with all my pertinacity in reading quartos, I could not wade through the tautology and impertinence of D'Ancarville's, though he has lately been here to draw a bronze I have of Ceres with a bull in her lap'. D'Hancarville's drawing was subsequently engraved for Richard Payne Knight's *Discourse on the Worship of Priapus* (1786), another antiquarian work that Walpole disparaged.[5] Payne Knight's patronage of d'Hancarville provides another reason for Walpole's unease, and similarly Gough's patronage became a pretext for critics of Schnebbelie and John Carter, who felt that their client status compromised their scholarship.

Collaboration was central to the research promoted by the Society of Antiquaries. First of all, correspondence, exchange of specimens and drawings, and oral debate in meetings were often indispensable to the Fellows' scholarly engagement with periods and questions in which evidence was fragmentary and problematic. Second, this type of network research required collaboration between members and non-members and thus crossed boundaries of social class almost as a matter of course. I will argue in Chapter 6 that explicit acknowledgment of class difference and the division of labour distinguished a growing number of antiquarian publications, especially those associated with Gough, who was Director of the Society from 1771–1797. This collective pursuit of national antiquities did not unsettle the hegemony of the landed gentry, but it was democratizing in the sense that it diffused the prestige of classical antiquity and made research the site of widening social interaction. My last chapter engages with the similarly collaborative scholarship of Henry Englefield, Vice President of the Society of Antiquaries during

[5] *Horace Walpole's Correspondence* XXXIII. 478–79 and n. 15; cp. *Walpoliana* II.26. Walpole set himself apart as a gentleman scholar, but Evans has pointed out that cracks were beginning to show in this model, suggesting that Walpole's 'approach' was 'too unscholarly for him to feel at home in the Society' (*History* 120). Walpole's solitary pursuit of scholarship put him at odds with new scholarly forms of collaboration.

Gough's tenure, who published extensively on both medieval and geological antiquities.

Even though his own historical research had the democratic tendency of casting doubt on royal prerogative, Walpole maintained, instead, an elitist vision of 'learning' as dependent on 'taste'. By portraying the Antiquarian Society as a bunch of squabbling old women he ridicules the collective adjudication of evidence in their meetings, designed to mirror the Baconian practice of the Royal Society. At the same time, he seems more sensitive to class distinction than the Antiquaries, as suggested not only by his distaste for d'Hancarville's practical pursuits but also by his sharp rejection of the supposed fifteenth-century manuscripts submitted to him by Thomas Chatterton in 1769. Unlike Walpole, Jeremiah Milles, then President of the Society of Antiquaries, embraced as authentic the forgeries of the 'marvellous' if self-educated and low-born Chatterton. Gough did not endorse Chatterton's 'Rowley poems', but he gave a greater role to self-educated 'practical antiquaries' on the intellectual front, even as he sought to maintain distinction on the social. Walpole, who was already disaffected by the time Gough launched the publication of *Archaeologia* in 1770, ridiculed these institutional efforts, at least in private letters to his friends—even as the quantity and quality of published research increased in the ensuing decades.[6] The reason may have been, in part, that Gough insisted on broad-based collaboration as essential to the progress of antiquarian 'science', as he called it in a 1784 obituary.

Specifically, Gough maintained the importance of technical skills such as architecture over against aesthetic judgment and literary ability, valued as credentials of gentility not only by Walpole but by many more orthodox Fellows of the Society. This scientific view accounts for his collaboration with Schnebbelie, Carter, and many other 'practical antiquaries'. His 1784 obituary of the Cambridge architect James Essex is typical in this regard:

> Those who have made Gothic architecture their study, as well as those who on a superficial view cannot help being struck with its stupendous efforts, will for ever regret the irreparable loss of this great master of that science. The repairs and improvements of King's College Chapel at Cambridge, of Ely and Lincoln minsters, planned and conducted by him, will be a lasting monument of his skill (*LA* VI.625n.).

By calling him a 'great master', Gough diverts attention from Essex's scanty publication record and emphasizes the built works instead. Gough also promoted the career of John Carter, another self-educated antiquary with

[6] On Walpole and Chatterton, see, among others, Russett, *Fictions and Fakes*, ch. 2; on Chatterton and Milles, see Evans, *History* 180–1; on Walpole's judgment of Gough, see Evans, *History* 137. On Gough and the increasing scientificity of antiquarianism, see further Sweet, 'Antiquaries and Antiquities' 189–91, and Smiles, *Eye Witness* 27–9, 53, 71.

architectural training; Carter went on to 'generate almost single-handed the concept of national heritage', in Rosemary Sweet's words (*Antiquaries* 293).

Although himself a conservative scion of the landed gentry, Gough espoused a view of antiquarianism that seems to anticipate the class-neutral rhetoric of modern science. History, he declared in an early publication, is 'not a mere narrative taken up at random and embellished with poetic diction, but a regular and elaborate inquiry into every ancient record and proof'.[7] Historians have made similar claims since the advent of historical writing. Gough differed in his insistence that technical expertise, independent of gentility, is required for disciplinary rigour. To this end he championed outsiders including Essex, Carter, and the publisher John Nichols, for all of whom he helped to secure election to the Society, as well as artists such as Schnebbelie, whose short career as an antiquarian draftsman forms the primary case study for Chapter 6. Similarly, Englefield employed Thomas Webster for his technical expertise, particularly in geological illustration, and ten years after their joint publication of *Description of the Isle of Wight* Webster became the first Englishman to declare himself a 'professional geologist'.

On the one hand, Gough's pursuit of colleagues from outside the dominant social group left traces that rendered antiquarian research, in fact, something less than class-neutral. These traces include meticulous visual documentation, always clearly signed by the delineator and the engraver, and discussions of working relationships in Gough's own publications and in others associated with the society. On the other hand, his agenda was not primarily a social one: his patronage might be viewed as simply a more gracious form of condescension than Walpole's contempt. Scientific antiquarianism, for Gough, depended on the technical input of an intellectually aspiring, economically dependent class. In this sense, modern science is not genuinely class-neutral either, though some would argue that the sequence of the authors' names on a scientific paper represents a system of apprenticeship rather than a set of class distinctions.[8] Gough, ardent medievalist that he was, might well have understood Carter and Schnebbelie as his apprentices. These knowledge workers themselves had no illusions of a threatening social mobility, but they did aim to build literary careers on the expertise they acquired while working for Gough and the Society. Collectively their experience corresponds more closely to the 'elaboration of intellectual strata', as described by Antonio Gramsci, than to explicit class conflict.[9]

[7] Gough, 'Introduction' ii.

[8] Although individual authorship is the norm in humanities scholarship, a comparison might be made here to the way in which even substantive contributions by research assistants are conventionally indicated in the acknowledgments or footnotes.

[9] In describing 'the 'organic' intellectuals which every new class creates alongside itself', Gramsci has in mind primarily the entrepreneurial culture of pre-Fascist Italy and its use of technical schools as an 'instrument' for achieving hegemony. It makes sense, however, to

The medievalism cultivated by this group also had a different set of implications that may have contributed more directly to the rise of middle-class culture. In its increasing historical and methodological rigour, the preservation of Gothic cathedrals—including documentation, advocacy, and historically faithful restoration—ultimately cancelled out the class interests of Gough and the landowners and clergymen he represented as Director of the Society. In this sense national heritage is at odds with private property, and by 1807 the preservation movement had successfully challenged the prerogative of bishops and chapters to make architectural decisions. Nietzsche had a more select readership in mind when he argued much later that the proper 'use of history for life' depends on a precise balance of the antiquarian, monumental, and critical impulses. But his formula helps to clarify the social relevance that allowed antiquarianism finally to live down its dry-as-dust reputation.

To some extent, the Society of Antiquaries, founded in 1707 by a librarian, an engraver, and a shoemaker, had always been socially inclusive in its composition and in its research programme. The Society's research prospectus of 1717 included 'the people' as one of its major headings, with subheadings including 'customs' and 'manufactures'.[10] But by the 1770s the majority of a much larger and more heavily aristocratic membership understood the Society's publications primarily as the benefits of membership in an exclusive literary society, notwithstanding Gough's efforts to focus increasingly on empirical studies and fieldwork.[11] His more socially conservative colleagues reminded him that many of his protégés were no true antiquaries. 'Before Mr Carter delivers his sentiments', wrote Samuel Denne in 1799, Carter should submit to the authority of Gough 'or some other person more conversant with books than himself'. Another correspondent of Gough's complains about Essex's unsophisticated prose: 'Mr Essex's quartos certainly contain many new and curious hints, and he certainly throws about his brick and mortar in a masterly manner—but . . . he has got into regions he has no map of, and gropes about wonderfully in the dark'.[12] When Gough nominated his intimate

understand ascendant 'social groups' in early industrial England, too, as seeking to 'multiply and narrow the various specializations' as part of the larger process by which, 'in the modern world[,] the category of intellectuals . . . has undergone an unprecedented expansion'. See 'On the Formation of the Intellectuals' 6, 10, 13.

[10] At the same time, there was an inherent tension in the Society's representation of antiquarianism as both literature and science, exemplified in the differing approaches of two successive presidents, Milles and Edward King. On the Society's incorporation in 1717, see further Sweet, *Antiquaries* 83–4. Sweet also gives a detailed account of Parliamentary intervention in architectural decisions concerning Westminster Abbey beginning in 1807 (293).

[11] An anonymous antiquary writing to the *Gentleman's Magazine* in 1802 complains that the Society of Antiquaries has been besieged by candidates 'who had no pretensions whatever to belong to that or any other literary society' ('Hints to the Society' 1181).

[12] *ILH* VI.772; *LA* VIII.610. The 'quartos' referred to here are apparently the three papers Essex read before the Society between 1775 and 1782, which were printed as pamphlets. For more on Essex's background and contributions, see Sweet, *Antiquaries* 258–60.

and frequent collaborator, and the Society's printer, John Nichols, as a candidate for fellowship in the Society in 1778, the President informed him that Nichols would not be acceptable. Nichols was finally admitted in 1810. Nichols himself, in his memoir on Gough, pointed out that as 'the independent master of an ample fortune, he was in all respects pre-eminently qualified for the labours of an Antiquary; the pain of whose researches can but rarely meet an adequate remuneration'.[13] Webster, who tried to make his fortune in geology in the 1820s, found that the same fiscal constraint shadowed the labours of the geologist.

Gough's collaboration with 'practical antiquaries' such as Schnebbelie contributed to make antiquarianism a more class-conscious, politicized, and commercial pursuit. In her richly documented account of these issues, Sweet highlights the ability of antiquarian discourse to cross class barriers and also points out that some less polished antiquaries, such as Joseph Strutt, forged ahead independently of the Society.[14] The fieldwork and the technical expertise increasingly required by antiquarianism eventually helped to precipitate the formation of professionalized history, art history, and archaeology, which had distanced themselves from amateur antiquarianism by the 1840s. Combining archaeological and geological analysis, Webster's case shows that the emergence of geology, too, owed something to antiquarianism. Through the early decades of the nineteenth century, antiquarian pursuits including Webster's helped to build up the marketplace for knowledge work, with its various avenues of professionalization, because of their intimate connection with the publishing industry and their dependence on fieldwork. The paradoxical combination of a pronounced class hierarchy, on the one hand, and the populist values associated with preservation, on the other, also helps to make antiquarianism an especially visible instance of the social dynamics of knowledge production. Gough's recognition of Schnebbelie, Carter, and others may have been intended to keep them in their place, but its very visibility illustrates the movement toward a more bureaucratic organization of knowledge, a movement that takes different forms in the history of the university and other learned bodies. The scholarly nature of these collaborations also sets them apart from the history of popularization, which formed another, quite

[13] *LA* VI.295. On Nichols's initial rejection, and on his recognition by other societies, including the more liberal Society of Antiquaries of Scotland, see Sweet, *Antiquaries* 114. In his prefatory essay to *Archaeologia* I Gough points out that *any* collection potentially 'supplies materials to those who have sagacity or leisure to extract from the common mass whatever may answer useful purposes. Here begins the province of the antiquary' ('Introduction' ii). 'Sagacity or leisure' suggests that, whatever the economic circumstances, antiquarian research is never mere fact-gathering.

[14] *Antiquaries* 60, 58. I am indebted to Sweet's pioneering discussion of this issue (57–66). Sweet more fully situates the careers of Strutt, an artist-antiquary who made a living from his publications, and others like him, in her splendid account of popular antiquities (309–43).

different point of intersection between the Society of Antiquaries and the literary marketplace, as shown in Sweet's account.

Walpole exaggerated when he charged antiquarian culture as a whole with compromising the prestige of classical antiquity. But a levelling force was latent in the antiquaries' catholic conception of 'ancient' periods and their 'antiquities'—terms that occur in the titles of ten essays in the volume reviewed by Walpole and refer to everything from Phoenicia to thirteenth-century France. This broad range did not, in fact, imply a devaluation of classical antiquity for most observers. It took a populist like John Carter to turn the neoclassical hierarchy on its head. But even Gough, who maintained a strong identity as a classicist throughout his medieval research, worried about the perception among visual artists that 'Grecian and Roman' subjects were 'more profitable' than 'Gothic and more domestic monuments' (*SM* I.i.9).[15] Carter and Schnebbelie, like Parkinson, d'Hancarville, and other knowledge workers, pursued a tenuous profit in the proliferation of antiquities, which sometimes did entail the explicit anticlassicism that Walpole deprecated.

Nietzsche, writing in 1873, sought to rescue Greek antiquity itself from the antiquarian fascination with the 'utensils of our ancestors' [*Urväter-Hausrat*] that he saw as afflicting historical study. In Nietzsche's conception of the antiquarian, monumental, and critical 'uses of history for life', critical historiography is reserved for those oppressed by a need to throw off the burden of the past, a new intellectual elite that seems quite removed from these eighteenth-century antiquaries. The practical antiquaries and their patrons came before the claims to objectivity put forward by nineteenth-century historicism, the claims that spurred Nietzsche's revolt. Yet their celebration of Gothic monuments is tempered by a rejection of earlier grand narratives, an implied criticism of the connoisseurs and philosophers who soared above material evidence in their contemplation of the past.

Gough's and Englefield's increasing dependence on fieldwork and the technical skills of others practically undermined their ability to promote a monumental view of history as underwriting the social hierarchy, even as they held to this view in theory. Nietzsche himself had no objection to such a conservative use of history, which he viewed as a form of popular antiquarianism deployed to keep 'inferior populations' [*minder begünstigte Geschlechter*] in their place; but he too turned a critical lens on grand narratives, asserting this prerogative on behalf of an intellectual elite. For Schnebbelie and Carter,

[15] Sweet points out a discrepancy between the more heavily classical content of the Society's meetings and the more heavily British content of the publications (*Antiquaries* 95). Susan Pearce's table of subjects shows that the 1785 volume of *Archaeologia* includes equal numbers of articles on Romano-British and medieval subjects (8)—as well as a record-high nine articles on prehistoric subjects (which attracted Walpole's notice). Pearce's figures, however, exclude the roughly 30% of articles dealing with 'material not from the British Isles'. See her 'Antiquaries and the Interpretation of Ancient Objects' 152.

however, it was precisely the scrutiny of ancestral utensils that allowed them a critical purchase on the past by creating a demand for knowledge work and exposing the tacit presentism of 'mere narrative' about the past.[16] John Ruskin turned a more monumental lens on Gothic architecture in *The Stones of Venice* (1853), rewriting the antiquaries' earlier recovery of the Gothic in the idiom of art history. Where Ruskin praised Gothic architecture for affording intellectual freedom to its workmen, the antiquaries pursued their own intellectual freedom by recovering and literally re-drafting Gothic monuments newly valued for their authenticity. The next two chapters explore that recovery as it extended beyond the architectural to the ethnographic and even geological history of Britain.

[16] Nietzsche, *On the Advantage* 20–2. In addition to the (then) new academic historicism embodied by Wilamowitz, Nietzsche is also seeking to cast off the moribund remains of antiquarianism. Nietzsche's text thus illustrates a sea-change in perceptions of antiquarianism in the space of eighty years, but his three categories of historical inquiry also provide a perennially useful analytic. The newer New Historicism revived a kind of Nietzschean criticism, as suggested by Stephen Greenblatt's critique of E. M. W. Tillyard. For Gough's dismissal of 'mere narrative', see, again, 'Introduction' ii.

6

Antiquarianism and the Science of Preservation: Jacob Schnebbelie, Richard Gough, and Gothic Antiquity

During the short course of his progress in science, it may with great truth be asserted, that no one laboured more intensely in his profession, or sought to gratify the enquirer with greater facility than Mr. Schnebbelie.
Review of Jacob Schnebbelie, *The Antiquaries' Museum* (1792)

6.1 GOTHIC MATERIALISM

The project of preserving medieval ecclesiastical architecture, which became increasingly central for British antiquarianism, relied less profoundly on local knowledge than did natural history or archaeological projects more remote in space and time. But the production of major publications on the subject depended strategically on draftsmen who could travel from site to site and investigate, excavate, climb scaffolding, record, and sometimes even restore the monuments that antiquaries wished to publicize. John Carter built his long and remarkable literary career on this kind of research-for-hire. Carter gained a fellowship in the Society of Antiquaries with the help of Richard Gough, but was ostracized shortly afterward because his strident activism precipitated a major conflict within the institution. Carter's attacks on historically unsound renovations at Salisbury and Durham Cathedrals, published in the *Gentleman's Magazine*, polarized the Society. With battle lines already clearly drawn, James Wyatt, the fashionable architect responsible for the renovations, was put up for election to the Society of Antiquaries by Sir Joseph Banks and other court insiders. When Wyatt—by now the king's architect as well—was elected a Fellow in 1797, Gough immediately resigned and left a small cadre of Gothic purists to soldier on.

Partly because of Gough's role at the *Gentleman's Magazine*, the debate continued for years, and one indignant Fellow wrote to the magazine in 1803 condemning Wyatt's party by neatly inverting their class distinction. Echoing an earlier complaint that such members constitute an 'indiscriminate rabble', this writer declares that 'unworthy members are admitted, to keep up the treasury and swell the number, though they know no more of the matter than the rusticks or journeymen to whom the British Museum used to be exhibited as a raree-shew, till the librarians were wearied out with doing nothing'. 'The few men of science among them may be frighted away', he warns, concluding that 'the Society is grown too *fashionable*, too *polite*, and too *rich*.'[1]

The so-called *querelle de Wyatt* encapsulates the social conflicts precipitated by the progress of scholarship on British antiquities. If Taste militated against too faithful an adherence to the original aesthetics of the medieval builders, then Learning (to adopt Joan Evans's terms) insisted that preservation was the paramount goal of antiquarian scholarship, even if preservation was at odds with current neoclassical style.[2] Carter was dismissed as a mere 'servant' soon after Wyatt's election in a symbolic displacement of the knowledge worker by the gentlemanly professional. Focusing primarily on Gough's collaboration with Jacob Schnebbelie (1760–92) rather than Carter, this chapter explores the ramifications of the increasingly refined division of labour within the study of British antiquities. The increasingly well-defined contributions of writers, draftsmen, engravers, and other specialists made the Gothic visible as antiquity while also giving rise to ambiguities of status, which ranged from the unstable opposition between 'science' and 'polite' knowledge to outright social conflicts like the one surrounding Wyatt's election.

After he resigned, Gough focused on his scholarship, incorporating some reflections on these conflicts in his next publication. His magnum opus, *Sepulchral Monuments of Great Britain*, was published in stages beginning in 1786, with the final volume of 150 engravings appearing in 1796. He added a fifth large folio volume to the set when he completed his 300-page introduction in 1799. In his preface to this volume he alludes to the dissensions in the Society of Antiquaries. Gough's account of the conflict is more decorous than those in the magazine, but his allusions are unmistakable partly because he elevates his draftsmen and engravers to the status of heroic antiquarian collaborators, displacing the rich and polite Fellows who had broken what

[1] See 'An Antiquary' 105–6 and his postscript (316). In a separate published reply to the original letter, Carter assumes that the phrase 'indiscriminate rabble' must have been intended for him, a divergence that illustrates the ambiguities of status arising from this conflict ('Publication of Cathedrals' 106–7). For the essay that prompted both replies, see the anonymous 'Hints to the Society of Antiquaries'. In addition to collaborating on many antiquarian projects with John Nichols, who was the Society's printer and publisher, Gough also contributed heavily to the *Gentleman's Magazine*, which Nichols edited.

[2] Evans, *History* 148–69; see also 207–13 on the *querelle de Wyatt* (citing *The Diary of Joseph Farington* among other primary sources).

he saw as their commitment to preservation. 'It were an invidious boast', he confesses, 'how little is owing to the assistance of my fellow labourers in the vineyard of antiquity' (*SM* II.i.5). Gough explains:

> The little success of invitation to communicate correct drawings of monuments serves but to prove how novel the subject is... While I congratulate myself in having contributed to preserve from decay so many of these beautiful remains, I congratulate my country that so many monuments of art have yet survived the decay of time and ruder devastation of ignorance, violence, malice, and accidents; [and] that... I have found a Schnebbelie, a Carter, and a Basire, to second my efforts, which, without their hand, would most imperfectly have fulfilled the task.

Gough goes on to specify the 'capricious alterations [by Wyatt primarily], in which all varieties of stile, all history of architecture as a science among us, is totally done away, and much historic knowledge besides' (7), all countenanced by a Society that, in his view, now 'counteract[s] the very object of its institution'.

Gough highlights two essential components of the science of preservation: the creation of a visual record and the critique of historically unfaithful alterations. Both are (or ought to be) informed by what is here termed the 'science' of architecture. The visual record, for Gough, is scientific in another sense as well, one more closely related to the ethnographic images and discourse we have seen in Parkinson and d'Hancarville: the sculpture in the cathedrals provides a record of medieval dress and material culture, as worthy of preservation as the architectural design itself. Both these senses of 'science' presuppose a materialist vision of history, or at least an emphasis on material artefacts—yet to modern eyes the visual record created by eighteenth-century antiquaries can seem idealizing. Maria Grazia Lolla acknowledges that 'the Society [of Antiquaries] pursued publishing as the best technology of preservation'. But the ultimate motive for these engravings, she argues, was to convey an aesthetic experience that rendered the monuments immaterial. Commenting on the presence of human figures, decorative borders, and compositional elements that are either superadded or not really visible from the assumed perspective, she argues that 'such artful compositions approximated a ritual of preservation. They celebrated the monuments; they were monuments to the monuments, but not the monuments themselves... at each step they make clear that they are *not* the objects they represent.'[3] This critique is reminiscent of recent commentary on d'Hancarville's two-dimensional renderings of Greek vase paintings.

[3] Lolla, 'Ceci n'est pas un monument' 19, 20. Lolla is concerned with *Vetusta Monumenta*, an ambitious series of engravings that also involved Gough and the same team of artists who worked on *Sepulchral Monuments*.

Even at the time, there was a perception that the decaying monuments themselves were neglected all the more once they had assumed an ideal form in print. Both Lolla and Rosemary Sweet comment on the stylistic differences between an earlier illustration (by William Stukeley) of the thirteenth-century Waltham Cross and a Schnebbelie/Basire image of the same subject *c.*1790: the later image aestheticizes its decay.[4] These images appear in *Vetusta Monumenta*, an apt source for comparisons because it was published serially from 1718 to 1906. Lolla argues that the later emphasis on admiration brings antiquarianism 'close to Romanticism', and William Blake's long apprenticeship under James Basire might be cited in support of this point. James Basire Sr (succeeded by his son James in 1802) was the semi-official engraver to the Society of Antiquaries, and Gough celebrates Basire's unprecedented accuracy in his 1799 preface.

Basire's work, however monumental it may be in itself, was simply urgent for Gough because the destruction of monuments had gained such momentum. Although he assigns the artists one collective 'hand', the hands of Basire and his workshop actually perform the separate function of ensuring that 'correct drawings' from the hands of Schnebbelie and others (though not from Gough's own 'fellow labourers') are transmitted without generational loss, ensuring that their accuracy reaches the printed page. For Gough, then, what is salient is the aesthetic restraint of the prints, as handwork that 'seconds' the historical grasp of architecture conveyed by his words—that may even stop the 'destroying angel' who 'now rides triumphant over a *third* cathedral, which is new modelling, under the uniform unvarying idea of a lengthened choir, bearing no proportion to the body or transepts' (*SM* II.i.7). The claims of preservation seem more robust in light of the fact that not just the picturesque hand of time but also the human agency of 'bishops, deans, and chapters' now perversely threaten to 'destroy their cathedrals'. Gough's allegiance with handwork and quotidian heroism reflects the institutional politics of the *querelle de Wyatt* as well as his scientific ambition.[5]

This allegiance also stems from a materialist view of history with more complicated relations to the wider politics of the 1790s. Gough carves out a scientific domain for antiquarianism by placing an unfashionable emphasis on material culture and its association with research methods being developed in

[4] For discussion of the two images, engraved for *Vetusta Monumenta* I and III, respectively, see Sweet, *Antiquaries* 448–49n., and Lolla, 'Ceci n'est pas un monument' 22. Sweet (305) and Lolla (19) also review period evidence of the conflict between illustration and preservation.

[5] The '*third* cathedral' in question is almost surely Durham, where Carter had observed Wyatt's alterations in 1795. It is not surprising that Gough would have a more empirical view of the accuracy of the prints than Lolla. It is, however, striking that the prints accommodate such a wide range of contemporary comment, with scholars such as Lolla and Myrone leaning more toward an ideological reading of the prints as representations and others, such as Smiles and Reeve, insisting on their evidentiary function as visual documentation.

France. This emphasis is qualified in the second volume (1799) by his inevitable reaction against the intervening events of the French Revolution, but this explicit ideological engagement only underscores the contradictions between Gough's methods and his political views. In the first volume (1786), he begins with praise of Bernard de Montfaucon, whose richly illustrated antiquarian volumes also inspired d'Hancarville and many others. Gough singles out Montfaucon's history of the French monarchy, signalling his own Church and King priorities, but then he turns with equal enthusiasm to current French research on popular antiquities by P. J. B. le Grand d'Aussy. Gough draws particular attention to the emphasis placed on visual documentation by both these French antiquaries. As a result, a materialist antiquarian method is juxtaposed sharply against the existing discipline of English history. In 1786, Le Grand d'Aussy had just published his *History of the Private Life of the French*, the broad survey of food and cuisine for which he is now best remembered; Gough also notes that Le Grand d'Aussy will soon publish his 'collection of drawings, copied from tombs, painted windows and fronts of churches, and other antient monuments, a complete history of dresses and fashions in all parts, and among all orders and ranks in France'. And it is only from these sources [i.e. material culture] that we can deduce such parts of the history of past times, and our acquaintance with those who lived before us' (*SM* I.i.1). What Gough did not know was that the new revolutionary government would sanction this project, which finally appeared in 1799. When he returns to the French in his second preface, he mingles much more qualified praise with horror at revolutionary anticlericalism. He does not mention Le Grand d'Aussy but concedes that a 'feeble attempt has been made, and favourably received by the National Assembly, to draw and engrave the "National Antiquities"' (II.i.5).[6]

The French get much less credit in the new preface for political reasons, but this wartime nationalism does not cancel out the essentially materialist line of critique that led Gough to challenge the stewardship of British bishops and antiquaries to begin with. Even in the context of widespread destruction of churches in France, Gough still feels as much anxiety about the imminent risk to ecclesiastical monuments in an increasingly establishmentarian and nationalist Britain. He even uses the same allegorical personification, 'destroying angel', to refer to anti-ecclesiastical violence in France (6) and alterations to British cathedrals (7). The first preface shows that the destruction and neglect were already well underway in the 1780s:

[6] The author of *Antiquités nationales*, Aubin-Louis Millin, receives credit only for producing more careful engravings of Montfaucon's original drawings. Le Grand d'Aussy's book appeared in 1799, one year before his death, and he read a paper from it at the Institut National on 25 February 1799, 'Mémoire sur les anciennes sépultures nationales'. Gough also 'boasts' of a related collection of French drawings that 'fell into [his] hands' in 1786 (*SM* II.i.6).

> In a few years more we shall have no foundation left for such a work. The two statues of Henry I and his queen, on the West door of Rochester cathedral, are crumbling away without having been drawn. The picture supposed of the Crusades in Canterbury cathedral has been whited over, and the arras in the antechamber of the House of Lords has never been examined with sufficient care to determine whether its subject be the same (I.i.3–4).

There are both ideological and methodological reasons, then, why Gough would seek to distinguish his work from a history:

> It is not an History of England that is here presented to the Public. After the number of histories already in print, what prospect of finding readers of a new one? I have neither the object, the plan, nor the method of an Historian. Our materials are different, and my plan adopts only what his excludes. Great events, great personages, great characters, good or bad, are all that he brings upon his stage... It [i.e. sepulture] is a picture of private mixed with public life. (4)

Gough notes that this distinction is 'borrowed' from the 'lively pen' of Le Grand d'Aussy and he reinforces his shared commitment to the empirical history of manners by describing his own fieldwork (5), a description resumed and amplified in the second preface (II.i.5, 9).

Gough's explicit politics rules out the possibility that his affiliation with material culture, and with artistic labour, results from any democratic programme. He critiques History, implicitly at least, not for its elitism but for its monumentality—here again Nietzsche's paradigm seems to square with the critical awareness accompanying Gough's early romantic historicism. As an archaeological method, his materialism is not necessarily political. Rather, as Stuart Piggott argues, archaeology as a discipline began to take form when antiquaries around 1800 turned their attention away from manuscripts and texts and toward material culture and built works. At the same time, such a turn in the 1790s is difficult to sever completely from the political unconscious in which materialism registers as sedition. This contradiction seems to be reflected in Gough's anxiety about anti-ecclesiastical violence, juxtaposed against his grudging recognition of the French Republic's state antiquarianism. David Simpson has described an English 'revolt against theory' in the 1790s, a movement that may help to explain Gough's scrupulous emphasis on his practice in *Sepulchral Monuments.* If his practice is not explicitly theory-driven, it does mirror his historiographic interest in 'all orders and ranks' because it relies on artists and artisans as heroic collaborators in a scientific process of rediscovery. Gough's science of preservation relies on the practical knowledge of draftsmen, engravers, and architects for its revisionist history. He hopes to preserve, both in visual and in material form, not only the ancient science of Gothic architecture but an inclusive picture of customs and manners in the Middle Ages. Gough's methodological preface, with its rich (if unconscious) analogies between his quotidian historical subject and his workmanlike collaborative method, practically invites political reading.[7]

Gough's remarks about preservation seem at times to presuppose the total extinction of medieval built works, either to mourn the monuments, as Lolla suggests, or, just as often, to encourage preservationist action. He sets out, monumentally enough, to imagine the future impact of his work:

> It is perhaps too weak a hope that this design may induce more attention in future to preserve what remains from immediate ruin, the defacement of the whitewasher's brush, the rude hands of school-boys and the vulgar, and various other unfavourable circumstances; or that this work should be taken up and improved by abler hands. It may be enough if it [i.e. *Sepulchral Monuments*] survives the national taste for such pursuits (I.i.5-6).

By turns either self-aggrandizing or exaggerating for rhetorical effect, Gough imagines that his book may in some future time be the only record remaining of numerous built works. This scenario may allude to the neoclassical topos of England in ruins that gained increasing currency in early nineteenth-century poetry.[8] Basire's unprecedentedly accurate engravings, Gough imagines, may serve 'as incitements to patronage, while the times are favourable; before the caprice, or the avarice, or the indifference, or false taste of Deans and Chapters, suffers and authorizes a havock, less rapid it may be, but not less ruinous, than that with which the axes and hammers of Reformation brake down the carved work of antient art' (6–7). In 1786 no-one could have imagined the violence against churches that followed in the wake of the French Revolution, but in his second preface Gough makes the connection explicit, referring to secular 'Iconoclasts' who 'revive the Puritanism' of the Reformation on both sides of the Channel, only with less gross violence in Britain (II.i.5).

The nineteenth-century Gothic Revival cast a long shadow that still obscures the path-breaking preservation work of Gough and his contemporaries, though the scholarship of Sweet and Sam Smiles has helped recently to bring it back into the light.[9] Part of the reason is that this earlier preservation movement did not yet have a public. Gough presents himself as a voice crying in the wilderness, yet the bewildering variety of dangers to Gothic architecture

[7] On the 'revolt against theory', see Simpson, *Romanticism, Nationalism,* ch. 2. I believe that Gough, like his conservative contemporaries, resisted theory as something continental and atheistical, in the manner described by Simpson. Simpson treats Baconianism more as a broad ideology than as a scientific problem, but there are scientific examples that may be more germane to Gough's situation. See, for example, Martin Rudwick's analysis of the first English translation of Georges Cuvier in *Bursting the Limits of Time* 596–9. On Baconianism and the archaeological turn, see, again, Piggott, *Ruins in a Landscape* 120.

[8] Examples include Anna Letitia Barbauld, *Eighteen Hundred and Eleven* (1812); Felicia Hemans, *Modern Greece* (1817); and Horace Smith, 'On a Stupendous Leg of Granite' (1818).

[9] Sweet, *Antiquaries*; Smiles, *Eye Witness*; and (especially relevant here) Smiles, 'Data, Documentation and Display'.

in his vision seems to prevent the formation of a clear rallying point for a broad public movement.

The secularism of the French Revolution threatens ecclesiastical tradition in France and may—from the point of view of the late 1790s, with their fears of invasion—seem to pose an imminent threat to English churches as well. Yet the model for revolutionary violence, the 'hammers of the Reformation', itself comes from England, invoking a familiar parallel between the regicides of Charles I and Louis XVI. Moreover, a French invasion is hardly needed, as long as misguided English deans and bishops 'sit at their ease' directing unsound alterations that they pay for by 'impoverishing the parochial clergy' (II.i.7). The immediate threat seems to emanate from an unlikely alliance between the clerical elite and the 'vulgar'. Whitewashing was a part of Wyatt's stylistic signature, so the 'whitewasher's brush' implicates the clerical authorities who employed him (some of whom belonged to the Society of Antiquaries) and equalizes their blunted historical sense with 'the rude hands of school-boys and the vulgar'. As with the debate over the 'rabble' in the *Gentleman's Magazine*, the levelling of class distinction seems unlikely to popularize the cause of preservation.

Along with these many contradictory sources of violence that seemed to threaten England's Gothic 'remains', neoclassical aesthetics itself—the inspiration behind Wyatt's 'decluttering' of cathedral interiors—became suspect. Although many untranslated Latin (and some Greek) quotations in Gough's writing signal his own debt to the classical tradition, he is willing to go considerably farther than Walpole in challenging its hegemony as a visual style. In solidarity with his artists or 'coadjutors', he concludes: 'the walk of fame for modern artists is not sufficiently enlarged. Emulous of excelling in History, Portrait, or Landscape, they overlook the unprofitable, though not less tasteful, walk of Antiquity, or, in Grecian and Roman, forget Gothic and more domestic monuments' (I.i.9).

6.2 'A TRUE PRACTICAL ANTIQUARY'

Gough's reflections on historiography in his 1799 preface help to situate the science of preservation both as a sociopolitical project and as a methodology for the work at hand, a vast compilation of engravings from medieval monuments with only loose internal principles of organization. Collaboration is one aspect of this method, but Gough goes beyond methodological concerns in his remarks on one collaborator in particular, Jacob Schnebbelie, who made more than half the drawings from which Basire prepared the engravings for *SM* II (1796). Schnebbelie, a self-educated draftsman who had abandoned his first career as a confectioner, visited at least three dozen churches in sixteen counties in the course of his work for *Sepulchral Monuments* alone. He died of rheumatic fever in 1792, at the age of thirty-one, while in the midst of this work. Gough commemorates him in the 1799 preface: 'In this I am to erect a monument to that able artist and antiquary Jacob Schnebbelie... Often have I indulged an unlimited confidence in him by himself; often I received instruction from his suggestions, when we visited many churches together... Let me not be accused of undue partiality when I say he was a true practical antiquary, nor of vanity when I add, we mutually instructed each other' (*SM* II.i.9). This epitaph, which continues at some length, makes an apt conclusion for the meditative preface to a monumental work on fifteenth-century funerary sculpture. It is not without condescension, but compares favourably to an established convention, among scientific writers, of acknowledging their draftsmen merely as 'tool[s] in [their] hands'.[10] If Gough was ever unsettled by the conflict in their working relationship, his eulogy does not show it. Nevertheless, it is suggestive that in posthumous references by Gough and others, Schnebbelie is granted the title 'Draftsman to the Society of Antiquaries' that he coveted during his lifetime.

The Society of Antiquaries library still holds a manuscript collection of 192 letters to and from Schnebbelie, more than three-quarters of them written by Schnebbelie to Gough during his work on *Sepulchral Monuments* between 1787 and 1792. These letters form the archival basis of this chapter. Their tone changes gradually as Schnebbelie gains scholarly expertise and enters into the kind of collaboration that Gough recalls in his preface. His increasing confidence eventually leads to a crisis, an eruption of the class conflict concealed beneath the smooth surface of Gough's prose. The first sign of a collaborative impulse, several months into the correspondence, comes at a moment when Schnebbelie 'begs leave to offer [his] conjecture' concerning a painting in Canterbury Cathedral, which he thinks has been misinterpreted by the antiquaries whom he has been reading.[11] By June of 1788 he is reporting his discovery in Winchester of 'figures as large as life finely painted on the wall in the N. Transept entirely covered over with a thick coat of plaister... having been hid for many years'. Schnebbelie adds, in a competitive spirit, 'Mr Carter has not seen them' (*Letters* 75). Gough valued Schnebbelie's work, and though he did not become a co-author in the way that Hamilton's collaborators sometimes did, Schnebbelie did eventually publish scholarly text with several of his own images that appeared in *Vetusta Monumenta* as well the *Antiquaries' Museum*. It seems possible that his competitive spirit might eventually

[10] I quote the Dutch anatomist Albinus, who thus describes Jan Wandelaar, the artist who drew and engraved the figures for Albinus's 1747 *Tables of the Human Bones*: 'he was instructed, directed, and as entirely ruled by me, as if he was a tool in my hands, and I made the figures myself'. See *Albinus on Anatomy* 18.

[11] *Letters from and to Mr. Jacob Schnebbelie,* Antiquarian Society MS 267, fol. 11. (Parenthetical citations of these *Letters*, by leaf number, refer to the recto unless otherwise noted.)

have led him to a prominent career like Carter's if he had lived longer. As Matthew Reeve puts it in the first (and so far the only) study of Schnebbelie's career, his completed work is still of 'exceptional importance' for its pioneering documentary approach.[12]

The correspondence differs in some important ways from the correspondence between d'Hancarville and Hamilton examined earlier in these pages. The stakes in d'Hancarville's case were larger: he continually sought investments in a long-term project—the *Collection of Etruscan, Greek, and Roman Antiquities*—over which he had editorial control, and he tried to engage Hamilton on the socially neutral terrain of connoisseurship, as a successful pretender to the status of a foreign savant on tour in Italy. Gough had authorial control of *Sepulchral Monuments*, and although he financed it privately he used the established institutional procedure of the Society of Antiquaries to engrave and publish the work. Schnebbelie was more like a contract employee than d'Hancarville, collecting payment for individual excursions to draw and examine monuments. In this case some of the patron's letters are preserved, and Gough's attitude is clearer than Hamilton's for that reason. From this handful of drafts (the fair copies did not enter the archive), it appears that Gough actively managed the details of Schnebbelie's fieldwork and also promoted his career by recommending him to other patrons.

Although increasingly ambitious, Schnebbelie's letters observe a comparatively strict decorum associated with class and status differences that he—unlike d'Hancarville—made no attempt to conceal, and which signified differently in a British and institutional context. Gough's published eulogy suggests that his relations in the field with Schnebbelie were collegial, somewhat like those of Hamilton and d'Hancarville at Paestum, and the MS correspondence tends to support this view. The explosion of conflict near the end of this correspondence suggests, however, that their collegiality depended on a delicate balance. Schnebbelie's drawings provided crucial evidence for Gough's claims concerning authenticity, but his potential contributions to *Sepulchral Monuments* were constrained by Gough's strong authorial role and institutional authority. The conflict between them—to be examined in detail in its chronological place—erupted when Schnebbelie tried to harness this institutional affiliation to his own book project, *The Antiquaries' Museum*, in 1791.

It seems fair to say that Schnebbelie worked feverishly during his short artistic career, which coincided with roughly the last five years of his life. There is no record of his artistic training, but the story of his career begins with a

[12] Reeve, 'Jacob Schnebbelie' 79. This chapter began life as a lecture given for the Johnson Society of the Central Region (US) on 15 April 2005. I have profited greatly from Reeve's work while revising and expanding this material.

fortuitous meeting on the grounds of the Earl of Leicester's estate, where Schnebbelie happened to be sketching. Leicester, president of the Society of Antiquaries from 1784 to 1811, was evidently so impressed with Schnebbelie's topographical and architectural draftsmanship that he recommended him to the Society at a moment when there was an increased demand for visual documentation.[13] The Society had officially appointed Carter as draftsman in March 1784, but Carter became offended over their employment of another draftsman, S. H. Grimm (who also drew the Tahitian implements for Stanfield Parkinson's publication of his brother's *Journal of a Voyage to the South Seas*, detailed in Chapter 1). Schnebbelie's earliest drawings for the Society, eventually engraved for *Vetusta Monumenta*, are dated in the middle of 1788, during the period of controversy with Carter (a conflict that clearly set the stage for the later one involving Schnebbelie).[14] By this time Schnebbelie had already made drawings of architectural sculpture for Gough at several locations, especially Canterbury.

There was a strong need for his services because Gough was working so actively on his own research while also pushing for an increased volume of publication by the Society in the form of prints for *Vetusta Monumenta*, plates and commentary for the series of English cathedrals, and scholarship for the journal *Archaeologia*. By the time of Schnebbelie's death in 1792 he had published nine drawings in *Vetusta Monumenta* as well as the first three numbers of his *Antiquaries' Museum*, for which he created twelve etchings, some aquatinted. Of the fourteen new engravings after Schnebbelie that appeared in the third volume of *Vetusta Monumenta* (1796), twelve had appeared as separate plates before his death, some of them with his own text. His largest single body of work, the seventy-six drawings selected and engraved for *Sepulchral Monuments*, only appeared posthumously in 1796. All but the *Antiquaries' Museum* prints were engraved in Basire's workshop. At his death Schnebbelie left numerous finished and unfinished drawings, some of which were coloured and sold by his widow and sons. Schnebbelie's eldest son, Robert, became an antiquarian draftsman in his own right.[15]

[13] John Bowyer Nichols claims that Schnebbelie had 'learned something of drawing under Paul Sandby' before he left the confectionery business to become a drawing master (*ILH* VIII.550). But in the absence of any direct evidence, it seems safer to assume with F. M. O'Donoghue and Rosie Dias that 'it is likely that he was self-taught' (*ODNB* 49.233). Schnebbelie's 'accidental' (or perhaps not so accidental) meeting with the Earl of Leicester is recorded in several period sources, including his obituary in the *Gentleman's Magazine* ('Schnebbelie' 189–90).

[14] On Carter's appointment and the ensuing conflicts see Evans, *History* 182, 191. The Society's cathedral series eventually became Carter's primary venue.

[15] Schnebbelie also published a small number of views as separate plates beginning in 1788, and he contributed to two other antiquarian books, most notably Nichols's *History and Antiquities of the County of Leicester*. In this chapter I am concerned primarily with the collaboration of Gough and Schnebbelie and limit my focus accordingly to the manuscript correspondence, *Vetusta Monumenta*, *Sepulchral Monuments*, and Schnebbelie's own book-length work, *The*

Schnebbelie's letters during the period 1787–88 record his visits to Canterbury and other cathedrals to produce documentation for Gough's treatment of fifteenth-century tomb architecture, the subject of his second volume. (Volume I, published in 1786, covers the eleventh through the fourteenth centuries.) This work was still in its beginning stages, so it was to Schnebbelie's benefit that Gough decided to publish some of the images in the second volume of *Vetusta Monumenta* (1789; 1790), which appeared long before Gough's own book was finished. Schnebbelie's first commission was Canterbury Cathedral, but since Gough saved these drawings for *Sepulchral Monuments*, it was his second major commission at Winchester Cathedral that made his reputation. These drawings of three fifteenth-century chantry chapels were lavishly engraved by Basire for *Vetusta Monumenta* and also gave Schnebbelie an entrée at the Royal Academy. Though it is less well-known, Schnebbelie's Canterbury work is quite fully documented in the letters, and the plates and text eventually published in 1796 still bear the imprint of his discoveries and his enthusiasm. This first group of letters includes the largest number of drafts in Gough's hand, containing detailed instructions as well as praise of Schnebbelie's zeal in all weathers (*Letters* 1, 8).[16] Over the course of several weeks in late summer 1787, Schnebbelie drew elevations and bird's-eye views of chapels, tombs, and tomb effigies, following the format established in Gough's first volume. His replies increasingly report his own independent discoveries, as when a painting over the tomb of the Black Prince proves to be 'so much defaced that I was obliged to clean it and put some varnish on by which means I discovered a crucifix that was not known before' (10).

The painting, included on a list of subjects assigned to him by Gough, proved to be more fully recoverable than Gough had imagined, and this archaeological pattern recurs throughout Schnebbelie's letters. He comments that parishioners are pleased with the restoration, anticipating a theme that runs throughout the body of contemporary comment on his work. Schnebbelie is often praised for his ability to uncover not only obscure areas of visible paintings but also earlier layers of wall and ceiling painting that had been forgotten. In the case of the Black Prince's canopy, he writes:

> I beg leave to give my conjecture as it is described in all who have wrote a description of the monument only as a painting of our Saviour with the four Evangelists at the corners—I think it rather represents the father—supporting the son—as the figure is seated on a rainbow with the hands held up on which rests

Antiquaries' Museum. At Schnebbelie's death Gough also retained many unpublished drawings by him, including those not selected for *Sepulchral Monuments*. These are now at the Bodleian Library, Oxford. Both the *ODNB* entry and Bowyer Nichols include a brief account of Robert Bremmel Schnebbelie.

[16] Reeve quotes a substantial sample of Gough's instructions ('Jacob Schnebbelie' 72).

> the upper part of the crucifix, the Ground being blue spotted with stars—represents the firmament, and on the drapery is a number of radiated circles with an inscription in each represented at large in the drawing.[17]

Though this tomb was too early to be included in Gough's second volume, he did publish engravings from Schnebbelie's drawings of seven fifteenth-century monuments in this cathedral. In one case Gough accepts as authoritative a detail in the tomb effigy of Henry IV, the figure's 'high shoulders', that he had originally challenged in his response to Schnebbelie's drawing (*Letters* 8; cp. *SM* II.31). Schnebbelie also notes that Archbishop Bourchier's monument 'being now cleaned and repaired will make three fine drawings' (11), and all three were engraved for the volume, showcasing Schnebbelie's meticulous recording of architectural detail (Plates CVI–CVII). The frequent references to restoration work suggest that scholarly interest in the history of the fabric was a novelty to cathedral officials, leading them—in some combination with Gough's authority and fees—to grant Schnebbelie a surprising degree of latitude for such work.[18]

By the time Schnebbelie and Gough set out for Winchester in May 1788, Schnebbelie had gained significant archaeological and artistic experience. Working either on his own or on commissions from Gough, and sometimes travelling with him, he recorded monuments in London, St Alban's, and parish churches in Kent during the winter and spring of 1787–88. He must have made the twenty-two-mile journey to St Alban's repeatedly, as six plates in *Sepulchral Monuments* are based on his drawings of four monuments in the cathedral there. He also published four individual prints of St Alban's Abbey, aquatinted by Francis Jukes, in 1788. In London, he executed the first of his commissions drawing artefacts belonging to Fellows of the Society of Antiquaries in the Society's rooms in Somerset House. One set of these, depicting a reliquary from the collection of Thomas Astle, appeared with the Winchester engravings in *Vetusta Monumenta* (II, Pl. LI–LII). He also undertook local archaeological expeditions to Westminster, where he observed ongoing work on the Chapel of Henry V at Gough's request (*Letters* 27–28), and to Clerkenwell Church (in hopes of obtaining a commission): 'Yesterday I received

[17] *Letters* 11. Schnebbelie's view is closer to recent consensus that this image represents the Trinity, but curiously the large father figure behind the crucified Christ actually looks more like the son than the father, with the familiar light brown beard, long hair, and soulful face. Schnebbelie refers to a similar image reproduced in Nichols's *Bibliotheca Topographica Britannica*, which he evidently carried with him. The monument was not engraved for *SM* I, but Gough does mention this painting on the underside of the canopy or tester (137), and may have intended to include the monument in a supplement.

[18] One notable exception to this rule would be Exeter Cathedral, which was governed from 1747 to 1784 by two successive deans who were also medievalists and presidents of the Society of Antiquaries. See Smiles, 'Data, Documentation and Display'. As Reeve notes, Schnebbelie brought in a local authority to corroborate his 'short-necked' representation of Henry IV ('Jacob Schnebbelie' 72).

information that Westons Monument in Clerkenwell Church was nearly taken down. Immediately attended the same and took the following account of what I found remaining' (29). At Clerkenwell (27 April 1788) Schnebbelie opened the coffin and sketched the remains under this monument. In Westminster Abbey he reports being unable to get 'a ladder long enough that [he] could make sketches of the coronation' (28)—another index of the bodily, and in this case dangerous, sense of 'knowledge work'.[19] In preparation for the Winchester journey, Gough secured a commission for Schnebbelie from the Council of the Society of Antiquaries, a change of status reflected in the relatively stiff and formal manner of Schnebbelie's first letter after the event (33); it also helps to explain his later resentment at being denied the right to use the title 'Draughtsman to the Society of Antiquaries'. All these experiences produced an artistic and antiquarian self-consciousness that is reflected in the variety of Schnebbelie's early work, in the style of his letters, and in visual details of the works themselves.

A newly elaborate signature in one of the engravings for *Vetusta Monumenta* (III. pl. IV) suggests the artistic ambition that such intense immersion in the Gothic must have fostered. The signature ('Schnebbelie Delin. 1788'), apparently carved on a fallen block of stone in the bottom right-hand corner of this image of the chancel and stalls in Chatham Church, Kent (Figure 6.1), lays claim to the monumentality of the subject. This slightly showy displacement of the delineator's signature from its ordinary place on the lower left, just outside the image, fits the pattern of 'romantic antiquarianism' described by art historians including Lolla and Martin Myrone. Myrone uses the term 'atmospheric depiction' to characterize a Schnebbelie image from the same series, and Lolla emphasizes the importance of the observer's sense of wonder as a thematic element of these compositions, a trope that could certainly be linked to Schnebbelie's incorporation of his own signature into the monument.

But romantic antiquarianism should not be regarded as a contradiction: if Schnebbelie participates in antiquity through 'aesthetic pleasure', he also insists on documenting the dilapidated state of these stalls (technically *sedilia*).[20] It would be a mistake to divorce ruin as a merely picturesque state from ruin as a material condition that demanded archaeological knowledge.

[19] One of Schnebbelie's successors in the line of recording cathedral interiors, C. A. Stothard, was killed when he fell from high scaffolding while sketching a stained-glass window in Devon in 1821. See the biographical headnote in Stothard, *Monumental Effigies* 17. Schnebbelie's letter includes a detailed pen and ink sketch of sculpture over the entry to the chapel. He was finally able to record details of the coronation (sculpted on the fascia at the north end of the chapel) in 1790 when there was a scaffold in place (*Letters* 115).

[20] Myrone, 'Society of Antiquaries and the Graphic Arts' 113; Lolla, 'Ceci n'est pas un monument' 22. The term *sedilia* (which I owe to Reeve) is not included among Schnebbelie's Latin quotations in this number, and as neither his co-contributor (Wells) nor the editor (Gough) inserted this term, it seems likely that it remained to be rediscovered by architectural history.

Myrone, commenting on the arrangement of Schnebbelie's images of Magdalen Hospital, Winchester (the plates just prior to Chatham Church in *Vetusta Monumenta* III), points out a 'fissure' between 'antiquarian exactitude and aesthetic pleasure'. The 'atmospheric depiction' of the chapel (Plates I–II), showing windows and vaulting in sharp relief against extremely dark shadows created by etching on the plate, is certainly more pictorial than the adjacent plate (III), a composite of 'plans, details, and exterior views juxtaposed without regard to pictorial space'. The Chatham plate, however, combines both types of representation on one sheet and the text that Schnebbelie supplies with both groups of images also undermines any fundamental distinction between the antiquarian and the artistic or romantic. The plate contains three figures, only the largest of which is reproduced here (Figure 6.1). Reeve's analysis best captures the effect of the whole: 'As though to transform his subject into a biological specimen, Schnebbelie produces an image of three panels' to

Fig. 6.1. *Chancel and Stalls in Chatham Church, Kent.* Engraving by James Basire from a drawing by Jacob Schnebbelie. *Vetusta Monumenta*, Vol. III (1796), Plate IV. The trompe l'oeil carved signature appears on the diagonal block in the lower right hand corner. Courtesy of the Division of Special Collections, Archives, and Rare Books, Ellis Library, University of Missouri.

illustrate the 'passage from destruction to discovery and from preservation to curiosity'. Another figure shows the end result, sculptured ornaments from the *sedilia* abstracted and set against a blank background: 'the lapidoptery presentation of these objects indicates that they . . . have been preserved by antiquarian intervention'.[21]

Schnebbelie's scholarly effort, inspired though it may be by the gloomy prospect of imminent destruction, is substantiated by the two pages of explanatory text included in the letterpress for each of these two suites of plates (I–III, IV–V). Following John Carter's lead, Schnebbelie later provided some commentary on his images in his own work, *The Antiquaries' Museum*, but very few numbers of *Vetusta Monumenta* included text by or even any reference to the artists, as the Society followed the more usual practice of assigning the letterpress to scholars (often Gough himself). The first note does not venture much in the way of interpretation, but Schnebbelie's contribution on Chatham Church is doubly unusual in this regard. He begins by pointing out that 'very few persons' knew of these stalls, as they had been covered with boards for an indefinite period until shortly before the demolition. Schnebbelie then reports: 'I fortunately got there in time to make a drawing of them, as likewise of the chancel, as it was taking down, which, I think, cannot be less than four hundred years old.' The date is significant because, as many scholars have noted, the stylistic development of Gothic architecture was only beginning to be understood in the late eighteenth century. Schnebbelie supports his date with references to Rochester Cathedral and other examples, though of these there are 'none that [he] recollect[s] so highly finished in the back ground and ceiling; the principal ornamenting of the others are bestowed on their fronts, pediments, and spandrils'.

Schnebbelie's signature observations here concern the remnants of the original structure. He carefully identifies traces of gilding and colour, and makes inferences concerning the now vanished canopies and front of the stalls: 'from six cavities in the arch above, I conclude it was decorated with finials, and must have been very elegant when perfect'.[22] Finally, he cites a

[21] Reeve, 'Jacob Schnebbelie' 81. Besides this end result (Figure 3), the Chatham plate (Plate 4) also includes an 'atmospheric' cutaway view (to continue borrowing Myrone's useful term) of the church as Figure 1. The largest of the three figures on this plate (Figure 2), which includes the carved signature, is the detail of the three stone seats (my Figure 6.1). Myrone adopts the distinction between 'exactitude' and 'pleasure' from Smiles, although I find it hard to see how this 'fissure', as self-conscious as it is, undermines the scholarship of a hired illustrator like Schnebbelie. As Smiles has shown, Gough promoted visual documentation precisely because it belonged more than ever to what Myrone calls the 'real work of the antiquarian' ('Society of Antiquaries and the Graphic Arts' 119).

[22] *Vetusta Monumenta*, vol. III, Plates IV–V, accompanying text, pp. 1–2. Schnebbelie also shows his attention to varieties of building stone in this text. On the vague understanding of chronology in Gothic architecture before 1800, see Smiles, 'Data, Documentation and Display' 82. Smiles quotes Gough's complaint that there was 'no system of Gothic architecture', but it is

recent *Gentleman's Magazine* essay on the use of this type of stall and gives a brief digest including Latin quotations from several missals used by the essay's author to argue that these groups of three seats were used by the priest and co-celebrants during pauses in the Mass. The second half of the letterpress with these two plates is a letter from David Wells, a Fellow of the Society and owner of one of the objects etched by Schnebbelie for *The Antiquaries' Museum*. Citing the same *Gentleman's Magazine* essay, Wells describes a field trip he took with Schnebbelie, two years after the Chatham drawing was made, to document the use of similar 'stone seats' in Westminster Abbey. Wells hints that 'a drawing from these seats would not be unworthy of the attention of the society'.[23]

The monuments of Cardinal Beaufort, Bishop Wainflete, and Bishop Fox in Winchester Cathedral (drawn June 1788) were the first to be published by the Society of Antiquaries and were probably Schnebbelie's most widely noted achievement during his lifetime. He and Gough reached Winchester by 30 May 1788 and must have spent some days examining the cathedral in the manner recalled by Gough in his eulogy.[24] Schnebbelie stayed on, and on 11 June he sent Gough three drawings from the cathedral, followed by an invoice for the Council of the Society (*Letters* 35–36). Schnebbelie's drawings of these monuments differ fundamentally from those discussed so far. The three principal engravings evidently depict large free-standing structures, exemplified here (Figure 6.2) by the monument of Bishop Wainflete (d. 1486), but in a technical sense at least they focus on a single architectural detail taken out of context (*Vetusta Monumenta* II. Plates XLV–VI, L). There are three supplementary plates (XLVII–XLIX) that illustrate details of the monuments, including the tomb effigies themselves, but unlike the numbers on Chatham Church and Magdalen Hospital, the Winchester material does not illustrate the place of the monuments in the church as a whole. Nor was the effort prompted by the ruin of the fabric—a much less likely, though not unheard of, fate for so large a structure.

As we have seen, Gough was pessimistic about the survival of these monuments, but here (as in *Sepulchral Monuments*) engraving is at least potentially a part of preservation, rather than a substitute. As his cross-references to the

worth noting that after Schnebbelie's death Gough was moved to credit him for moving in this direction: 'Mr Schnebbelie was not contented with drawing the remains of antiquity: his close pursuits had made him proficient in the study of our National Antiquities, and a judge of the different styles of Gothic Architecture and Monuments' *(Antiquaries' Museum* vii).

[23] *VM* III. Plates IV–V, accompanying text, p. 6. Wells's account also helps to explain Schnebbelie's feeling that he was entitled to call himself 'Draughtsman to the Society of Antiquaries'. See also the letter by 'Indagator' (*Gentleman's Magazine* 57).

[24] On 30 May 1788, Samuel Pegge wrote to Hayman Rooke: 'Gough and Nichols have gone to Winchester with a draughtsman, to copy two grand monuments there.' Antiquarian Society MS 891/2 (Rooke Papers), fol. 39. The high reputation of these drawings is attested by the fact that they were exhibited at the Royal Academy in 1789 and singled out for special mention in the review of *The Antiquaries' Museum* quoted in my epigraph, 'The Antiquaries Museum'.

Fig. 6.2. *Monument of Bishop Wainflete, Winchester Cathedral.* Engraving by James Basire from a drawing by Jacob Schnebbelie. *Vetusta Monumenta*, Vol. II (1790), Plate XLVI. Courtesy of the Division of Special Collections, Archives, and Rare Books, Ellis Library, University of Missouri.

Winchester material attest (*SM* II.147–49, 298–300), these six plates derive their visual style as well as their science of preservation from his programme for *Sepulchral Monuments*, rather than from the programme of exclusively visual preservation pursued in parts of *Vetusta Monumenta*.[25] Like some of the engravings in d'Hancarville's *Etruscan, Greek, and Roman Antiquities*, the decontextualized presentation of these details makes them monuments in their own right. Unlike Grecian urns, however, these containers really are architectural in their scale, as Schnebbelie emphasizes by showing them with open doorways (on the right in Figure 6.2).

Although monumental, these engravings are more documentary than pictorial in their approach. It is a curious coincidence that the original drawings in question secured Schnebbelie his only opportunity to exhibit at the Royal Academy. More painterly works such as the images of Magdalen Chapel would have been more likely candidates for exhibition—as Schnebbelie himself recognized—but the historical accuracy of these drawings from Winchester played the decisive role. Schnebbelie wrote to Gough in March 1789, as he was finishing these drawings for the engraver, to say that he had met the president of the Royal Academy and the royal architect at Somerset House and had shown them the drawings:

> Sir Joshua Reynolds and Sir William Chambers have seen the drawings of the Winchester Monuments and were pleased to express very great approbation. Sr Wm thought it a pity that such beautiful pieces of Gothic Architecture should be so neglected as little notice to be taken of them while our students were sent abroad to study the dull Greetian—Sir Joshua said they were the most elegant things of the kind he ever saw, he was sorry he had not seen them before, particularly the figure of Cardinale Beaufort. he having finished his picture for Boydel being the Cardinal's death from Shakespeare and had given him a long beard when by the figure on his tomb he had none Sir Joshua said they would be entitled to a proper place in the next exhibition. (*Letters* 63–63v.)

He goes on to ask Gough for permission to exhibit the drawings of Beaufort's and Wainflete's monuments as well as 'the two inside views of Magdalen Chapel' at the Academy's annual exhibition two months later (64). When Gough sent Schnebbelie four of the Canterbury drawings to show to Reynolds on a return visit, Schnebbelie repaid the favour by promoting *Sepulchral Monuments*, which Reynolds did not yet know: 'he thought it a great and very useful work particularly for the artists who studied or painted British

[25] One of the more notable examples is the six views of Cowdray House (*VM* Plates XXXIII–VII), taken in 1782–83, shortly before its destruction, by S. H. Grimm, 'of whom they were fortunately purchased by the Society a few months before his death' in 1794 (13). Lolla includes the Eleanor Crosses, drawn by Schnebbelie (*VM* III. Plates XII–XVII), in this category of antiquities whose physical decay is promoted, or at least not hindered, by engraving ('Ceci n'est pas un monument' 22). The west tower of Hereford Cathedral collapsed in 1786.

Historical pictures', offering to purchase it for the Academy (65). The painting alluded to above, *The Death of Cardinal Beaufort* (based on *2 Henry VI* III. iii), was Reynolds's own contribution to the Shakespeare Gallery, John Boydell's massive commercial effort to 'establish an English School of Historical Painting'.[26] History painting did not become central to Reynolds's own (at that point very advanced) career, but for Schnebbelie, any role in history painting, even as a source of better raw material, was professionally promising.

In the event, faithful documentation proved incompatible with the aesthetics of history painting, and Schnebbelie's work has only now begun to attract the attention of art historians. Smiles refers to *Sepulchral Monuments* in his account of this incompatibility, drawing evidence from the career of an initially even more obscure contributor to the first volume, Basire's apprentice William Blake. Smiles points out that the plates in question (from Westminster) are signed by Basire as both draftsman and engraver: 'as Basire's apprentice Blake's name would not have been recorded'. Art historians uncovered his authorship only as they became interested in 'the mature artist's creative abilities'. Even the most distinguished antiquarian draftsmen 'have dropped into relative obscurity... by dint of the lowly status associated with such work'. Smiles points out that—contrary to the hope of the Antiquaries themselves—few artists treated antiquities in a scientific spirit, and few antiquaries were professional artists. One exception was C. A. Stothard, one of Schnebbelie's successors in the recording of Gothic architecture. Smiles observes that Stothard 'is better considered as an independent scholar as well as an artist and engraver', partly because he was a Fellow of the Society of Antiquaries, like George Vertue and John Carter earlier, and not a 'menial' such as Basire or Schnebbelie (or the youthful Blake). In a scholarly context this term seems prejudicial, especially given that it repeats—whether consciously or not—Stothard's own contempt for Gough's illustrators.[27]

[26] Boydell, *Catalogue of the Pictures* v. Reynolds evidently made good on his offer to purchase Gough's book, as the Royal Academy's online catalogue entry for *Sepulchral Monuments* cites the Council minutes that document a resolution to purchase Vol. I on 19 March 1789, just a week after his meeting with Schnebbelie. For Reeve's account of the episode, see 'Jacob Schnebbelie' 73–4.

[27] Smiles, 'Art and Antiquity' 126, 133. Smiles's use of 'menial' is not intended to be pejorative, and he notes that 'the numerous exchanges taking place between artists and antiquaries in this period are still subject to critical neglect' (126). Smiles's own earlier scholarship has done much to remedy this neglect, but this recent essay seems ambivalent on the aesthetic value of antiquarian illustration. Here he makes quite a strong distinction between creative work, or a 'poetic apprehension of antiquity' (141), and commissioned documentation of monuments (126), a distinction that seems indebted to the civic humanist division of artists into the classes of public men and mere mechanics. This division is certainly operating in Stothard's own account of *Sepulchral Monuments*.

My recovery of Schnebbelie's contributions aims to show that he, too, was 'an independent scholar as well as an artist'—a view supported by Reeve's study of his unpublished Salisbury drawings, which provide evidence of 'Schnebbelie's own analytical and archaeological bent'.[28] One proof of Schnebbelie's scholarship is that Reeve, a historian of medieval rather than eighteenth-century art, is the first modern scholar to have devoted substantial effort to Schnebbelie's work. As in the cases of Parkinson, Fabris, Piaggio, d'Hancarville, and Webster (see Chapter 7), exclusion from the learned societies is not just a negative limit on knowledge work, but an enabling condition.

By the end of 1789, Schnebbelie had six mouths to feed, and when he did attend meetings of the Society he was more concerned to secure commissions than to secure membership. He began attending meetings in the autumn after his sojourn in Winchester. The record of these meetings shows us another aspect of his pragmatic and in many ways symbiotic collaboration with Gough. Gough became disillusioned over the Society's lack of scholarly rigour and even contemplated resigning in 1780. Although he did not really resign until much later, his attendance at meetings became 'erratic'; sending Schnebbelie to meetings as his deputy allowed Gough to keep his distance while still receiving detailed reports, several of which are included in the archive. Schnebbelie makes allusions in more than one of these reports to the rationale that Gough must have supplied—namely, viewing the meetings as an opportunity to secure commissions—and that Schnebbelie later came to question.[29] On some of these occasions he exhibited drawings that had either been commissioned by the Society or were being considered by them for purchase. Though Gough was not always a popular director, his influence was clearly crucial, and at the same time the members of the Society's Council seem to have been critical consumers of work recommended to them. In February 1790 Schnebbelie wrote: 'if you intend proposing any of the drawings I made in Lincolnshire to the council I shall be much obliged to you to let me know which they are that I may be prepared to answer any question put to me' (*Letters* 111). Other letters from the period (103–04) indicate that Gough successfully proposed Schnebbelie's drawings of Holy Sepulchres at two provincial parish churches, which duly appeared in *Vetusta Monumenta* III. One of these (*VM* III. pl. XXXII) belongs to the much larger group of drawings done in Lincolnshire the previous summer, and another from this group (III. pl. VI)

[28] Reeve, *Thirteenth-Century Wall Painting* 62.

[29] See *Letters* 59 (30 Jan 1789), which also provides a particularly detailed report, and *Letters* 142 for his later doubts. Schnebbelie attended meetings almost weekly in November–December 1789 (97–102). Gough's 'increasingly erratic' attendance is noted by Sweet, 'Incorporated Society' 85. See also Pearce, 'Antiquaries and the Interpretation of Ancient Objects', on the conduct of the Society's meetings. Schnebbelie announced the birth of his fourth child, a girl, in January 1790 (*Letters* 108v.). A draft in Gough's hand from this year shows that he was actively promoting Schnebbelie to other patrons (*Letters* 71).

had been acquired earlier because of its connection to Bishop Wainflete, whose monument Schnebbelie had drawn for the Society.[30]

6.3 'PRIVATE MIXED WITH PUBLIC LIFE'

Schnebbelie's next major commission, at Salisbury in late summer 1789, resulted in nine plates in *Sepulchral Monuments*, including the headpiece to the volume (*SM* II.ii.1). It also got him involved for the first time in the politics—as well as the science—of preservation. Most histories of the Gothic Revival cover Wyatt's renovation of the cathedral (1789–91), which Reeve calls 'a major media event'. Michael J. Lewis, for example, gives a fairly dispassionate survey both of Wyatt's importance for the early Gothic revival and of the historical inaccuracies that earned him the epithet 'the destroyer'.[31] A great deal of commentary on Wyatt appeared in print, both during and after his lifetime, but Schnebbelie's letters give us an especially intimate glimpse of Wyatt's methods and of the role of local authorities who employed him. They help to contextualize the published polemics of Gough, Carter, and others, and the images that he and Basire produced illustrate the specific architectural features and aesthetic values that were at stake in the debate, which often lost itself in generalities. The first letter in the series simply states, 'Mr Wyatt is in town I have been to his house near a dozen times but have not been able to get sight of him' (*Letters* 81).

The monuments in Salisbury needed to be recorded before Wyatt's renovation work was too far advanced—if anything the need was more urgent in this case because the destruction seemed not just imminent but arbitrary. Schnebbelie did make contact on one occasion, when he learned that 'the whole ceiling of the choir would be made of one uniform colour and all the painting thereon would be totally destroyed, this Mr Wyatt informed of the day before I left' (*Letters* 85). Wyatt may well have avoided Schnebbelie after this encounter, which probably took place on the eve of one of his earlier visits to Salisbury, only partially recorded in the archive. But he evidently gave him permission to enter, because *Sepulchral Monuments* includes several engravings after Schnebbelie of two chantry chapels that Wyatt's workmen had already torn down by the time of his fateful visit on 24 August:

[30] Schnebbelie's main subjects in Lincolnshire were in Tattershall [Tateshale] Castle and the associated Holy Trinity Collegiate Church, monuments from which are depicted on six plates in *SM* II (LXII–VI, XCVIII–CXVII). Wainflete supervised the construction of the church beginning in 1472. The group of subjects from this location is especially varied, including chimney pieces from the castle and tombs of two powerful fifteenth-century women.

[31] Reeve, 'Jacob Schnebbelie' 76; Lewis, *Gothic Revival* 41–2.

> I arrived here last Saturday morning but too late to make a sketch of the stalls they being down as likewise the tomb of – Hungerford whose body was this morning discovered . . . The Bishop . . . wishes very much to have some conversation with you concerning the improvements of the Cathedrale and other matters he told this morning a paragraph had appeared in the Gen. Eveng. Post of Saturday signed Victor at which he was much hurt and thought it must come from you which makes him the more regret his not seeing you as he could have given substantial reasons for the taking down the Chapels etc. he always was a lover of antiquities & wished to preserve them as much as possible but here was a case of necessity in taken them down in order to preserve the main fabric (84).

The engravings and published text reveal more concerning these monumental chapels and the controversial decision to remove them. Schnebbelie's last visit to Salisbury, where he made sketches of the doomed ceiling paintings between 15 September and 5 October, is recorded only in manuscript (*Letters* 85–89). His high price of ten guineas for these sketches suggests the difficulty and the volume of work involved. Gough must have paid considerably more for the finished drawings (delivered March 1790) and perhaps more still for the 'proofs' delivered in November 1791 (155). Reeve suggests that Gough intended them for publication in *Vetusta Monumenta*, which would explain the preparation of proofs for the engraver. In the event, they were not published until 2008, when Reeve reproduced eight of the finished drawings (in colour) and four of the preliminary sketches in his valuable study of wall paintings in the cathedral. Schnebbelie quotes Bishop Barrington as saying that 'it would be impossible to preserve "the paintings" because "it was intirely opposite to Mr Wyatt's plan to make the whole of one uniform colour and the expence of the repairing them too great"' (90). Gough's severe condemnation of Wyatt in print suggests that he was determined to give no quarter despite his knowledge of the mitigating circumstances that Schnebbelie impartially reports in these letters. In *Sepulchral Monuments* he is more concerned with the destruction of the chantry chapels, since both are fifteenth-century works that fall under the purview of the second volume, than with the ceiling paintings. Although posterity has largely seconded Gough (along with Carter, John Milner, and other critics) in his condemnation of Wyatt, at least one recent authority states that 'decisions to amputate the nave at Hereford, and to remove the chantry chapels at Salisbury, had been taken before Wyatt even arrived on the scene'.[32]

Schnebbelie's richly detailed drawings of the two chapels provide excellent ammunition for Gough's elegiac polemic. The case of Salisbury bears out some of his more pointed claims in the preface concerning the Antiquaries' betrayal of their trust and institutional vandalism within the church (*SM* II.i.7). By

[32] Crook, *John Carter* 35. For the paintings, see Reeve, *Thirteenth-Century Wall Painting*, Plates 20–21, 24–25, and colour Plates IV–XI.

selecting Schnebbelie's unusually soft, atmospheric rendering of Bishop Beauchamp's chapel as the vignette or headpiece for the entire fifteenth century (II.ii.1), Gough draws attention to the Salisbury portions of the main text as having special relevance to the general claims that frame the work. His accounts of the two destroyed chapels—Beauchamp's and that of Robert, Lord Hungerford, both adjoining the Lady Chapel—stand out for their use of the past tense, their reliance on data from Schnebbelie's letters, and their ethnographic detail. The narrative of Beauchamp's chapel is certainly the more polemical of the two, driven not only by preservationist outrage and feudal nostalgia but also by the special veneration Gough reserves for Bishop Beauchamp's contribution to Gothic architecture. No authentic record of the chapel's interior remained. Since the Schnebbelie/Basire vignette was only a reconstruction, it was banished from the scientific narrative of the main text, which is accompanied by only one plate, an austere ground plan from Walpole's collection (Pl. CI).

Gough's empirical description resembles that of many other entries, except for his sustained use of past tense—a rare enough occurrence to mark the difference between 'decaying' and 'destroyed': 'on the fascia of this beautiful but decaying chapel was the mitre: also on the ceiling of oak with a rich stone cornice' (*SM* II.271). A simple grammatical change like this signifies powerfully in the context of a long and often dry narrative. It also contradicts the sophisticated understanding of romantic antiquarianism as being more invested in aestheticized decay than real preservation. Gough's lament continues:

> I have been the more particular in describing this building, as this beautiful memorial of its munificent founder, has since been sacrificed, with the corresponding chapel of the Hungerford family, to modern *taste* and *improvement*, and the highly finished niches and other carvings are transported out of their place to form a modern altar piece, and contribute to decorate different parts of the chancel, to the utter confusion of all style and chronology in architecture (271).

Although the architectural importance of these chapels is in direct proportion to the wealth and power of the bishop and the baron who endowed them, Gough makes a particular effort here to show how these chapels, in their splendour, also 'illustrate' the quotidian 'history of families, manners, habits, and arts', as promised in the subtitle of his book. Beauchamp's architectural activity constitutes one such worldly, if not especially humble, pursuit. Gough points out that in addition to this chapel, built during his lifetime, Beauchamp designed the great hall of his palace and served as master and surveyor of the works when Edward IV rebuilt St George's Chapel at Windsor. (This was the position Wyatt enjoyed under George III.) For testimony of the bishop's dedication both to architecture and to lay piety, Gough appeals to an English inscription in his honour in St George's Chapel, which is reproduced in black letter in Gough's text: 'this nervous simple inscription was carved and painted, and having survived to the present time, was new blacked in the late repair of the chapel' (271–72). He also quotes Edward as praising Beauchamp's daily attention to the rebuilding work, adding dolefully that 'other labourers have entered into his harvest, thinking they could make a better use of his materials, and raise themselves a name from darkening his' (273). Beauchamp also secured the canonization of St Osmund, a favourite saint both of the nobility and of the populace—and one whose memory is now disregarded along with Beauchamp's own: 'the bodies and resting places of St. Osmund, bishop Beauchamp, and Robert, Lord Hungerford... have suffered in our time a violation too gross to pass unnoticed' (272). It is curious that Gough concludes this condemnation by returning to the secular idiom of taste: 'No genuine judge of the Gothic style of Architecture but laments the demolition of the Beauchamp chapel, which we have endeavoured to preserve a faint resemblance of in the vignette to this century'.[33]

Gough's description of the Hungerford chapel is much less polemical, and was composed somewhat earlier, before Wyatt's continued activity had raised preservationist anxieties to their peak. Another difference is that Schnebbelie had a chance at least to sketch this chapel before it was torn down, and the section includes an interior view (Figure 6.3) in the same style as his vignette of the Beauchamp chapel (Figure 6.4). (In 1803 Carter cited this particular image for proof that the removal of the chapel was not, in fact, structurally necessary.) Instead of polemicizing, this particularly long description (*SM* II.186–91, plus three plates) focuses on details of the chapel, and aims in particular to preserve the record of customs and manners represented by wall paintings in the chapel.

Schnebbelie's detailed rendering (Pl. LXXI–II) provides a visual record of these paintings as they appeared on the eve of their destruction, and Gough's description presents an unusually rich intertext. In lieu of the obtrusive moralizing prompted by Beauchamp's chapel, Gough here merely notes that the paintings had been 'vanishing apace' (186), and that the chapel had been used most recently 'as a cellar by one of the vergers' (191). The best-preserved of these paintings, clearly visible in Schnebbelie's interior view of the chapel as well as a detailed separate plate, shows a Dance of Death motif. Gough's

[33] In the next sentence we find Gough's most direct reference to Wyatt: 'But while we regret the injudicious havoc making in our finest cathedrals by one architect, we are happy to find what strict attention has been paid to the preservation of the fine specimens in the conventual church of St Bartholomew in Smithfield by another' (*SM* II.272). For Carter's 'professional' argument against removing the chapels, see 'Pursuits (LXIV)' 737. Gough himself supplies a footnote to document the 'gross violation' of the bodies in Salisbury, referencing 'Gent Mag LIX.873 and 1194'. For an account of the (then) recent renovation of St George's Chapel, Windsor, see *Vetusta Monumenta* III, pl. VII–IX. Gough's evident religious feeling in such passages may have fed speculation that he was a crypto-Catholic.

Fig. 6.3. *View of Hungerford chapel lately adjoining the North side of Salisbury Cathedral.* Engraving by James Basire from a drawing by Jacob Schnebbelie. *Sepulchral Monuments of Great Britain,* Vol. II (1796), Plate 71. Yale Center for British Art, Paul Mellon Collection.

detailed account of the human figure's dress, which can be verified point by point from the engraving, leads to the identification of this figure as the 'amorous esquire' (188). Fragments of other paintings suggest additional images of Death confronting other stock characters, including a 'baron or a pilgrim', as suggested by a fragmentary pair of shoes (not shown here) in a very different style from those of the esquire (189).

Gough depends heavily on Schnebbelie's work here, along with an earlier engraving, and he draws in another collaborator, 'Mr Wansey of Salisbury, FAS', whose interpretation of the paintings he quotes at length (*SM* II.187–89). Such use of correspondence is a vital antiquarian strategy for this period, as Sweet and other historians have shown. Gough gives his own elaborate description of the figure's dress (187), recalling the methodological emphasis on material culture, following Le Grand d'Aussy, in his preface. Wansey, for his part, reports the views of 'common people' on this painting and then goes on to supply its forgotten affiliation with the 'moral tale, or morality, as it was called in antient time'. In other words, the fifteenth-century painting was copied from older ones, which in turn recorded a type of 'spiritual

Fig. 6.4. *Inside new [reconstruction] of Bishop Beauchamp's Chapel, Salisbury Cathedral.* Engraving by James Basire from a drawing by Jacob Schnebbelie. *Sepulchral Monuments of Great Britain,* Vol. II (1796), part 2, p. 1 (headpiece to main text). Yale Center for British Art, Paul Mellon Collection.

masquerade' performed in churches much earlier; Gough takes the *terminus ad quem* for the iconography back to 1313 in a learned footnote (188n.). He quotes Wansey's somewhat different description of the figure's costume and his comparison of the two styles of shoe, thus building a history of customs and manners both for the mid-fifteenth-century moment of the painting and for the early fourteenth-century era of the iconography and the practice behind it.

Schnebbelie's interior view (Figure 6.3) sharpens the contrast between the Hungerford and Beauchamp Chapels. His view emphasizes this wall painting, located by Gough on the south wall of the chapel, thereby reinforcing the narrative. This interior view resembles the vignette of the Beauchamp chapel (Figure 6.4) in its perspective and in the architectural detail represented. In both views we see the central supporting arch in the middle ground and windows on two sides, somewhat more perpendicular in Beauchamp's slightly later chapel. The Hungerford plate, however, shows the dilapidated state of the paintings, fallen carved work littering the floor, and the tomb and shrine in disarray; the Beauchamp reconstruction, by contrast, shows the elaborate carving intact, along with an ornate vaulted ceiling and, most crucially, the tomb and altar in their places. The third, detailed Hungerford plate, showing the 'amorous esquire' painting flanked by two others (LXXII), ably captures the comparatively modern style of these paintings, with their realistic details of dress and a domestic interior. Gough notes more than once the 'large as life' scale of the paintings, underscoring the magnitude of their loss as well as the general opulence of these chapels, which is another major concern here and throughout the volume.

Despite his appeal for accurate historical judgments of Gothic architecture in the Beauchamp text, these Salisbury descriptions tend to favour the ethnographic aspect of Gough's archaeological project over its architectural aspect. The opulence of the chapels, too, is presented as an aspect of social history. The common feature linking these descriptions with other entries in the volume is a strong emphasis on heraldry. Heraldry was perhaps the most canonical subject for antiquaries and Gough presents it as a place where architecture intersects with 'the history of families' in the form of carved and painted heraldic devices. The most opulent monuments, tending to represent the most powerful aristocratic families, require the most space in the text—a two-page list of arms and their devices in the case of the Hungerford Chapel (*SM* II.189–91). The first plate in the sequence includes details of this heraldic carving, but it also gives the standard elevation and bird's-eye view (LXX), so it is surprising that Gough's text does not mention any other architectural features. His use of the past tense again suggests the chapel's destruction as a possible reason, although in this case there is a better visual record. Schnebbelie's elevation, a north-east view of the chapel, atypically sets the structure against a blank background in 'lapidoptery' fashion (to use Reeve's term), perhaps to allude to its amputation from the cathedral. The slightly earlier iron chapel or 'cage' of Hungerford's father, Walter, however, was not demolished but only moved to another part of the church during renovation work in 1779, and Gough gives some architectural description in this case (159). His descriptions of the Winchester monuments, however, include much more architectural detail.

Gough's accounts of the more elaborate chapels typically feature a discussion of their endowments, another aspect of social history. These descriptions resemble an ethnographic narrative of potlatch, with founders vying with each other to endow their chapels more richly. In this context, Bishop Beauchamp sounds more like Robert Browning's Roman bishop than the good English prelate: 'He gives all his lands and tenements in England, to the endowment of a chantry of four chaplains, to celebrate in said chapel' (273). Hungerford's widow, Margaret, invested two manors, three messuages, 508 acres of agricultural land, and other property to endow their chapel and 'a chantry of two priests to say and pray divine service for ever within it' (186).

For Gough and Schnebbelie, architectural history meant reconstructing the lived experience of the monuments more than it meant devising or applying any system of periodization. The importance of funerary customs for this approach accounts for the intense interest in human remains discovered by antiquaries and innovators alike. Perhaps the most famous image of Gough is the widely reproduced cartoon by Thomas Rowlandson, *Death and the Antiquaries* (1816), which recalls a rumour that Gough surreptitiously removed a finger from the body of Edward I when it was exhumed in Westminster Abbey in 1774. Rowlandson's connection between antiquarianism and the Dance of Death is especially apt in the case of the Hungerford chapel, the site of the painting detailed by Gough and Schnebbelie and also of unusually well-preserved remains, brought to light to prepare for the removal of the chapel. Gough does not name Schnebbelie in this account, but his date of 24 August 1789 coincides with the letter in which Schnebbelie informed him that the chapel was down and that Hungerford's 'body was this morning discovered' (*Letters* 84). The most striking part of the detailed description, which Gough owes to Schnebbelie's letter in whole or in part, is that 'the coffin was very dry, and had not the least smell, and the skeleton very entire, except the right foot, and some of the flesh remained under the upper ribs' (*SM* II.191). Gough gives the dean and bishop credit for their 'decency' on the occasion, a much more conciliatory gesture than the invective prompted later by the Beauchamp chapel. He pays particular attention to the way the body is positioned and to the grave clothes; these are both major themes of his introduction to the volume as well. The discovery of the body in Schnebbelie's presence gave Gough an occasion to explore his own ethnographic interest in funeral rites. This is one reason why the description reads so differently from the account of Wainflete's monument in Winchester (Figure 6.2), where Gough concentrates on 'the lighter and richer style' that distinguishes this 'second specimen of the improvement of Gothic architecture' from Cardinal Beaufort's monument.[34] At the same time, the elaborate fan vaulting and lofty pinnacles of both these chapels make them visually more spectacular than the Salisbury chapels.

[34] *VM* II, pl. XLVI–IX, p. 1 of accompanying text (signed R. G.). Here and in the text on Cardinal Beaufort, Gough makes basic observations on the period style, drawing on Walpole's *Anecdotes of Painting* for support. He describes the structure and ornaments of each chapel in some detail, glossing two particularly fine Schnebbelie/Basire plates of architectural detail (XLVIII–IX), but devotes the majority of his pages to the two prelates' political and ecclesiastical careers, especially that of Beaufort. This biographical attention to their public lives, via textual sources, is much closer to the grand historical narrative that he rejects in the preface to *Sepulchral Monuments* (I.i.4). In fact many of the texts in this work, too, indulge in public history; the biographies of Henry IV (*SM* II.31–4) and Henry V (esp. 59–62) form an interesting contrast in this regard.

In the context of the other fifteenth-century Salisbury monuments, the account of Hungerford's remains appears as part of the quotidian history—along with dress, lay piety, architectural practice, and other customs—for which, as Gough declares in his preface, these monuments provide the best, and in some cases the only, available source material. The upheaval associated with Wyatt's renovations was a kind of archaeology *manqué*, and the exhumations that it required promoted the ethnographic part of antiquarianism as much as the associated destruction stimulated architectural history and preservation. The main text of *Sepulchral Monuments* includes no images of human remains as striking as the engraving after Henry Emlyn of the body of Edward IV, discovered during renovation work in St George's Chapel and published in *Vetusta Monumenta* (III. pl. VII–IX). The composition of this skeleton and attendant artefacts deserves to be compared with Nolli's famous engraving of the Trebbia tomb in d'Hancarville (see Figure 5.4 above), which for some scholars represents the primal scene of archaeology. Another remarkable image in this vein appears in the introduction to *Sepulchral Monuments*, an engraving after S. H. Grimm of the remains of young St Hugh in Lincoln Cathedral, supposedly the remains of a boy crucified by Jews in the thirteenth century (II.lxviii–lxxiv).[35]

The preparation of bodies for burial is a major concern of both the first (1786) and the second (1799) introductions, the subject of rich scholarly narratives touching on last rites, methods of embalming, the treatment of various organs, and much more. Gough cites many 'instances of extraordinary preservation' (I.lxxvi–lxxxii), drawing particularly on saints' lives, which helps to explain his interest in Hungerford's remains. His larger body of evidence—largely textual—comes from numerous Greek and Roman sources, as well prehistoric English barrows and travel narratives ranging from Cook's voyages to John Filson's *The Discovery of Kentucky*.[36] In this mode his writing is closer to Octavien de Guasco's comparative discourse on ancient and primitive cultures than to d'Hancarville's art historical idiom.

Schnebbelie's archaeological bent manifested itself in his investigation of the subsurface of cathedral walls, of the structure of landscape, and (in at least one case) of Romano-British finds that he encountered on his travels to and from the architectural sites. On his departure from Salisbury in October 1789 he wrote to Gough that he was returning to London with 'the following pieces

[35] The long narrative concerning St Hugh provides a grim instance of Gough's credulity (always the bugbear of antiquaries) and anti-Semitism. Having witnessed the opening of the child-size coffin near the altar of the cathedral, he concludes: 'It is scarcely possible henceforth to doubt the truth of such acts' (II.lxxiv).

[36] The section on 'sepulture and its rites' is the longest in the second introduction (i–ccviii; cp. I.i–lxxvi). Both introductions also have substantial sections on 'habits'—clothing, hairstyle, ornament, and so forth (I.cxxiv–cxciv; II.ccviii–ccxxx). Gough cites the official account of Cook's third voyage for an example of 'extraordinary preservation' in Tahiti (I.lxxxi)

of antiquity found in the chalk clifts on the east side of Winchester[:] a Spur . . . a fibula, two buckles and a Roman coin which Mr Milner says you may either exhibit them to the Society or send them to Mr Urban' (*Letters* 89). Milner was one of Gough's allies in the Society of Antiquaries and another of Schnebbelie's patrons. He was with Schnebbelie at Winchester as well, and the finds came from new construction at a site where Schnebbelie had made drawings the previous summer, Magdalen Hospital. Gough chose to exhibit them, and he included Schnebbelie's drawing of these finds, with Milner's identification of the Roman tombs in which they were discovered, as an appendix to the plates of Magdalen's Hospital (*VM* III, pl. I–III, pp. 13–14).

Milner also supplied text for the first number of Schnebbelie's *Antiquaries' Museum*, which was in preparation by the end of 1789 and appeared in January 1791. For one of the plates in this number Schnebbelie drew again on the work he did at Winchester, publishing one of his discoveries originally announced in a letter to Gough in June 1788. In this letter (partially quoted above) he reports 'having discovered some figures large as life finely painted on the wall in the N. Transept entirely covered over with a thick coat of plaister and in a small chapel near them . . . all finely painted in a very free bold stile . . . and having been hid for many years. Mr Carter has not seen them the height of the figures are above three feet. Mr Milner has just seen them and is much surprised they are so perfect and will give an account of the place etc' (*Letters* 75). Milner's account also confirms Schnebbelie's archaeological curiosity and imagination, which left a strong imprint on the account of Salisbury and other parts of *Sepulchral Monuments.*

6.4 *THE ANTIQUARIES' MUSEUM*

The Antiquaries' Museum, though unfinished, stands out for its detailed representations of little-known monuments and for its substantial letterpress descriptions. In these respects it resembles a low-budget *Vetusta Monumenta* much more than similar-sounding volumes of topographical prints such as Paul Sandby's *Virtuosi's Museum* (1778). Schnebbelie's work does include some topographical views of buildings, like Sandby's, but the Winchester plate—showing a detail of one of the wall paintings from the cathedral—is more typical. Gough's influence was key: he introduced Schnebbelie to Milner in Winchester, but he also declined to purchase these particular drawings, which Schnebbelie offered in the 1788 letter. The images therefore contributed to his own stock, while the scholarly authority and prose of Milner provided a template for the work as a whole, which Schnebbelie advertised as pairing his hand-coloured plates with texts 'by Gentlemen of the first Literary Abilities'.

The published text shows that Milner admired and encouraged Schnebbelie's scholarship from the first: 'his insatiable curiosity and active ingenuity, to which the writer of this was witness, led him to make many discoveries in that inexhaustible mine of antiquarian researches [i.e. the cathedral], which had escaped the attention of former investigators'.[37] Milner, a Catholic priest, sounds the note of decline that we also hear in Gough's account of the Salisbury chapels, noting that the north transept, 'the chief scene of his discoveries', was only being used to store building materials, though the paintings and other remains provided evidence of former sacred uses. Milner praises Schnebbelie's keen attention to this neglected evidence:

> Though the subjects in some parts of these paintings were sufficiently discernible to the curious observer, yet Mr. Schnebbelie has the merit of bringing many more of these to light, and of rendering others much more visible and certain. The most singular discovery, however, in the opinion of the writer, was that of a double coat of painting in this part of the Cathedral, containing a series of subjects, one underneath the other, which had evidently been painted at different periods, and of which the undermost was far the freshest and the most perfect. (*AM* 2)

The plate (Figure 6.5) shows two figures from this earlier (twelfth-century) layer, taken from a group of saints, though these two 'seem to represent some holy persons of the ancient law, by the philacteries which they bear'. An added note by Gough correctly identifies these figures as belonging to a series of Old Testament prophets.

Throughout this period Schnebbelie was consciously competing against John Carter. Carter, twelve years his senior, was more established. Not only was Carter the first (and, as it turned out, the only) draftsman officially appointed by the Society of Antiquaries, he also launched his own serial publication of antiquarian etchings in 1780. By the time Schnebbelie made contact with Gough in 1786—stepping into the breach left by Carter's conflict with the Society—Carter had collected the first fourteen numbers of *Specimens of the Ancient Sculpture and Painting Now Remaining in this Kingdom* into a volume published with an impressive list of subscribers. That same year Carter launched another series, *Views of Ancient Buildings in England* (1786–93), this time writing all the text himself. Schnebbelie, perhaps consciously emulating this act of scholarly self-assertion, chose to write some of his own text for *The Antiquaries' Museum*. Carter was on his mind as early as 1788, as noted in his letter on the Winchester wall paintings. Not coincidentally, Carter had drawn other wall paintings in the cathedral, in St Mary's Chapel on the east end, in 1784, and published them in *Specimens* with an explanatory letter by the same

[37] 'Proposals for publishing a new work' (*Letters* 137); *Antiquaries' Museum* I, pl. III, accompanying text, p. 1. The latter is cited parenthetically in the text hereafter as *AM*; each essay is paginated separately. [The title page engraved in 1800 lacks the apostrophe following 'Antiquaries', but as it appears in the other versions I have retained it here.]

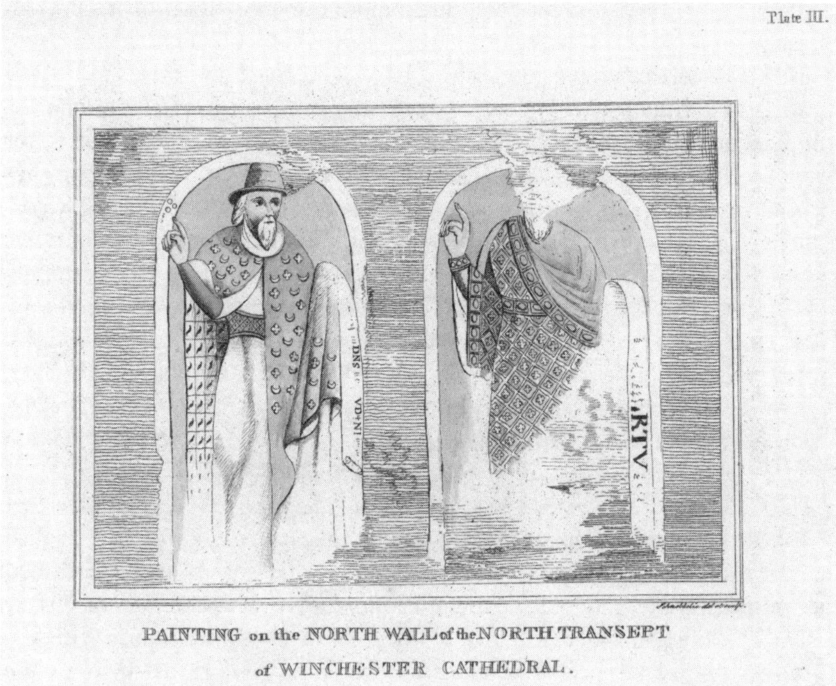

Fig. 6.5. *Wall Paintings, North Transept, Winchester Cathedral.* Hand-coloured etching by Jacob Schnebbelie from his own drawing. *The Antiquaries' Museum* (1791), Number I, Plate III. Yale Center for British Art, Paul Mellon Collection.

patron, John Milner. By 1790 Carter was working for the Society again, recording monuments in St Stephen's Chapel (Westminster Palace) for what became the Cathedral Series, a major venue for his work.

Schnebbelie was literally following in Carter's footsteps that summer when he recorded the chapel of Henry V in Westminster Abbey for *Sepulchral Monuments.* Carter first drew this monument in 1786, but Schnebbelie's letter to Gough shows that Carter was examining it closely again in preparation for the four detailed etchings of its sculptures that he published in *Specimens* (1791–92): 'I found Mr Carter had been there the day before and had permission to have a scaffold erected for him' (*Letters* 115). Schnebbelie made good use of the scaffold for the plates in *Sepulchral Monuments* (XXV–VI). He also betrayed his knowledge of his competitor's work, quoting Carter's suggestion of a 'swelling or wen' on one of the figures—'under the Chin of Henry (Vth as he supposes)'—and dismissing it as 'nothing but whitewash which run down the face (when the figures inside were whitewashed probably just before the coronation of his present Majesty)'.[38]

In the last year of his life, Schnebbelie continued to travel and make drawings for Gough and Nichols, but his main focus was *The Antiquaries' Museum*. 'It would be the making of me', he writes, in a letter asking for Nichols's help with publication (*Letters* 121v.). The present chapter has focused primarily on Schnebbelie's work for *Sepulchral Monuments*, a gorgeous and encyclopedic book that offers ample evidence of the collaborative knowledge work that is my subject in *Sciences of Antiquity*. It is appropriate now to close the discussion of Schnebbelie with a glance at *The Antiquaries' Museum*, not least because it represents a new kind of artist-led collaborative project. The full title is a good place to start: *The Antiquaries' Museum, Illustrating the Antient Architecture, Painting, and Sculpture, of Great Britain, from the Time of the Saxons to the Introduction of the Grecian and Roman Architecture by Inigo Jones in the Reign of King James I.*

This title gestures toward the narrative of rise, progress, and decline that British antiquaries adopted from classical antiquarianism in order to establish their own discipline, a narrative that promoted what Susan Pearce has called the 'typological method' of 'focusing on comparative characteristics of surface design'.[39] Gough decided to close his history of Gothic antiquity with the fifteenth century because it produced work so 'advanced' as to merit the designation 'Gothic architecture', a term that he self-consciously proposes as a neologism (*SM* II.i.1). Following Walpole, he decries the 'barbarous mixture' or 'mongrel species' of architecture that he dates from the later years of Henry VIII and that was consolidated—as Schnebbelie's title explicitly registers—by the classicizing innovations of Inigo Jones (*SM* I.civ).

In his *Gentleman's Magazine* column, Carter is still more strident in rejecting the (late-arriving) classical influence in English architecture, but even in his scholarly work he does not attempt a systematic chronology of style for the canonical period preceding Jones. In the introduction to *Specimens of Ancient Sculpture and Painting*, Carter disavows any intention of presenting a 'complete history' of monuments in the 'chronological order' familiar from works on classical and even French antiquity. He also hints at a reason, suggesting that British antiquities, unlike those of Greece and Rome, are nowhere near the point of being 'exhausted'. Like Carter's *Specimens*, *The Antiquaries' Museum* is not a systematic or chronological work and it does not

[38] Schnebbelie goes on to say sarcastically that the whitewash under the chin must have 'hung like a drop it may be said with equal propriety as several of the figures have Blisters raised by the Damp and then the sun shining on them that they have pimpled faces' (115v.). For the passage in Carter, see *Specimens* II.34. The plates and commentary appeared in four successive numbers of the work, and the relatively brief text (Carter's own) appears on pp. 34, 36, 39, and 45–6 of the second volume (published in book form in 1794). For Carter's Winchester wall paintings, see *Specimens* I.40–1. For Gough's account of this chapel, see *SM* II.59–71.

[39] Pearce, 'Antiquaries and the Interpretation of Ancient Objects' 159; cp. Reeve, 'Jacob Schnebbelie' 81.

address the problems of periodization and definition that Milner finally began to solve by delineating three stages of 'pointed architecture'.[40] The overarching rise and progress narrative forms an important subtext for *The Antiquaries' Museum* inasmuch as it justifies the importance of the specimens. At the same time, the antiquarian method, by virtue of the traditional distinction between history and antiquities, exempts any individual project of documentation from direct involvement in a grand narrative.

The 1000-year period defined by these antiquaries as a privileged national past affords a broad scope to documentary projects such as *The Antiquaries' Museum*, and the miscellaneous nature of Schnebbelie's book is typical for the genre rather than a contingent effect of its largely posthumous construction. As Gough explains in the biographical memoir that prefaces the posthumous volume, 'just before his death, he had completed the third number of it, and left behind him several drawings intended to form the first volume. These materials falling into the hands of his friends Mr. Gough and Mr. Nichols, they in justice to his memory and to the publick, determined to complete it. The several subjects were illustrated with accounts by Mr. Gough, Mr. Hawkins, Mr. Milner, Dr. Pegge, Mr. Nichols, and other friends' (*AM* iii). Besides these twelve numbers, comprising four plates each, there are several added prints from the *Gentleman's Magazine* and elsewhere, making a total of fifty-five plates. Two of the texts can be clearly attributed to Schnebbelie himself, and a few of the unsigned texts may be his, though the majority are certainly by Gough, with the others supplying one or two each. Unsurprisingly, churches are the most common subject (twenty-seven plates), with the style and point of view ranging from exterior and interior views to architectural detail, carving, wall paintings, inscriptions, and even a few monuments engraved in the grand style of Gough's volumes. Apart from the latter three (12.2, 12.3, 14.1), however, these plates are etched, in some cases aquatinted and/or hand-coloured, and they range in quality from highly finished to fragmentary.

Gough informs us that Schnebbelie 'had early in 1791 made himself master of the art of aquatinting', and he certainly etched the completed numbers himself. The rest have no engraver's signature, but many were probably etched by another hand. Several of the exterior views are topographical, not only views of churches but also of other built works including a medieval manor, a park, and a tower. Topographical drawing remained part of Schnebbelie's repertoire; in his last letter from Salisbury we find him wishing for time to capture the 'most beautiful picturesque scenes' of the Devil's Punch Bowl (Surrey) on the way back to town (*Letters* 89v.). Topographical pursuits had

[40] For Carter, see *Specimens* I. ii; on Milner's contribution, see Lewis, *Gothic Revival* 48. Carter echoes Gough's view of neoclassicism as 'barbarous' in many of his essays. Gough believed that Schnebbelie was, in fact, at work on a systematic history of Gothic architecture at the time of his death.

taken him to the Earl of Leicester's estate in 1786, where his whole antiquarian career had been launched by a chance meeting with the proprietor.

This commercial dimension of Schnebbelie's art links him to the popularizing antiquarian publications of the day, such as *The Virtuosi's Museum* and the especially influential publications of Francis Grose.[41] Schnebbelie's concept for his *Museum*, quite faithfully completed by Gough and Nichols, was to intersperse these views among a larger number of scientific representations influenced by his more scholarly work with Gough. In addition to objects from churches, this group of subjects includes ten illustrations from a MS life of St Guthlac in the Cottonian Library, painted glass, vestments, devotional figures, seals, and 'antient keys'.

The first subject, St Sepulchre Church in Northampton, is one of nine Northamptonshire churches included in the *Museum*. Schnebbelie's career itinerary in Northamptonshire (1789–91) included one of the Eleanor Crosses (Hardingstone) and three churches for *Sepulchral Monuments*. He also passed through many times on his way to commissions from Nichols in Leicestershire and to other points north. The geographical focus on Northamptonshire reflects these opportunities to gather notes and observations there, and the particular appeal of St Sepulchre may have been that it differs so sharply from the later Gothic monuments to which Schnebbelie was most frequently assigned. In his commentary he emphasizes the round plan and early date of the church, both of which link it historically to the first Crusade. Although the first recorded reference is 1127, Schnebbelie adduces architectural evidence that the church 'must have been built soon after, if not prior to, the first crusade', a supposition confirmed by the presently accepted date of 1099.[42]

The interior view (Figure 6.6), in spite of technical constraints, shows some of the distinctive features now associated with the Norman period, including the variously shaped capitals and bases of the columns. Schnebbelie draws attention to some 'curious piece[s] of very rude sculpture' in the church included in his second plate (not shown here), using language commonly linked with what was then still termed 'Saxon' design in learned and popular texts alike. At the same time etching is a 'ruder' technique than line engraving, for which Schnebbelie had neither the time nor the money. The contrast between Norman and Gothic is thus repeated by the contrast between Schnebbelie's etching and Basire's lavish engraving. The admiring figure in the foreground marks a related difference, since this topographical convention, still accepted in William Stukeley's time, was banished from antiquarian

[41] On Grose and popularization, see Bending, 'Every Man Is Naturally an Antiquarian'.

[42] *AM* 1, pl. I–II, p. 2. Schnebbelie did not have access to the nineteenth-century (and current) conjecture that Earl Simon de St Liz, who also built Northampton Castle, founded the church in thanksgiving after returning successfully from the first crusade. One late-nineteenth-century account concurs with his date on purely stylistic grounds, however. See Cox and Serjeantson, *History of the Church of the Holy Sepulchre* 23.

Fig. 6.6. *St Sepulchre's Church, Northampton*. Etching by Jacob Schnebbelie from his own drawing. *The Antiquaries' Museum* (1791), Number I, Plate I. Yale Center for British Art, Paul Mellon Collection.

images produced under Gough's supervision. The heavy shadows and dramatic shaft of light at the rear are reminiscent of the interior view of Magdalen Chapel (*VM* III, pl. I–II), except that the effect, in the absence of engraved architectural details, is less spectacular.

The first two plates, with Schnebbelie's commentary, make a good test for Reeve's assessment of him as a 'seasoned amateur' in the field of antiquarianism.

Schnebbelie's archaeological bent and keen eye do not make up for the lack of book-learning apparent in his work, and so 'seasoned amateur' seems reasonably apt in the context of modern usage. In the context of late eighteenth-century usage, however, it is noteworthy that Schnebbelie was a *more* seasoned amateur than most of the Fellows of the Society of Antiquaries. In this context, the term 'amateur' is in no way pejorative, except that too many of the Fellows (in Gough's view) were lukewarm in their love of antiquities. Neither these nor most of the truly enthusiastic antiquaries would have had as much field experience as Schnebbelie amassed in just five years, so 'seasoned' is a very important qualifier in Reeve's formulation. Moreover, Gough and other commentators did not scruple to refer to antiquarianism as Schnebbelie's 'profession', which further complicates Reeve's implicit distinction between the amateur and the professional.

My emphasis in this chapter has been on knowledge gained in the field, not as a professional qualification in the modern sense but as an indispensable contribution to the empirical knowledge of antiquity as it emerged at the high point of copper engraving technology—slightly 'before the era... [of] true mass production' of images, as Arnold and Bending have noted.[43] Schnebbelie's work was knowledge work in the sense that the accuracy of state-of-the-art representations of Gothic architecture depended on him, and others like him, for the expertise they gained while making observations in the field, in lieu of formal training or learned guidance. In this way he resembles the other knowledge workers detailed in this study, with their mixed motives of intellectual curiosity, economic need, and social ambition, and this group in turn is a common factor among the sciences of antiquity.

Schnebbelie did not have the resources to give his own work the documentary richness of *Sepulchral Monuments* or the Cathedral Series. And even these state-of-the-art works were soon superseded: their success, as sweet observes 'is perhaps best seen in the rapidity with which [their approach] appeared to be outdated' (*Antiquaries* 275). Schnebbelie initially hoped to find an engraver for his project, in part surely because he knew that from a documentary or scientific point of view engraving was the best medium, and in part perhaps to increase potential profits. Many popularizing works, such as those of Francis Grose, also featured engravings. Carter, on the other hand, found etching the best medium for producing a large volume of scholarly antiquarian prints in the context of a one-man operation, and Schnebbelie had some exposure to etching (aquatint in particular) through his collaboration with Francis Jukes and his later involvement with James Moore's *Monastic Remains*. Schnebbelie first mentions the project in a hasty, unpunctuated letter to Gough in January 1790: 'understanding there is a council to be held next week I shall beg the favor of you to ask permission to have my work dedicated to the society I shall send you in time

[43] *Tracing Architecture* 1. For Reeve's formulation, see 'Jacob Schnebbelie' 82.

specimens of the plates the letterpress Mr Nichols has now in hand the proof sheets I shall esteem favour if you will oblige me to look over and anything amiss to strike out' (*Letters* 104). Schnebbelie's next move was to try to secure a loan from Nichols to hire an engraver, assuring him that 'it would be the making of me' (121v.). Although Gough 'was pleased to say he thought my intended work would make me more known than what I had done for the Society of Antiquaries' (121)—high praise indeed—Nichols apparently did not come through, perhaps because he knew that £100 would not suffice. From Gough's memoir we know that Schnebbelie spend the ensuing months mastering 'the art of aquatinting', and all these circumstances together explain the additional year's delay between the initial letter to Gough and the publication of the first number in January 1791.[44]

Along with the title page and texts, Nichols also printed a proposal (20 January 1791) to promote future numbers of the work, which Schnebbelie submitted to the Council of the Society of Antiquaries for consideration. Having taken such care beforehand to secure the approval of Gough and 'difft Gentlemen with whom I have conversed on the subject' (*Letters* 121), Schnebbelie was outraged by the negative reaction communicated by the Vice President of the Society, Owen Salusbury Brereton, who was presiding: 'Nothing can equal my astonishment—Mr Brereton . . . having declared from the chair that I was not draftsman to the Society of Antiquaries . . . it was to you sir I thought I was indebted for that appointment and little did I think I was imposed on . . . when I laid before them the intended title page they approved of it at least you told me so. Believe me sir, it's more than I can bear' (*Letters* 46). The draft of Gough's reply—the most legible and complete of those included in the archive—says that he cannot remember having seen the proposed dedication and also argues that Schnebbelie never was officially appointed by the society (143). Schnebbelie's reply of 5 February challenges Gough sharply on both points and responds with equal alacrity to his paternalism: 'It is very true that the manner in which I am paid by the Society proves I am not the draftsman but a collector and disposer of drawings to the Society, a situation which for the future I must beg leave to decline. After the reprimand I recd from the Chair you surely sir cannot expect to see me in my place as you are pleased to call it. Your letter proves I have no place in the Society' (141–42). Rather than a fellowship in the society (which John Carter was gratified to receive in 1795), Schnebbelie's disappointed reaction suggests

[44] The printed proposal, included with the archive (*Letters* 137), indicates that he spent some of the intervening time thinking about how to market the work. One line specifies, 'No Money to be paid but on Delivery of the Numbers', indicating that he would not collect subscription money in advance, as Carter (and d'Hancarville) had done. The advertised price of five shillings a number is relatively high.

that he envisioned a salaried position as the fruit of his efforts submitting piecework for *Sepulchral Monuments* and *Vetusta Monumenta*.[45]

As Reeve points out, this dramatic conflict remains 'one of the problems in Schnebbelie's career that cannot be readily answered', in part because it was resolved so fully and so quickly—yet few traces remain of how the resolution was achieved. It is a compelling problem, in part because of Schnebbelie's eloquence in making these charges and because of the efforts made by Gough to rebut them. The controversy is significant for my purposes because it shows that the working relationship between Gough and Schnebbelie was an unfamiliar kind of relationship in which social decorum—at least in this instance—gave place to the claims of collaboration and a certain kind of professionalism. I have presented similar kinds of conflicts in the other chapters—another such conflict, between Thomas Webster and the Geological Society, is still to come—but this episode of Schnebbelie's dedication is a particularly vivid case of social ambiguity resulting from an intellectual division of labour. Empirically, the main issue is Schnebbelie's first letter concerning the project, in which, as we have seen, he asks Gough to vet his proofs and bring the project to the attention of the Council.

This evidence from Gough's own archive speaks against his claim that 'I do not recollect . . . any such title' (*Letters* 143). Part of the confusion is that Schnebbelie's main concern was to secure permission to dedicate his work to the Society, while their main concern was to stop him from using the title 'Draftsman to the Society of Antiquaries'. In reply to Gough's letter, which insinuates that the title was not included in the material presented the previous year, Schnebbelie pointedly quotes the earliest version of the dedication, which he claims was formulated in exactly the same words: 'When I gave you the MS Title requesting the favour of you to ask for permission to dedicate the work to the society and to know whether they had any objection to the title—you told me the council had given me permission and that they had no objection to the title where was written the following words and in the same manner . . . ' (141–141v.). The 1790 letter does not explicitly refer to the title, but it must have been present in the proofs that Schnebbelie submitted, and it seems clear that he genuinely believed he was authorized to use it.[46]

The disagreement over Schnebbelie's title cannot be resolved—though it is hard not to sympathize with him—but it does shed light on the ambiguities surrounding research-for-hire at this moment in the evolution of the learned

[45] Fol. 46, which is undated, is out of place in the archive. Actually it is Schnebbelie's reply to Gough's reply (not extant) to Fol. 140. 141–2 is Schnebbelie's reply to the first three-quarters of 143, written on 4 February; the last quarter of 143, drafted 11 February, is Gough's final reply, promising to make everything right. The learned societies did generally have salaried officers, although the stipend paid to the secretaries of the Royal and Antiquarian Societies (for instance) was held to be nominal. For more on this issue see my next chapter on Thomas Webster, the secretary of the Geological Society. Reeve's view is that Schnebbelie may have been hoping for admission to fellowship rather than a stipend ('Jacob Schnebbelie' 74).

[46] Schnebbelie also insists that the title has some intrinsic significance, reminding Gough that he had introduced him as 'Draughtsman &c' to Milner and others, 'which I look on as a confirmation of that empty name as you are pleased to style it'.

societies and of European scholarship in general. Both Gough and Schnebbelie review the original circumstances of Schnebbelie's appointment and the economic terms on which he operates within the Society. The first letter of this sequence, written four days after the proposals were printed, already shows some anxiety over the council's delay in approving the dedication, which Schnebbelie had reason to count on (*Letters* 140). The next letter, reporting the 'cut' he received from Brereton at the next council meeting, is undated and incorrectly numbered (46), though it necessarily falls between 24 January and 4 February 1791. This gap makes it impossible to reconstruct a complete sequence of events, but it is clear that Gough did not intervene on Schnebbelie's behalf in this period. Schnebbelie charges Gough with enabling him in his belief that he was Draftsman to the Society of Antiquaries—'little did I think I was imposed on' (46)—and perhaps, as Reeve suggests, Gough knowingly led him on to serve his own purposes. Schnebbelie's next concern has to do with status difference: 'I have devoted my time to them been moderate in my charges & perhaps too humble in deportment I must say I little expected so severe a cut.' By pointing out that he gave up his work as drawing master at Westminster School for them, Schnebbelie seems to suggest that he expected not only financial gain but some level of social recognition—or at least a status not inferior to that of a drawing master. He closes appropriately by requesting payment on outstanding bills.

The debate that follows turns on the question whether a 'stipend' or salary should be attached to the kind of work that Schnebbelie is doing, which has implications relative not only to these outstanding bills but also to the conduct of a learned society and scientific work. Gough's reply of 4 February alludes to a council meeting of that morning, which is probably the one in question. He points out that the Society had resolved never again to make an exclusive appointment like Carter's because of the problems it caused, and notes that even Basire is not officially Engraver to the Society (*Letters* 143–143v.). All this is true, but Schnebbelie may or may not have been advised of it as Gough claims here. Gough's second argument is more to the point: he is not Draftsman to the Society because 'the manner in which you are paid for your drawings proves the contrary'—but this mistake about the title has not prevented him from being 'continually employed', a phrase Schnebbelie uses in his own letter. Gough's closing argument—that Schnebbelie's own 'favourite pursuit' is best served by using standards of merit rather than a sinecure to promote the best visual documentation—is also of particular interest because it enlists Schnebbelie on the side of science, in effect compensating him for his loss of status. Schnebbelie's reply acknowledges these objections in part: 'At the first Council I attended I was informed I was draughtsman to the Society but not to have the stipend allowed to the former ones (their names being still on the books) and this merely on account of their improper conduct—I agreed to attend the Society on these terms' (141).[47]

Having recognized that he is merely an object of patronage to the Society, Schnebbelie concedes the point that 'Draughtsman to the Society of Antiquaries' is an 'assumed title': 'at the compleating of first Vol. I shall entirely cancel the present title page and give an engraved one—and rest more on the merit of the ensuing plates than any patronage' (*Letters* 141v.). Schnebbelie bitterly laments his 'loss of time', having attended 'night after night & not had a single drawing to make' (142). The conclusion suggests not only that his arrangement with Gough concerning the meetings may have been less mutual than Gough promised. It also suggests, more fundamentally, that by reducing his status the Council has effectively challenged his right to attend meetings at all.[48] In Schnebbelie's mind, at least, the title and the salary (or prospect of a salary) associated with it signify a kind of scientific entitlement that to some degree compensates for his social exclusion from fellowship on the basis of class. (The cases of Vertue and Carter show that this exclusion was not absolute, but Carter in particular is an exception that proves the rule.)

Gough tacitly recognized this logic when he urged Schnebbelie to accept whatever fiscal arrangement would best serve the interests of science, but he was either unable or unwilling to act on the fundamental problem. He does not appear to have insisted that Schnebbelie continue attending meetings on his behalf, and at least some of their interaction in 1791 focused on *The Antiquaries' Museum*, which is the subject of the next letter in the archive (144) following a three-month gap. Gough did reply to Schnebbelie's last letter on the subject of the dedication, promising on 11 February that 'everything will be set right at the Council on Thursday next'. The scrawled draft, which Gough appended to the draft of his previous letter (143), is not entirely legible, but it appears that Leicester had promised to preside at the next council meeting and reconsider Schnebbelie's use of the title in his printed proposals.[49]

[47] Gough had written: 'You concur eve[r] in your favourite pursuit with a Socy who make every exertion to preserve its objects and for that very reason do not confine themselves to the efforts of a single artist but are ready to patronize all who [join and plan?] with equal merit' (*Letters* 143v.). Schnebbelie may have expected that he would receive the 'stipend' once the other names (Carter's and Thomas Underwood's) were removed from the books. According to Reeve's account of the dedication affair, Gough may have 'exercised his authority and led an aspiring young artist to believe that he held a prestigious position that he did not' actually hold ('Jacob Schnebbelie' 74).

[48] 'Perhaps were I to attend the Society as usual I may be called before the next Council and told there is no order for my attendance' (*Letters* 142). Schnebbelie's spirited rejection of patronage in favour of merit recalls Samuel Johnson's famous critique of patronage in the preface to his *Dictionary*.

[49] A final ironic twist, which goes unmentioned in the archive, is that Schnebbelie actually is identified by the contested title on the title page of a work to which he contributed some etchings, James Moore's *Monastic Remains* (1791). Moore, FAS, must have held a more liberal view than the council's concerning Schnebbelie's use of the title. The irony is that if Schnebbelie had not

Whatever the final decision may have been, it was poetic justice in effect, since the contested title appeared widely in print only after Schnebbelie's death, just one year after this final meeting (21 February 1792). An obituary in the *Gentleman's Magazine* announced that he had died of a rheumatic fever occasioned 'by too intense an application to his professional engagements'. This obituary refers to Schnebbelie as 'Draughtsman to the Society of Antiquaries', as do the first published review of the *Museum* (1792) and the posthumous edition of the completed work (1800). The last twenty letters in the archive, from Schnebbelie's widow, Caroline, to Gough and Nichols, document the rapid social decline of a family once on the verge of respectability. The Society took up a collection for the widow and spent £50 to set her up with a print shop, which sputtered and eventually failed.[50] After years of financial struggle, Caroline Schnebbelie wrote a letter in 1800 indicating that Nichols had sent her proof of the biographical memoir prefixed to *The Antiquaries' Museum*. In her reply she begs Nichols to delete a phrase that she says will derail the pending admission of her youngest son to Christ's Hospital: 'My maternal feelings have been torturd to the extreme there needs not the addition of publishing it to the world' (*Letters* 191). The published text, though Nichols may have changed it, still shows the source of Caroline's objection: 'his surviving family', Nichols reports, 'received a handsome relief from the Society to which he was draughtsman' (*AM* viii). Her anxiety that this financial disclosure might disqualify her son seems reasonable, since Christ's Hospital was a charity school.[51] In the course of the 1790s she issued several more numbers of *The Antiquaries' Museum* (*Letters* 175) and collected

been so scrupulous about the title page of his own book, the whole issue might have gone unnoticed. Then again, his use of the title in this publication—assuming that the title page was printed after the second council meeting—could prove that the council did reverse its decision and authorize him to use the title after all.

50 For the obituary (almost surely by Gough), see 'Schnebbelie'. The last group of letters begins with a letter to Gough from the engraver Barak Longmate (*Letters* 168). A letter from Gough shows that some money remained in trust even after £50 had been paid out to stock the shop in October 1792 (172). My evidence that the shop was a print shop is conjectural, based on Longmate's involvement and on the family's continued efforts to earn income by colouring and selling prints after the shop had failed (175, 180, 182, 184–7). Apparently Caroline Schnebbelie tried at least once more with the shop after losing her first lease in June 1794 (175). As of December 1795 the shop was earning no more than one shilling per day (182).

51 The youngest boy, born a few days after his father's death, had been placed in an institution because she could not support him (191). According to Longmate, a Dr Layard initially secured places in grammar school for the two older boys (168), but they both went on to a series of apprenticeships disrupted by financial and disciplinary troubles. Placing them is a frequent topic of her letters to Gough, who evidently helped a good deal (175, 177, 180). Jacob Christopher Schnebbelie, his father's namesake, ran away to sea in the middle of his apprenticeship as a stationer (179) but eventually found success in his father's original career of culinary arts (188). See the 1805 edition of W. A. Henderson's *The Housekeeper's Instructor*, 'revised and considerably improved' by Schnebbelie Jr, with his engraved portrait as frontispiece. Robert Bremmel Schnebbelie, as mentioned above, followed his father's second career as antiquarian draftsman. Another boy and a girl died when the family was visited by illness in 1791 (133, 136).

a number of unpublished drawings of Schnebbelie's from his patrons, so they could be printed (sometimes coloured) and sold (181, 182, 184).

6.5 EPILOGUE: JOHN CARTER AND THE SCIENCE OF PRESERVATION

John Carter was another 'practical antiquary' who was rebuffed by the Society of Antiquaries. This 'cut', as Schnebbelie would have called it, or series of cuts, was more dramatic, since Carter had just been elected to fellowship, and the conflict developed into a pitched battle that ended with Gough's resignation. This story, connected with the *querelle de Wyatt*, is better known, and I have already given a brief outline. In the course of this conflict, Carter was condemned as a 'servant' trying to 'dictate to his masters', but Gough continued to support him until they both lost the cause. Gough was driven by his anxiety that 'all history of architecture as a science among us, is totally done away', and he recognized Carter's activism as the antidote to this forgetting.[52] If James Essex was a practicing 'master of this science', as Gough put it in his obituary of the Cambridge architect, Carter's architectural training provided a scientific basis for the documentation and defence of medieval antiquities. Gough and Carter belonged to a small but vocal minority of antiquaries who invoked the name of science in their critique of modernizing interventions by Wyatt and other 'improvers' of medieval built works.

The science of preservation, therefore, recognizes medieval building practices as architecture and demands an empirically documented authenticity. 'Science' in a more modern sense can also be invoked to explain why Carter was the only antiquary to earn a prominent place in histories of the Gothic Revival from A. W. N. Pugin to the present day. Reeve, for example, points to the 'earliest scale drawings produced by artists such as John Carter shortly after 1800, which represent the first truly scientific and analytic approach to the study of medieval monuments'. The scientific standard in this case is that of archaeology, which privileges physical over textual evidence. J. Mordaunt Crook is more skeptical of Carter's architectural understanding but still praises his 'dogmatic commitment to Gothic archaeology'. Noting that Pugin, the architect most strongly identified with the Gothic Revival, 'acknowledged Carter's role as proto-prophet', Crook declares: 'Carter played John the Baptist to Pugin's Christ'.[53]

[52] *SM* II.i.7. On the condemnation of Carter by Lord Leicester, Sir Joseph Banks, and others, see *The Diary of Joseph Farington*, quoted in Crook, *John Carter* 27. Carter was made a full Fellow of the Society of Antiquaries in 1795 in recognition of the quality of his work in the cathedral series (*John Carter* 11).

[53] Reeve, 'Jacob Schnebbelie' 80; Crook, *John Carter* 62, 66.

Carter brought the project of preservation fully into the public sphere in his writing for the *Gentleman's Magazine* from 1798–1817. From the series title, *Pursuits of Architectural Innovation,* to the dramatic content and pseudonymous signature, 'An Architect', Carter's essays show him to be a skilful polemicist, which Gough was not and Schnebbelie did not try to be. He shared both Gough's antiquarian zeal and Schnebbelie's gift for meticulous visual documentation. Carter's antiquarian publications, perhaps especially his volumes for the Antiquaries' Cathedral Series and *The Ancient Architecture of England,* have enjoyed continuing recognition, while his relatively small body of built works—including St Peter's Roman Catholic Chapel (1792), built for Milner's congregation in Winchester and dismissed by Crook as 'jejune'—has not. Carter's magazine persona appeals rhetorically to both his practical and his theoretical expertise, but his patient legwork provided the real foundation for his incendiary investigative journalism.

Carter travelled all over England and visited the sites to document both the true architecture of antiquity and the 'ravages of innovation'. His commissions from the Society of Antiquaries after 1790 put him in a position to observe closely over a long period of time, most famously at Durham Cathedral, where his published exposés had the effect of halting Wyatt's planned improvements in mid-stride, and where he is fondly remembered especially for saving the Galilee Chapel at the west end. In these essays Carter skillfully sets off 'relics' that may yet be saved against the antiquities that have already fallen victim to 'the destroyer': 'Salisbury's and Lichfield's high-altar screens are no more; Durham's endures still'. Another arresting juxtaposition pits County Durham's tough Carboniferous sandstone against Italian marble, 'our "grit-stones" portraying an Edward and Eleonora' versus 'your armless Venuses, your legless Herculeses'.[54]

Then as now, the word 'science' has many valences, and its currency in the circle of artists and antiquaries around Gough has both synchronic and diachronic ramifications. John Pinkerton described 'National History' in 1788 as 'the only science neglected' in Britain. Edward King, briefly President of the Society of Antiquaries in 1784, defined antiquarianism in more Baconian terms as a 'science of antiquities' in a published address to the Society. King promoted Carter's initial appointment and emphasized the collection of drawings as essential to antiquarian science. Modern archaeological assessments of the Society of Antiquaries in this era, however, tend to cite John Frere's short, descriptive 1797 paper on flint weapons found at Hoxne as a 'landmark'

[54] Carter, 'Durham Cathedral' 402, 231. For a good overview of Carter's work at Durham see Sweet, *Antiquaries* 289–90; Sweet also reproduces Carter's elevation of the West Front from the Cathedral Series (Pl. 35). While later antiquaries such as John Britton carried on in the domain of documentation—and popularization, from which Carter refrained—Carter's journalistic work was unparalleled.

achievement because it situates artefacts in relation to the surrounding stratigraphy.[55] From this point of view, artefacts are paramount. Although the work of Gough and Schnebbelie has an archaeological aspect (as in the case of the Hungerford remains in Salisbury), they are more apt to think of architecture as their science. The scientific element in architectural history is apparent from Lewis's claim that the Gothic Revival proper begins only with the early Victorian recognition of Gothic architecture as a system of structural engineering. Gough's archaeological work, in part because it was conducted indirectly through Schnebbelie and other knowledge workers, depends too heavily on the written record to satisfy the stringent archaeological criteria of historians such as Piggott.[56]

The issue of science arises in yet another way from the collaborative production of text and image exemplified by *Sepulchral Monuments.* Smiles has argued that Gough's sponsorship of architectural illustration stems from his realization that any accurate 'system' would depend on 'proper visual documentation of medieval architecture', on the principle of 'absolute fidelity' to the object already established in 'the illustration of anatomy and natural history'.[57] Thomas Webster's fieldwork on the Isle of Wight in 1811–12, featured in Chapter 7, further illustrates this kinship between the study of medieval architecture and natural history.

The epigraph to this chapter, from a review of *The Antiquaries' Museum*, praises Schnebbelie for his dedication to his 'profession' of 'science'. The idea of the professional scientist is so thoroughly naturalized today that this language applied to Schnebbelie requires translation into a historical idiom. In this case and in others studied here, I prefer the equally anachronistic 'knowledge worker'.[58] Schnebbelie's reviewer implicitly recognizes both his mental and his physical exertions, and both were necessary to the larger agenda advanced by Gough and his associates in the Society of Antiquaries. Schnebbelie and other paid knowledge workers supplied the empirical evidence that allowed history to become more than 'mere narrative', as Gough demanded. Schnebbelie's contributions helped to make antiquarian scholarship more inclusive, collaborative, empirical, and professional in practice, long before the study of the past was deliberately reorganized along these lines. Like

[55] Pinkerton ['Philistor'], 'Letters to the People' 126; *Speech Delivered by Edward King* 5–6. Evans calls Frere's paper 'a landmark in the development of prehistory' (*History* 203); on Frere, see also Pearce, 'Antiquaries and the Interpretation of Ancient Objects' 158. On the importance of stratigraphy more generally, see Briggs, 'Prehistory in the Nineteenth Century'.

[56] Lewis, *Gothic Revival* 49–50; Piggott, *Ruins in a Landscape*, ch. 6

[57] Smiles, 'Data, Documentation and Display' 91, 94.

[58] 'Antiquaries Museum' 282. For one recent treatment of the element of labour in the history of science, see Harding, *Sciences from Below*, ch. 4.

Carter, though for a shorter time, he gathered evidence that helped to support the growth of historic preservation. According to his *Gentleman's Magazine* obituary, Schnebbelie had more work in hand at the time of his death that would have developed the archaeology of medieval dress as a key to the chronology of the period as a whole. The emphasis on material culture in *Sepulchral Monuments* might well have influenced him to pursue such a broad scientific ambition. According to this enthusiastic obituary (probably by Gough), Schnebbelie's ability to record Gothic architecture had reached a new peak in the last months of his life, when he produced drawings 'worthy' of the 'most perfect' medieval monument, King's College Chapel, Cambridge.[59]

Reeve's 2008 recovery of Schnebbelie's work on the Salisbury Cathedral ceiling paintings stands as the most impressive testimony to this scientific merit: his 'scientific' sketches and drawings, Reeve argues, 'offer us remarkably consistent and accurate data for interpreting the cycle' of medieval paintings, the originals of which cannot be recovered. Given the technical expertise that Schnebbelie acquired in plotting all aspects of cathedral design, it is curious that neither he nor Gough show much interest in the medieval artisans—knowledge workers in another sense—who originally plotted and executed these designs.[60] That territory was left for Ruskin to explore.

Schnebbelie and Carter paved the way for artist-antiquaries such as Charles Stothard, 'an independent scholar as well as an artist' (to repeat Smiles's words) who produced all the images and texts for his *Monumental Effigies of Great Britain*. Stothard's critique of *Sepulchral Monuments* betrays an inclination to repress his debt to these predecessors. The necessary fieldwork, he wrote, could not be undertaken by 'the unfeeling hand of a labourer; and can it be supposed that a mere draftsman, employed upon a work of which he is not the proprietor, will take upon himself the disagreeable and unprofitable task of clearing the surface of a subject, which his employer will probably never see or

[59] The obituary refers to Schnebbelie's planned work as 'Antique Dresses since the Reign of William the Conqueror'. See 'Schnebbelie' 189–90, and cp. the memoir in which Gough credits Schnebbelie with a systematic knowledge of Gothic architecture (*AM* vi). The archive shows that as of 22 October 1791 Schnebbelie was shortly to go to Cambridge on Gough's request (*Letters* 154). The drawings of King's College Chapel later provided a source of income to his family (184).

[60] Reeve, *Thirteenth-Century Wall Painting* 64. Reeve's assessment is worth quoting more fully: 'Schnebbelie's work anticipates the first application of a scientific method of scaled recording practised by his own rival and successor as draughtsman to the Society of Antiquaries, John Carter'. Reeve notes that Schnebbelie's original sketches, as opposed to the finished drawings, also 'remain valuable evidence for assessing the style of the original paintings' (66). Carter prefers to personify the architect of a cathedral in terms of a solitary romantic genius, and he goes so far as to ridicule Christopher Wren's more accurate perception that cathedrals were built by 'strolling companies of mendicant builders and artificers'. See Carter 'Pursuits (XXXV)' 416. Although, as Smiles points out in the case of Exeter, records of medieval workmanship were available, little was known and in at least one instance it seems that the anonymity of these artisans suits Gough's nationalist agenda (*SM* I.cxv).

examine?'[61] The accusation is patently false, but Stothard's explicit class-consciousness is a noteworthy variation on the theme of going beneath the surface, a trope of scientific progress that we saw Schnebbelie himself deploying earlier in his criticism of Carter.

Notwithstanding Stothard's and other critical views, Gough achieved something more than old-fashioned patronage in his collaboration with Schnebbelie. The Shakespearean George Steevens exposed what he saw as their scientific pretensions by using an elite connoisseurial tactic: he planted a forged Saxon inscription in Gough's way that Schnebbelie laboriously copied before Steevens exposed the hoax. Yet Gough's patronage of Schnebbelie looks somewhat less feudal, and more professional, next to Thomas Pennant's 1793 memoir of his relationship with his draftsman: 'In the spring of this year I acquired that treasure, Moses Griffith, . . . a good and faithful servant, and able artist'.[62] In other respects, too, Gough was at least an enlightened patron, whose orthodox politics perhaps allowed him to feel secure in this social role. He proposed to 'enlarge' the artist's 'walk of fame' (*SM* I.i.9) and promoted this expansion in practice. His patronage of and collaboration with artists is of a piece with the sympathy he expresses with rank-and-file clergy forced to bear the burden of fashionable and expensive renovations to the cathedral in their diocese.

However much Gough may have wanted to maintain the civic humanist fiction of reciprocal exchange, his collaboration with 'practical antiquaries' such as Schnebbelie helped to promote the politicized, class-conscious, and commercial aspects of antiquarianism, and with it the formation of more specialized disciplines and a more separate popular domain. The fracas over Schnebbelie's dedication was one of numerous occasions on which the Society of Antiquaries as a body repudiated any affiliation with commerce.[63] To the anti-establishment satirist John Wolcot, the antiquarian enterprise of Gough and Nichols was suspect because of the way in which it used the commercial press to market a combination of supposedly disinterested research and high-minded nationalism. Wolcot would have been delighted to get his hands on Gough's poems, which—perhaps luckily for Gough—remained unpublished during his lifetime: 'I glow! I burn! I feel transferr'd/Leland's, Camden's,

[61] Stothard further argues that 'had Mr Gough been draughtsman sufficient to have executed his own drawings, he might have avoided the innumerable mistakes' that Stothard attributes to his illustrations (*Monumental Effigies* 5).

[62] Nichols describes Steevens's hoax as a 'deep-laid trap into which that excellent artist Jacob Schnebbelie was instantly ensnared, as appears by his beautiful facsimile of the supposed relick' (*ILH* V.430). For the anecdote on Griffith, see Pennant, *Literary Life* 10. Gough's attitude is much more respectful in the published prose on his draftsman (quoted above), though he too strikes a somewhat paternalistic note in a private letter on Schnebbelie's death that Nichols published much later (*LA* VI.321).

[63] Pearce, 'Antiquaries and the Interpretation of Ancient Objects' 148; Evans, *History* 164–6.

searching spirit:/I dare the task! May I inherit/The smallest radiance of their name./My Country calls. With joy I see/Britannia rise again to fame'.[64]

Carter's medievalist activism, though equally chauvinistic, was more effective because of its populism: by using his professional qualifications as an architect to dismiss inauthentic restoration as 'mock Science', he ultimately persuaded the public that Britain's architectural legacy should be in the hands of publicly accountable professionals rather than those of the clergy and aristocracy. In this respect, then, the literary marketplace—*pace* Wolcot—served the cause of preservation well. The division of labour in eighteenth-century knowledge work, although it was created to facilitate the exploitation of knowledge workers, eventually empowered a new class of professionals, and in the case of architectural antiquities it also encouraged an increasingly democratic paradigm of national heritage.

For Carter, as for most historians of antiquarianism today, the national part of this equation bears a much stronger emphasis than the democratic part. Only Nietzsche might have seen it the other way around, as 'monumental history' tainted by populism. In Carter's insular vocabulary, Gothic architecture was a 'national science', and even the most passing historical reference to British antiquarianism is likely to allude to nationalism as the enabling ideological framework. The present case study of Gough and Schnebbelie places the emphasis on the division of labour in antiquarian research rather than pursuing a further elaboration of the period's cultural nationalism. This local narrative may also serve to complicate the relationship between antiquarianism and nationalism. Gough and Schnebbelie contributed to the proliferation of antiquities that is charted in this study as a whole. By drawing attention to non-classical antiquities, and refining the gradations of antiquity, Gough and his party challenged a hegemonic system of values according to which the national temper was decidedly neoclassical rather than Gothic.

The charges of Catholicism that were hurled indiscriminately at Protestants such as Gough and Carter as well as actual Catholics in the group—along with much other non-sectarian abuse—clearly indicate that the pursuit of an empirically documented Gothic antiquity was more troubling than reassuring to national identity.[65] The ambivalence of establishment figures such as Walpole and William Chambers underscores this instability: it is surprising to hear Chambers, the royal architect whose Somerset House (1776–1801) is an exemplar of neoclassicism, quoted in Schnebbelie's letter as lamenting that British art 'students were sent abroad to study the dull Greetian' (*Letters* 63). I have argued that Gothic antiquarianism was suspect as well for its materialism,

[64] For Wolcot's squibs against Gough and Nichols, see especially *A Benevolent Epistle to Sylvanus Urban, Esq.* For Gough's poetry, see *LA* VI.341.

[65] Even a modern scholarly work, Crook's *John Carter* (58), refers to Gough as a Catholic, showing how forceful and persistent this suspicion was.

its attention to built works and material culture. Sir Henry Englefield, who launched the Antiquaries' Cathedral Series, was a professed Catholic who was openly charged with Papism and adultery. This kind of pressure probably influenced his scholarly drift toward a deeper and more stable body of national antiquity, the geological matrix of the very building stones that the antiquaries learned to distinguish in their work on the cathedrals—as we shall see in the next chapter.

7

'The Whole of This Coast Is Composed of Ruins': Thomas Webster's Fieldwork on the Isle of Wight

7.1 GEOLOGY AND ANTIQUITY

One of the first researchers hired explicitly to do geological fieldwork, Thomas Webster, began with a vocabulary that evokes tourism more than science. In May 1811, he reported to his employer, Sir Henry Englefield, that 'the whole of the country between [the Undercliff] and the sea, had, apparently, at some remote period, been one immense mass of ruin, though now covered with woods, cornfields, and villas. It was impossible to view this scene, and to contrast its present with its former state, without feelings of interest'. This aesthetic discourse on a landscape, eloquent as it is, would be unremarkable for its time were it not that Webster (1772–1844) was the first person to refer to himself as a 'professional geologist'.[1] His major geological publications rely upon the picturesque and the sublime, as well as an antiquarian discourse of ruin, to establish analogies for geological processes and geological time. As Ralph O'Connor has recently argued, these and other literary techniques were essential to geological writing at least through the middle of the nineteenth century.[2] Webster's research and publications illustrate very clearly the continued importance of aesthetic categories even as geology parted ways with romanticism and conjectural history. A close affiliation with archaeology enriches his geology as well.

This chapter returns to the conjunction of natural history and antiquities by way of a pairing first noticed long ago by the critic Robert Aubin in an essay entitled 'Grottoes, Geology, and the Gothic Revival'. Webster's project links

[1] Englefield and Webster, *Description of... the Isle of Wight* 129. Cited parenthetically in the text hereafter. Webster identifies himself as 'Thomas Webster, F. G. S. &c. Professional Geologist, and Lecturer on Geology', in his *Proposal for Publishing (by Subscription) A Description of the Isle of Wight*, a manuscript source (British Library) recently discovered by Hugh Torrens.

[2] *Earth on Show* 241–5.

geology and the Gothic in a more scientific manner than the eighteenth-century poets whose atmospheric descriptions of rocks prompt Aubin's analysis.[3] Webster published his earliest geological observations by commission in Englefield's *Description of the Isle of Wight* (1816), a work that held together historic architecture and picturesque landscape with scientific natural history even as it accommodated the more specific disciplinary idiom of the new science. Englefield, like Banks and Hamilton, actively pursued natural history alongside his antiquarian research, and recorded both natural history—particularly geology—and antiquities in his journal when he visited the Isle of Wight in 1801. Englefield published a few of his observations on the island's strata at the time, but when he returned to his notes a decade later he realized that what he had observed belonged to an emerging science that had just established itself under the name 'geology'.[4]

Englefield hired Webster when he decided to make a scholarly plate book out of his journal. Webster was not only, like Schnebbelie, a topographical and architectural draftsman with a reputation for scrupulous accuracy. He was also close to the new science, having created illustrations for England's first course of public lectures in geology, given by Humphry Davy in 1805.[5] On the island, Webster drew several medieval monuments, some in a pleasing state of ruin, to accompany those Englefield himself had recorded. He then went on to make a whole series of drawings of geological sites and developed such penetrating original observations that Englefield credited him as a co-author, effectively launching Webster's geological career.

As an architectural draftsman, Webster was especially well equipped to translate Englefield's antiquarian idiom into a more geological register. He took the commonplace image of geological 'ruin' and used it to establish the historicity of geological processes. By attending to a new set of ancient ruins, this chapter shows how strongly the traditional language and research questions of antiquarianism informed a newer discipline, geology, in the very process of its emergence as a professional specialization. In his ambitious concluding letter, Webster lays out the sequence of geological events that formed the Isle of Wight in deliberately antiquarian terms, 'restoring (to use the language of the antiquary) this series of strata, which, though now in ruins, was probably much more entire in some former condition of the earth' (*Description* 201). Like a picturesque tourist, Webster declares that 'the whole of this coast is composed of ruins', but he literalizes the conventional figure by identifying the massive fault that runs through the chalk on this part of the coast (137–38). In his part of the volume, Englefield offers detailed

[3] Aubin, 'Grottoes, Geology'. For parallels in later, self-consciously scientific studies of both the Gothic and geology, see Buchanan, 'Science and Sensibility'.

[4] *Description* i. On the usage question, see Dean, 'The Word "Geology."'

[5] *Humphry Davy on Geology* xxxii–iii and 156n.13.

empirical reconstructions of the stages of building evident in early medieval built works on the island, both ruined and entire, and the volume as a whole vividly develops the many ramifications of this analogy between two kinds of ruin. Webster's key contribution, still acknowledged by geologists today, was to recognize the larger fold structure into which the island and opposite coast were incorporated. 'Ruin' (from the Latin *ruere*, to fall), which had been applied figuratively to geological features for generations, signified a more literal identity between human and natural structures in this context.

As a professional draftsman doing research-for-hire for Englefield, the Geological Society, and other patrons, Webster brought to his geological work a different intellectual background and different objectives from those of the paying members of the Geological and other learned societies. The *Description* includes numerous engravings made from Webster's drawings and Englefield also attributes the entire geological section of text to him. While the book was in production, Webster published a more technical version as a long article in the *Transactions of the Geological Society* and soon became the Society's first salaried officer. Historians of geology concur with his self-identification—the first on record—as a 'professional geologist'.[6] Webster's unique abilities and his socio-economic position visibly shape both the 1814 article and his extensive contributions to Englefield's book. As a knowledge worker whose various fields of expertise—architecture, agriculture, drawing—were called for on different occasions by different employers, Webster was apt to draw on diverse forms of local history for his geological essays, just as he took notice of geology and antiquities simultaneously when working as a draftsman on the Isle of Wight.[7]

Englefield (1752–1822) was a long-time vice president of the Society of Antiquaries as well as a Fellow of the Royal Society and had correspondingly wide-ranging interests, though he published only two short geological essays.[8]

[6] Webster continues to be of interest to geologists because, as Kirk has put it, 'he made major contributions both to the understanding of Jurassic and Cretaceous stratigraphy, and to the interpretation of the structure of the area [southeast], which was well ahead of its time'. See her 'Thomas Webster (1772–1844)' 311–12.

[7] The biographical literature on Webster is limited, though there is no shortage of primary materials, including numerous manuscript letters, some still unpublished, and a manuscript autobiography housed at the Royal Institution, utilized so far only by Ironmonger in 'The Royal Institution and the Teaching of Science'. Other sources on Webster's life and career include Kirk as well as articles by Challinor and Edwards, most of which consist largely of the texts of Webster's letters. Some 46 of these have now been published, together with 124 letters to him. Discussions of his geological work include Rudwick, *Bursting the Limits of Time* 512–21, and Torrens, 'Arthur Aikin's Mineralogical Survey' 138–9.

[8] The scholarship on Englefield is even more limited. Apart from the *ODNB* entry by Bernard Nurse, there are only passing references in Sweet's *Antiquaries* and other histories of antiquarianism. Englefield's essay on flints embedded in chalk and a brief continuation appeared in the *Transactions of the Linnean Society* (1802) and arose from his observations on the Isle of Wight. Englefield published occasionally in the *Philosophical Transactions*, more regularly in the

Though not a strictly geological qualification, Webster's architectural knowledge would have appealed to Englefield, who supervised the Society of Antiquaries' programme of publishing engravings of the English cathedrals. At the same time Webster's growing geological expertise made him an ideal contributor who could confirm and develop the connections between human and geological antiquity implied by the title of Englefield's book. Martin Rudwick has recently argued that the emergence of primary-source-oriented antiquarianism or 'erudite history' strongly influenced the evidentiary turn in geohistorical science.[9] The appearance of path-breaking geological research in a fashionable antiquarian work of local history shows how this symbiosis continued to operate in the nineteenth century.

By comparing Webster's text and images with Englefield's in the context of local history, we may see that the relationship between natural history and antiquity persisted in the era of advancing professionalization. Webster sought to capitalize on his influential research on the Isle of Wight and his position in the Geological Society to reinvent himself—albeit with limited success—as a professional geologist.[10] Webster collaborated with Englefield in developing a geological sense of 'ruin' that set the stage for this institutional self-fashioning, which reflects a broader reorganization of intellectual life. Englefield's antiquarian vocabulary provides the medium required by Webster to express the historicity of these geological ruins and facilitates his reconstruction of the stratigraphic sequence. Although Englefield's work on the cathedrals shared in the nationalist spirit infused into the early Gothic revival by Gough, his work with Webster also transcended the patriotism commonly associated with romantic discourse about ruins by expanding the geographic and temporal scope of 'ruin'. Georges Cuvier referred to himself as 'a new type of Antiquary' at roughly the same time, and as Alexandrina Buchanan has shown, elaborate analogies between the Gothic and geology remained

Archaeologia, and wrote the commentary for six numbers of the Society of Antiquaries' cathedral series. The work with which this chapter is mainly concerned, the *Description*, was reviewed by John Playfair in the *Edinburgh Review*, and anonymously in the *Augustan Review*, which places great (pejorative) emphasis on Englefield's Catholicism. There is a brief but useful discussion of the *Description* in Freeman, *Victorians and the Prehistoric* 14–17.

[9] Rudwick, *Bursting the Limits of Time* 182.

[10] As Edwards demonstrates, Webster was still struggling financially even after he was appointed the first professor of geology at University College, London, in 1841. See Edwards, 'Thomas Webster (Circa 1772–1844)' 469.

common even as specializations like palaeontology and 'ecclesiology' emerged in the mid-nineteenth century.[11]

7.2 LOCAL HISTORY, FIELDWORK, AND COLLABORATION

Stuart Piggott's influential antithesis between 'romance' and empirical science has been a touchstone for my larger argument and it is especially relevant to the kind of geological history that Webster and Englefield produced. The *Description* displays the new empiricism that, in Piggott's view, infused the English county histories of the period 1790–1820, displacing the 'irrational beliefs of Gothick romance' that had corrupted antiquarian research in the mid-eighteenth century and renewing the rigour of early Baconian investigators such as Robert Plot and Edward Llwyd. Piggott particularly praises Richard Colt Hoare's *Ancient Wiltshire*, published from 1812–21 under the epigraph 'we speak from facts, not theory'.[12] Webster's use of 'ruin', however, signals a debt not only to 'fact' but also to 'theory' and history.

Piggott's archaeological bias leads him to distinguish excavation and material culture categorically from manuscripts and archival sources of evidence and he even suggests that earth science went into decline between Robert Hooke and John Whitehurst because of the way in which texts came to dominate historical study.[13] Piggott seems somewhat too ready to dismiss textual and archival research, and Rudwick, by contrast, has emphasized its importance for geology. In fact both archaeological and philological research began to display new levels of empirical rigour in the later eighteenth century, as richly illustrated by the journal *Archaeologia*, founded in 1770 by the Society of Antiquaries. A variety of early contributions to the journal, including Englefield's, register these changing standards of evidence and authentication. As a leading preservationist in the Society, Englefield was closely associated with Gough, the founding editor of the journal. As editor of the

[11] Cuvier, *Recherches sur les ossemens fossiles* (1812), quoted in Buchanan, 'Science and Sensibility' 178n.10. Buchanan argues that an emergent architectural history in the mid-nineteenth century deliberately drew on geology and paleontology in an effort to establish scientific approaches transcending the impressionism and patriotic sentiments of earlier studies of medieval built works. Her brief case studies include numerous authors who were prominent in both fields, including Sir James Hall, William Whewell, and Robert Willis. On Webster and Cuvier, see Challinor, 'Some Correspondence of Thomas Webster [II]' 168. On ruins and nationalism, see Janowitz, *England's Ruins* 1–19, 118.

[12] Piggott, *Ruins in a Landscape* 120, 127–8.

[13] *Ruins in a Landscape* 117.

cathedral series, Englefield carried on a productive collaboration with John Carter that depended on structural investigation, excavation, and other kinds of field evidence. Piggott's emphasis on the new county histories helps to explain the connection between this kind of archaeological work and the new geology. Geological fieldwork *is* local history, and it makes sense that Englefield's concern with local archaeological evidence in particular would have led him to stray into the precincts of 'that part of natural science lately called Geology', as he terms it in his preface (*Description* i). The *Description* incorporated the new science into a local history framework that preserved the older synthesis of natural history and antiquarianism.

Englefield's unexpected detour into geology was—or might have been—the making of Webster's career. Webster's Isle of Wight research made him a recognized authority and led to some further publication, but not to the financial success that he not unreasonably expected.[14] Though he worked closely with Gough, Englefield's intellectual interests ranged widely, like those of Banks and Hamilton, his senior colleagues in the Society of Dilettanti. In addition to the cathedral series, Englefield published on astronomy and had his collection of ancient Greek vases engraved for publication after serving as secretary of the Dilettanti for many years. When he noted in 1811 that 'that part of natural science' might nonetheless be beyond his purview, this recognition was by no means inevitable, and even so he might have employed Webster merely as a fact-gatherer or ghostwriter. In the event, Englefield explicitly credited Webster as co-author on the title page and cited Webster as an authority repeatedly in his own part of the book.[15] By contrast, Carter and Schnebbelie received credit for their illustrations but only rarely for their research (though both eventually published their own books). Some of the reasons for this difference are peculiar to Webster and Englefield themselves, but the moment can also be understood historically as one in which the illustrator no longer needs the direct supervision of the naturalist-antiquary—as in Maria Toscano's eighteenth-century scenario—but rather becomes credible as an independent witness.[16]

[14] Kirk, 'Thomas Webster (1772–1844)' 313, 315.

[15] Here is the full text of the title page: 'A Description of the Principal Picturesque Beauties, Antiquities, and Geological Phoenomena of the Isle of Wight. By Sir Henry Englefield, Bart. With Additional Observations on the Strata of the Island, and Their Continuation in the Adjacent Parts of Dorsetshire. By Thomas Webster, Esq. Illustrated by Maps and Numerous Engravings by W. and G. Cooke, from Original Drawings by Sir H. Englefield and T. Webster. London: Printed by William Bulmer and Co. Cleveland-Row, St James's, for Payne and Foss, 88 Pall-Mall. 1816.' No information is available concerning Webster's financial arrangement with Englefield. Hugh Torrens has suggested (pers. comm.) that newly discovered Webster letters at Southampton University may shed some light on this question.

[16] Toscano, 'Figure of the Naturalist-Antiquary' 227.

Webster, like Carter, was trained as an architect and architectural draftsman, but his professional engagements had led him to natural science rather than antiquities. Webster designed the lecture hall of the Royal Institution in Albemarle Street and in 1805 he supplied panoramic paintings for Humphry Davy's pioneering geology lectures in that theatre.[17] While at the Royal Institution he came into contact with Englefield and with Arthur Aikin, who was with Davy one of the founders of the Geological Society. Hugh Torrens has suggested that Aikin probably gave Webster his first exposure to geological fieldwork. In 1812 Webster became draftsman to the Society and keeper of the museum. As the Society's first salaried officer, Webster embraced geology both as a vocation and as a plausible professional path. In 1826 he elected to pursue that path outside the Geological Society when they refused him a raise and promotion after fourteen months of negotiations. Edward Ironmonger has shown how Webster became committed to making knowledge public during his time at the Royal Institution, and this commitment also influenced him to return to lecturing in the 1820s.[18]

Englefield highlights Webster's expertise in his preface to the *Description*. Webster's research on the Isle of Wight led to two publications in his own name in the second volume of the Society's *Transactions*, the very substantial essay 'On the freshwater formations in the Isle of Wight' and a shorter piece on a new species of fossil Alcyonium (both published with Englefield's consent before the book appeared). Webster's attested geological expertise helped to justify the publication of Englefield's book, making it more competitive in a rather saturated corner of the literary marketplace. Englefield acknowledges as much in his preface, citing Webster's expertise, along with his own originality, to justify the publication in 1816 of observations made fifteen years before. Englefield warms gradually to Webster's praise, noting at first only that his own manuscript work from the turn of the century, especially the geological observations, still has not been 'superseded by the labours of later travellers', and therefore merits publication:

> As I felt my own strength and activity unequal to the exertions required, I requested Mr. Webster, who was not unacquainted with natural history, and who, to the talents of an expert draftsman, added the more valuable qualities of most scrupulous accuracy and patient investigation; to revisit those parts of the Island as I had examined imperfectly [. . .] and to extend his researches into that part of Dorsetshire where it was evident that the great chalk range of the Isle of Wight reappeared, after being cut off by the sea for near 12 miles (*Description* ii).

[17] *Humphry Davy on Geology* 156n.13. O'Connor reproduces a slightly later version in 'Facts and Fancies' (Fig. 1).

[18] Torrens, 'Arthur Aikin's Mineralogical Survey' 138–9; Edwards, 'Thomas Webster (Circa 1772–1844)' 472n.58–59; Ironmonger, 'The Royal Institution' 139–43.

Having begun in this somewhat patronizing tone, Englefield goes on to credit Webster with 'unparalleled' discoveries and declares that Webster's observations 'constitute the most valuable part of the book', fully deserving of the explicit authorial credit reflected in the title page and contents (Webster wrote just over half of it). Finally Englefield notes that he has given 'the whole of [Webster's] researches in his own words' and recommends Webster's 'masterly descriptions' to his readers (iii), having prepared them carefully to accept the authority of a mere draftsman.[19]

Although Webster's 'On the freshwater formations' has justifiably become the standard for geological readers, it is important to read this paper side by side with the co-authored book, both to recover the antiquarian spirit that motivated Webster's research and to analyse the changes that Webster made to adapt his work to a geological audience. Both works display a variety of literary and visual conventions. The *Description*, as denoted by its price of ten guineas, occupied the niche of the antiquarian plate book. In this work he does not aspire to the archaeological rigour extolled by Piggott in the case of Hoare's *Ancient Wiltshire*, but does evince a well-schooled scientific curiosity in both archaeological and geological matters, and the overall difference in style and content between Englefield's and Webster's parts of the book is less than one might imagine. The most important difference is Webster's ability to enumerate and correlate the strata (and observe their dip), which he does in verbal as well as tabular and cartographic form. Webster also made extensive observations by boat, which, as he notes in his first letter, was essential to reveal 'the true dip of the strata' (120). These features, and his concluding table ('Order of the Upper Strata') in particular, anticipate the larger scale of Webster's conclusions in their final form in the *Transactions*, where he applies his observations on the local history of the Isle of Wight and the corresponding parts of Dorset to the chalk formation as a whole. Both these habits of correlation and comparison and Webster's simultaneous attention, following Englefield, to geology, antiquities, and the picturesque, encourage the conceptual movement from local observations to regional structures. Both antiquarianism and picturesque theory promote attention to recurring visual patterns, relative dating, and geological situation.

[19] It is a bit surprising that Englefield does not mention Webster's credentials as a geological draftsman. In addition to his work for Davy's lectures, Webster had contributed illustrations to John Pinkerton's *Petralogy: A Treatise of Rocks* (1811). Pinkerton was, like Englefield, primarily an antiquary. Englefield might not have known of Webster's work for Aikin and couldn't have known that many of these illustrations would eventually be published in Roderick Murchison's *The Silurian System*, as shown by Torrens in 'Arthur Aikin's Mineralogical Survey'. Webster also 'draft[ed] the topographical base' of G. B. Greenough's 1819 *Geological Map of England and Wales*, according to Wendy Kirk, 'Thomas Webster' 312.

7.3 'A SORT OF ARCH OR VAULT'

Webster himself makes one particularly explicit analogy between archaeological and geological forms:

> In connecting together, and as it were restoring (to use the language of the antiquary) this series of strata, which, though now in ruins, was probably much more entire in some former condition of the earth, a singular conclusion presents itself to the imagination [. . .] with nearly the same certainty which we feel in putting together the fragments of an ancient temple. The conclusion to which I allude is, that the chalk of the middle range, together with the clay of Alum Bay and the inferior strata, once formed a sort of arch or vault, by which it joined to the horizontal strata of the hills of St Catherine and Shanklin. (*Description* 201)

This passage clearly illustrates the difference between the older antiquarian premise of continuity between natural and human history, still defended by Whitehurst in 1786, and the self-conscious use of historical references as not continuous with but analogous to pre-human events whose antiquity is of a higher order of magnitude. Rudwick has noted that this passage and the accompanying section focus attention on geohistorical sequence in place of conjectural causes, and especially on the astonishingly recent date of this 'arch' or fold structure.[20]

The antiquarian context of Webster's analogy thus merits careful attention. The 'denudating causes' responsible for removing the top of Webster's imagined arch (Figure 7.1) are directly analogous to the weathering that has removed and obscured the vestiges of beacons formerly set atop early medieval structures along the coast, and are linked in a more general way to the poor state of preservation of many of the region's architectural antiquities. Buchanan points out that the technique of visual reconstruction itself was practised increasingly by both architectural historians and palaeontologists with a consciousness of such heuristic analogies.[21]

Englefield encouraged Webster to return to the Isle of Wight in the summer of 1812, primarily to extend his correlation of strata between the Isle of Wight and the Dorset coast west of Purbeck. Twenty of the *Description*'s fifty plates are devoted to the Dorset coast, and four of them depict an antiquarian site that Englefield gave Webster particular instructions to examine in 1812, because it had not been illustrated before (*Description* 191).

Webster's writing on this site, St Adhelm's (now Aldhelm's) Chapel, beautifully exemplifies his stylistic amalgamation of antiquarian, geological, and picturesque elements. The chapel resides amidst the 'grand' and 'immense'

[20] Rudwick, *Bursting the Limits of Time* 517.

[21] Buchanan, 'Science and Sensibility' 173–5.

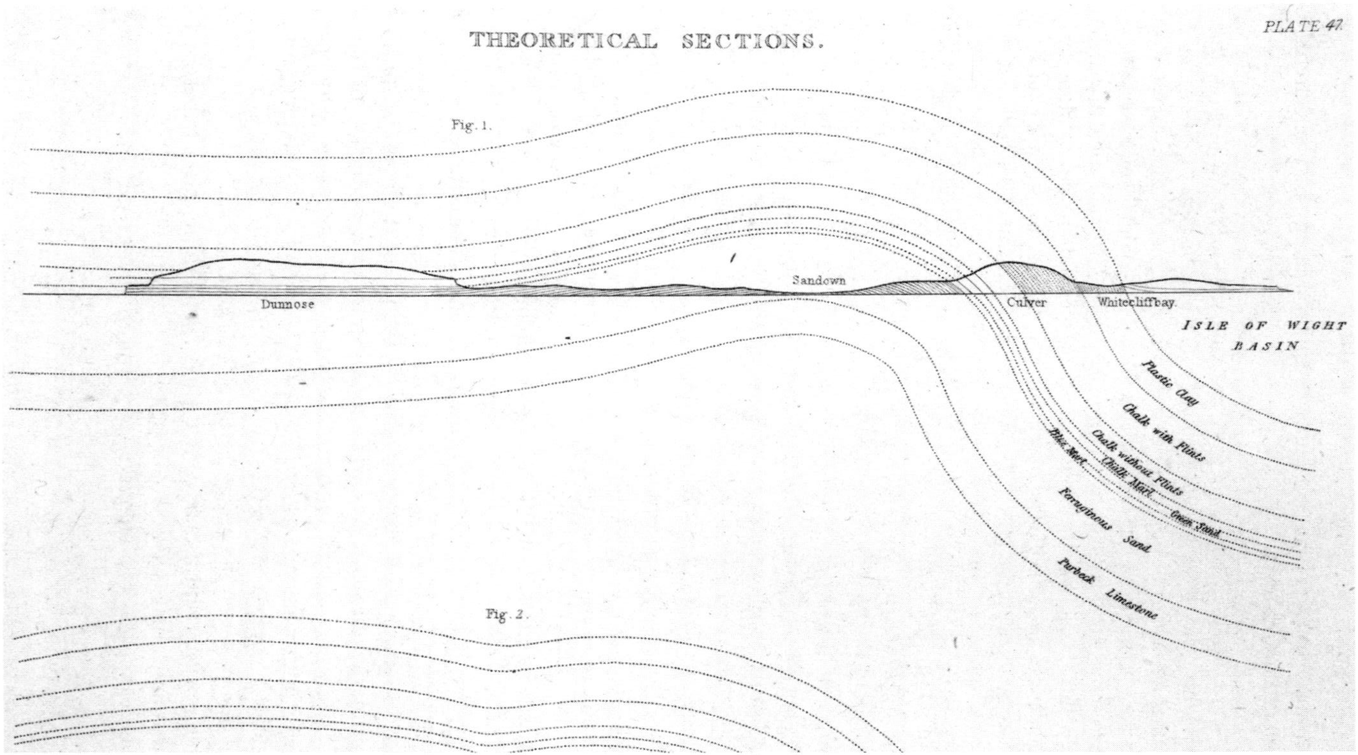

Fig. 7.1. *Theoretical Section* or 'ruined arch'. Engraving by W. and G. Cooke from a drawing by Thomas Webster. *Description of the Isle of Wight* (1816), Plate LXVII, Figure 1. Linda Hall Library of Science, Engineering, and Technology, Kansas City, Missouri.

Fig. 7.2. *Interior of St Adhelm's Chapel.* Engraving by W. and G. Cooke from a drawing by Thomas Webster. *Description of the Isle of Wight* (1816), Plate XLIII. Linda Hall Library of Science, Engineering, and Technology, Kansas City, Missouri.

ruins of the two lowest strata in Webster's column, not exposed at all on the Isle of Wight:

> St. Adhelm's Head, [. . .] whose dangerous rocks have been so often fatal to mariners, is a vertical section of the oolite, and of the argillaceous strata below it. [. . . D]eep fissures in the upper part of the cliff predict the impending ruin of many places, and the agitation of the sea for above a mile from the shore, shows the nature of its rocky bottom, the remains, no doubt, of land which has gradually been covered by the ocean. (188–89)

The chapel itself (Figure 7.2), displaying formal properties of 'the simplest kind of Saxon architecture' (189), bears marks of such 'high antiquity' that the unusual forms of the arches supporting the structure initially appear to Webster as a result of deformation over time ('high antiquity' is another phrase that recurs in Webster's geological descriptions). On closer inspection he realizes that these arches are not ruined, but were originally 'composed of portions of circles having their centres a little above the springing of the arch, which gives it something of the Moorish character, a circumstance rare, though not unique, in Saxon architecture' (xxiv; cp. 190).[22] The cross vaults supported by these arches also made an impression on Webster with their 'extremely elegant' simplicity, and in his conclusion he relies on the term 'vault' (205) to underscore the three-dimensional character of his hypothetical ruined arch or fold structure.

Architectural history also informs the mood of the engraving and the prose on this site. The recent discovery by workmen of a cylindrical base on the roof of the chapel (presumably used to hold a warning beacon for vessels) especially fires Webster's imagination. Echoing Englefield's description of a similar structure on the Isle of Wight, Webster imagines the 'religious inhabitant' of this 'chauntry', as he calls it, praying for the safety of distressed mariners as he shuddered at the 'dismal scene' visible through the structure's one window (190). Webster's initial description of the cliff, then, evokes the perceptions of this early medieval anchorite, while the oddly melancholy contemplation of the lone fisherman lounging in the converted chapel—now a storehouse—preserves the historical mood and at the same time provides a visual echo of the labouring figures featuring so prominently in many of Englefield's drawings (engraved as Plates I–XII) and in other illustrations of Webster's as well.

These figures must be understood as both picturesque and antiquarian elements. In some cases, where the figures indicate the scale of the cliffs, they are geological elements as well. In fact, the picturesque convention of rural labourers or banditti dwarfed by craggy landscapes, deriving primarily from the landscapes of Salvator Rosa, is likewise an important art historical legacy for geological representation.[23] The antiquarian aspect of these figures derives less from their scale than from their costume and their illustration of customs and manners. While the mood of the lounging figure may evoke the melancholy life of Webster's imagined anchorite, the baskets and other fishing gear speak clearly of this occupant's worldly vocation. Many of the illustrations, by both Englefield (Figure 7.3) and Webster (Figure 7.4), feature fishermen, their boats, and their equipment, though they are rarely mentioned in the text. Webster mentions them twice as sources of data (pp. 147, 156) and he also includes quarrymen and their technology in two plates (e.g. Figure 7.5), citing their methods more than once for the light they shed on the stratigraphy or lithology of a particular locale.

Many of these engravings incorporate ethnographic material richly reminiscent of the artistic exploration of customs and manners that features so prominently in the work of Banks and Parkinson, Hamilton and Fabris, d'Hancarville, and Gough and Schnebbelie. Webster's quarrymen directly

[22] On the patriotic anachronism of post-Conquest 'Saxon' architecture, see Sweet, *Antiquaries* 251.

[23] Gilpin, 'A Poem on Landscape Painting' (appendix), *Three Essays* 20–1; Bermingham, *Landscape and Ideology* 81–3.

Fig. 7.3. *Freshwater Cave, I.W.* Engraving by W. and G. Cooke from a drawing by Henry Englefield. *Description of the Isle of Wight* (1816), Plate VI. Linda Hall Library of Science, Engineering, and Technology, Kansas City, Missouri.

Fig. 7.4. *Handfast Point in Dorsetshire.* Engraving by W. and G. Cooke from a drawing by Thomas Webster. *Description of the Isle of Wight* (1816), Plate XXVII. Linda Hall Library of Science, Engineering, and Technology, Kansas City, Missouri.

Fig. 7.5. *Tilly Whim Quarry*. Engraving by W. and G. Cooke from a drawing by Thomas Webster. *Description of the Isle of Wight* (1816), Plate XXXIII. Linda Hall Library of Science, Engineering, and Technology, Kansas City, Missouri.

recall the quarrymen in Fabris's engravings (Figure 2.4) in *Campi Phlegraei*, a work cited by Englefield in the *Description* (79) and one likely to have inspired his project with Webster. Figure 7.4 also resembles a spectacular Fabris plate entitled 'View from the Sea of the Promontory call'd Capo dell'Arco in the island of Ventotiene' (*CP* XXXIV), in which the boats in the foreground similarly set off an imposing cliff face with strongly marked stratification. Figures 7.2 and 7.3 might also recall, in this context, Gough's vivid defence of the value of dress, and material culture more broadly, for the study of the past (*SM* I.i.1) and Schnebbelie's close attention to sartorial detail in his drawings of tomb effigies. The closely parallel emphasis on native costume and implements in Parkinson's ethnographic drawings, made in the very different arena of Pacific voyaging, points to the still under-recognized importance of antiquarianism as a framework for early ethnography, which in its broadest sense also includes the labouring figures in Englefield's *Description*. Closer to home, the 'costume books' or visual studies of dress among the labouring classes of Great Britain may have inspired Webster as well.[24]

[24] On the British labouring-class 'costume books' (his term), see Smiles, *Eye Witness* 77–110.

Like Parkinson and Fabris, Webster had to resist picturesque convention to portray his object of study more accurately, and his figures become more than merely conventional in this light. At Culver Cliff, for example, he reports that his point of view 'was chosen rather with a view to shew the real dip of the strata than as the most picturesque' (*Description* xi) and the same logic is implicit in many of the views, though at times the two objectives coincide (as in Figure 7.4). Unlike Parkinson and Schnebbelie, Webster did not die young of the hazards attending his profession, although he might easily enough have been swept off the stump of rock on which he sat for two hours, becoming soaked to the skin, while taking both his spectacular view of the Needles (XXV) and the view of Alum Bay that was included on the volume's only colour plate (XIX). In defending Webster's scientific merits, Englefield rightly insisted that 'in no instance has accuracy been sacrificed to the effect of the engraving' (iii), but effect was nonetheless crucial to Webster's art.

Though he was far from uncritical, Webster used picturesque conventions quite freely both in his prose and in his drawings—as we have seen in the case of the lounging fisherman—when they did not interfere with a stated scientific objective. Lulworth Cove in Dorset, for example, 'unit[es] great picturesque beauty with singularly instructive geological phenomena' (xxiii), while the columns of erosion-resistant chalk off Handfast Point (Figure 7.4) 'render it somewhat dangerous to pass, but add to the picturesque effect of the place' (xviii). Englefield, however, is more apt to note technical 'defects' in the island's picturesque beauty, which result in part from a deficiency in landed property; but he too waxes eloquent on its sublime scenes such as Alum Bay: 'the Needle rocks rising out of the blue waters, continue the cliff in idea, beyond its present boundary, and give an awful impression of the stormy ages which have gradually devoured its enormous mass [. . .] all is wild ruin' (81). For both Englefield and Webster, the unique geology of the island provided a new framework for the analogy between the ruins of nature (such as the Needles) and the ruins of time, rejuvenating the conventional aesthetic discourse associated with each type.

7.4 KNOWLEDGE WORK AMONG THE RUINS

Englefield's portion of the book is in conventional chapters, unlike Webster's letters, which appear more or less in their original form, in part surely to reinforce the authenticity of the fieldwork reported in them. Englefield devotes his longest chapter to picturesque beauties, but in doing so he creates a fairly consistent line of demarcation between the aesthetics of landscape and the science of antiquities, which (like geology and other more objective domains) is reserved for a separate chapter. His chapter on the picturesque is full of allusions to and borrowings from the classic authors of picturesque theory and travel: William Gilpin, Richard Payne Knight, Sir Uvedale Price, and others. Though he names none of these precursors explicitly, he shares the technical vocabulary of 'vistas', 'perspectives', and 'terminations' (*Description* 56) with Gilpin and his disdain for the landscape designs of Capability Brown (87–88) with Price and Knight. Webster, by contrast, uses the terms of the picturesque in more colloquial and empirical ways. One interesting and probably unintended effect of the epistolary form is Webster's synthesis of the three elements at least partially segregated by Englefield, a synthesis that shows how often geology and the picturesque, and/or antiquarianism, reinforce, rather than conflict, with each other.

Englefield displays his substantial expertise in architectural history in the chapter on antiquities, but it is far from being a dominant theme in the volume. There is a curious emphasis on the island's deficiencies that unites all the chapters and helps to explain, in my view, his delay in publishing the book as well as the need for Webster's contribution to justify its publication. Englefield continually reminds his readers that one of his objectives is to

Fig. 7.6. *Yaverland Church, I.W.* Engraving by W. and G. Cooke from a drawing by Henry Englefield. *Description of the Isle of Wight* (1816), Plate IX. Linda Hall Library of Science, Engineering, and Technology, Kansas City, Missouri.

demystify the Isle of Wight, to correct the enthusiasm of previous accounts and bestow praise on the landscape and antiquities only where it is due—in some cases on sites that previous authors have not noticed. 'So much has been within these few years written on the beauties of the Isle of Wight, that it may appear presumptuous to rank it lower in a picturesque view than these authors have uniformly placed it' (55). The interior of the island, Englefield observes, is 'as destitute of beauty, as any tract of the same extent in England'; even when its historic buildings are taken into consideration, 'scarcely a subject for the pencil can be found' (58).

The chapter on antiquities thus begins in very unpromising fashion: 'The remains of antiquity in the Isle of Wight are very few in number, and inconsiderable in point of size and beauty' (89). Englefield finds only about a dozen structures deserving of extended analysis, primarily churches, and five buildings are engraved from his drawings. One of these is Yaverland Church (Figure 7.6), which provides the occasion for some of Englefield's most detailed architectural history and also anticipates Webster's careful attention to the arches of St Adhelm's Chapel. The arch here displays the 'very rude Saxon zig-zag ornaments' that repeatedly draw Englefield's attention (101), and he notes that this arch 'is the only entrance into the Chancel, from the Nave of the Church. So solid a wall of separation, and so small an opening, are not common, and mark an high antiquity' (viii). The broken staircase to the left of the arch 'appears coeval with the wall, and it is a very curious and perhaps singular remnant of antiquity', recalling the reading desks attached to the choirs 'of the earliest Christian churches' (102). This analysis displays Englefield's knowledge of ecclesiastical history, but the reasoning is also closely analogous to the reconstruction of stratigraphic sequence and deformation adumbrated in his geological discussion and elaborated fully in Webster's letters. At the same time, Webster's geological observations confirm Englefield's feeling that the island—although not insignificant in a picturesque or an antiquarian view—is most curious for its geology, which ultimately justifies the publication of his work.

Another building noticed by Englefield resembles St Adhelm's Chapel in function, if not structure: a small freestanding turret on St Catherine's Hill, with an octagonal upper story containing windows suitable for the projection of a light. Archaeological evidence suggests that this lighthouse was manned by a hermit who lived in a cell close by, and for Englefield, 'standing on this airy summit, it is impossible not to picture to the mind the venerable inhabitant of this cell lifted almost out of the habitable world, and only recalled to it by the friendly care of his charitable lamp' (94). The 'religious inhabitant' of St Adhelm's imagined by Webster (190) arises from the same intimate relationship between coastal geology and navigation. These figures clearly mark the affinity between antiquarian local history and the new 'geohistorical science' (to borrow Rudwick's term). Just as Englefield consults both archaeological and manuscript evidence to reconstruct the religious history of the island, so too Webster takes an archival approach to the strata in his geological reconstruction.

This motif of marine piety is an essential feature of the local history of almost any rocky coastal landscape. Charlotte Smith's learned local poem *Beachy Head* (1807) provides a good analogue to both Englefield and Webster in its rendition of the legendary Parson Darby as a hermit residing in a cave under Beachy Head, piously pursuing his vocation of saving drowning mariners until he and a portion of the cliff are consumed by the raging sea. Webster's essay 'On the freshwater formations', which extends his observations eastward along the coast, tells of the narrow escape of Parson Darby's successor, the then-current vicar of East Dean, who leaped to safety across a widening fissure just before a 300-foot by 80-foot mass of the chalk plunged into the Channel. Smith's interest in the natural history of Beachy Head certainly extends to the qualities of the chalk, though she does not pursue a structural understanding of the formation as Webster does. Webster, even in this more formal paper, is far from indifferent to the sublime, whether in its coastal manifestations or in the form of 'mysteries on which all the speculations of geologists have not thrown any certain light'.[25]

The picturesque, the sublime, and the elegiac theme of ruin all converge in the tales of shipwreck that belong to the description of many sites of geological interest. Webster happened to make his hard-won sketch of the Needles the day after the wreck of the *Pomone*, a fifty-gun frigate coming from Persia (*Description* 161). Webster mentions several other shipwrecks (cp. xvii, xix–xx, 191) and Englefield is moved by his discovery of an impromptu burial ground, stemming from a 1782 shipwreck, to include a quite powerful original verse elegy (61), the only piece of his *Description of the Isle of Wight* ever to be reprinted.[26] The *Pomone* provides an especially compelling case because Webster actually witnessed the aftermath of the wreck, which sets the tone for his description of the Needles and reinforces the difficulties of his own practice as a draftsman and geologist. The letter in question (VI) begins with a tone of heightened interest and wonder, conditioned by the astonishment that Englefield and his predecessors reserve for Alum Bay: on the day described in this letter, Webster took his views both of the colourful clay cliffs of Alum Bay (Pl. XIX.3) and of the Needles (Pl. XXV) from the same rock off the western tip of the island (xv; 163). He points out that the shipwreck inevitably

[25] Webster, 'On the Freshwater Formations' 191–92n., 245. Charlotte Smith's description of the Isle of Wight in her First novel, *Emmeline* (1788), resembles Englefield more than Webster in its literary polish. See *Emmeline* 302–3. For a sustained comparison of Smith and Webster see Heringman, '"Very vain is Science' proudest boast"'.

[26] The poem, with a paragraph of Englefield's prose, was included in an account of a new technology (the 'Cliff Waggon') for rescuing 'shipwrecked persons' in an 1834 *Saturday Magazine* article entitled 'Preservation of Life from Shipwreck'.

reinforced the natural wonder of this landscape: 'Viewing this scene as merely picturesque, and independently of the feelings of regret naturally excited by the loss of so valuable a ship, it was one of the grandest I have ever witnessed. [. . .] such a wreck, on a spot so extraordinary, formed a combination, which, though not strictly accordant with the rigid laws of picturesque composition, yet was in nature highly sublime' (161). Webster notes that the wreck—together with the spectacle of feverish salvage work—had attracted many other 'spectators, in pleasure yachts and other boats' (162); ironically the same swell that precipitated the wreck now 'made it almost impossible to approach the *Pomone* without danger, but increased very much the picturesque effect of the groupes'.

An unusually large portion of this letter is devoted to aesthetic response (the clay cliffs, too, turn out to be more sublime than picturesque) (158). It also records Webster's important conclusion that the same strata not only run all the way across the island but also reappear (albeit horizontally) at Hordwell Cliff in Hampshire (159–60). Webster's willingness to view the coast from the water—which Englefield apparently had not done—was a prerequisite for the accuracy and comprehensiveness of his observations, as he indicated already in his first letter to Englefield (120). Webster also added a lengthy commentary on the aesthetic appeal of the coast as viewed from the Needles in the 'Explanation of the Plates' included in the front matter to the volume, a rhapsody on its 'romantic' character that merits attention in its own right:

> The wonderfully coloured cliffs of Alum Bay, the lofty and towering chalk precipices of Scratchell's Bay, of the most dazzling whiteness and the most elegant forms, the magnitude and singularity of the spiry insulated masses, which seem at every instant to be shifting their situations, and give a mazy perplexity to the place, the screaming noise of the aquatic birds, the agitation of the sea, and the rapidity of the tide, occasioning not unfrequently a slight degree of danger, all these circumstances combine to raise in the mind unusual emotions, and to give to the scene a character highly singular and even romantic (xvi).

Webster's literary skill in managing the sublime may well owe something to his time with Davy at the Royal Institution, since Davy's geology lectures rely heavily on the sublime as well.[27] The experience of the shipwreck infuses Webster's complex use of 'ruin', which already carries moral and aesthetic connotations, with a sense of catastrophe that enriches his new geological usage even as he sets it apart as geological.

Webster's own adventures by boat had a different purpose from those of the 'spectators' who flocked to the wreck, but his geological purpose did not rule

[27] *Humphry Davy on Geology* 12–13, 77–8, and *passim*. The spectacular plate described here (Pl. XXV) is reproduced in Freeman, *Victorians and the Prehistoric* 15; Kirk, 'Thomas Webster (1772–1844)' 311; and Rudwick, *Bursting the Limits of Time* 515.

out aesthetic pleasure, nor did it preclude his gathering more antiquarian data. The wrecked vessel, for example, 'afforded [him] a scale' for measuring the height of the Needles, which appear deceptively small from a distance (161). When the tide was out he ventured closer, not to view the wreck, but simply to find a station for his drawing, which he finally did, appropriately, on a ruin, 'the stump of the lofty spire of chalk which chiefly gave the name [Needles] to these rocks, and which being worn through by the waves, fell down in 1764' (163). Webster reports that he had about two hours on this six-foot wide rock to make his sketches, 'though completely wetted through by the spray', before the tide began to come in again. In this case Webster draws on an antiquarian work, Sir Richard Worsley's *History of the Isle of Wight* (1781), to historicize the geological monument.[28]

7.5 'THE LANGUAGE OF THE ANTIQUARY'

Worsley (1751–1805) was a significant figure in the antiquarian network that extended from Hamilton in Naples to Englefield in London. Worsley's involvement with the Society of Antiquaries and with the exchange of specimens and collections reflects some of the social and material changes that took place in the network over time. The persistence of the same network over fifty years, however, also creates continuity between the labours of d'Hancarville and Fabris and those of Webster. Worsley, who was for a time the British envoy in Venice, shared Hamilton and d'Hancarville's fascination with ancient cameos and intaglios. In 1791 he purchased Hamilton's second gem collection, notable especially for its 'very fine' cameos (*V & V* 192). Whereas Worsley figures as a historian of the island in Webster's prose, in Englefield's part of the *Description* he appears as a landowner and a connoisseur, particularly of classical antiquities. There is evidence that Englefield looked to Hamilton, too, primarily as a model of connoisseurship. When he hired Henry Moses to engrave his collection of ancient vases in 1819, Hamilton and d'Hancarville's vase book must have loomed large as a precedent.[29] When Englefield

[28] Worsley's book, referred to elsewhere by both Webster and Englefield, includes an engraving of the rock when it was still standing. In addition to Worsley's history, other histories of the Isle of Wight referred to in the *Description* include books by Henry Windham (1793), the Rev. Richard Warner (1795), and especially Charles Tomkins (1796).

[29] *Vases from the Collection of Sir Henry Englefield.* Moses's designs are influenced in many ways by the designs in *AEGR*. An 1848 reprint by H. G. Bohn adds twelve unpublished plates, suggesting that a second volume was in progress at the time of Englefield's death. His descriptions include the names of other collectors from whom he acquired these vases, thereby helping to locate Hamilton, Worsley, and himself within a much larger network of collectors (such as Sir John Coghill) who bought and sold from each other.

cites *Campi Phlegraei* in the *Description*, he points out a resemblance between the colour plate of Webster's Alum Bay drawings (*Description* XIX) and Fabris's depictions of multi-coloured strata (though not of clay) around Naples (79). He qualifies the resemblance as 'picturesque' rather than geological: there is 'nothing similar' to these colourful clay cliffs even in England, but there is a visual parallel to the plate in Fabris's painstakingly naturalistic representations of colour. At the same time Englefield implies a structural parallel between Hamilton's enlightened patronage of Fabris and his own relationship with Webster, praising him in terms reminiscent of the preface to *Campi Phlegraei*.[30] Webster's palette in the colour plate resembles that of Fabris and his staging of boats in coastal views (as in Figure 7.4) also recalls Fabris's style of composition (*CP* XXX, XXXIV). Unlike Fabris, however, Webster used his talent for scientific visualization to move beyond the patronage system to a salaried position with the Geological Society—still largely a gentlemen's society, but more specialized than the Society of Antiquaries in 1812.

Many episodes in Webster's letters document the intersection of human and geological antiquity, and specifically the utility of antiquarianism as a disciplinary framework for geological observation and induction. He recognizes this kinship most explicitly in the passage cited above comparing the Isle of Wight [now Hampshire] Basin to an 'arch or vault', in which he self-consciously decides 'to use the language of the antiquary' (*Description* 201). His observations concerning building stone sometimes prove crucial for determining the order of the strata, and his longest antiquarian digression concerning St Adhelm's Chapel is followed by an account of a more strictly geological antiquity, the 'coal money' found at Kimmeridge, just west of St Adhelm's Head. Having already described the 'Kimmeridge coal' [oil shale] as an effective but foul-smelling local fuel, Webster now touches on specimens that bear obvious signs of human workmanship and appear to be 'some ancient sort of token money' (191).[31] In exploring the coast between Luccombe and Bonchurch on the east side of the island, where Englefield had been bewildered by a massive landslip, Webster uses the language of the antiquary more routinely—the word 'ruin' appears ten times in Letter III. Although the specimens in this case are natural, the practice of the antiquary echoes in Webster's assurance that he 'climbed over the ruins' and 'searched, among the fallen masses, for some specimens of what you had seen' (133).

[30] Hamilton's statement that the Fabris plates display 'as much taste as exactness' (*CP* 5–6) echoes in several of Englefield's passages in praise of Webster. Although the geological histories of the two places are entirely dissimilar, one of Fabris's colour studies does document the production of alum in the colourful deposits of the Solfaterra (*CP* XXV), and the same mineral is connected with the colours of the clay strata in Alum Bay.

[31] Later found to be the central disks from Romano-British armlets. See Calkin, 'Kimmeridge Coal-Money'.

Webster's account amplifies the idea of sublime confusion from Englefield's description of this site (cp. 38), and yet his recognition that 'the whole of this coast is composed of ruins' (138) facilitates some effective geological reasoning: based on his observation of the dip in the remaining horizontal mass, he identifies the massive fault running through the chalk on this part of the coast (137). Along the same stretch of coast, while noting the location of a sandstone bed that was evidently used for 'carved work' in the island's Gothic churches (141), Webster observes 'singular forms in relief' that turn out to be natural (141–43): a new species of fossil Alcyonium.[32] In successive letters the antiquarian framework continues to facilitate the inductive process while also leaving room occasionally for bolder conjectures (e.g. 150) that come to fruition in the final letters reconstituting the chalk formation across south-eastern England as a whole. These two letters (XI–XII), though maintaining a strict insistence on 'deductions from the facts themselves' (198), also defend the antiquarian mode of reasoning by analogy. Webster acknowledges here that 'the fabric, whose ruins we have been contemplating, is of a nature too vast to be completely conceived as a whole', but he maintains that nonetheless 'the mode at least' of 'great changes' may be 'inferred' from the geological fragments 'by a reasoning not unphilosophical' (215–16).

Renewed comparison suggests that the elegiac tone is more pervasive in Englefield, perhaps mirroring his disciplinary affiliation with antiquarianism rather than geology. At the same time, Englefield is deeply invested in an ideology of the picturesque derived from the privilege attached to landed property, an ideology that might be presumed to have a weaker hold on Webster.[33] As a Catholic and an ardent medievalist, Englefield finds cause to lament the decline of architectural and moral standards in both his picturesque and antiquarian chapters. For Webster, 'ruin' as a geological metaphor—apart from the theme of shipwreck—does not carry these connotations.

Like other writers of picturesque tours and prospect poems throughout the eighteenth century, Englefield looks first to gentlemen's seats for instances of the picturesque.[34] He is disappointed both in the number and the quality of great houses with large landscaped grounds (*Description* 87). Worsley's family seat is one exception, but in spite of Worsley's good taste in classical antiquities, he allowed the property to be spoiled (in Englefield's view) by the celebrity landscape designer Capability Brown (88; 104). To make matters worse, at least two other villas on the island are in a Gothic revival style, one of them designed by James Wyatt (57), the architect who became the arch-enemy of Englefield, Gough, and other medievalists by carrying out radical renovations

[32] Detailed in Webster, 'On Some New Varieties of Fossil Alcyonia'.

[33] For Webster's controversial project of educating 'mechanics' at the Royal Institution see Ironmonger, 'The Royal Institution' 139–43.

[34] See further Barrell, *Idea of Landscape*, ch. 1.

in Salisbury (1787–93) and other cathedrals.[35] Englefield's contempt for this shallow Gothicism is readily apparent as he laments 'the infelicity of attempts to imitate the style of our despised ancestors, whose works, while modern science strives in vain to rival, modern insolence brands with the appellations of dark and barbarous' (58).

Although lacking in picturesque credentials, Worsley's property, Appuldurcombe, gets high marks from Englefield for its fine collection of paintings and classical sculpture. More important, Worsley is responsible for the restoration of the hermitage and lighthouse on St Catherine's Hill that prompts one of Englefield's most enthusiastic flights in his chapter on antiquities (94–95). Worsley's antiquarian labours thus benefit both Englefield's and Webster's portions of the book, if in rather different ways. Englefield identifies with him as a gentleman landowner and classically inspired antiquary, but dismisses the decadent modern architecture associated with his class in favour of 'Saxon' ecclesiastical buildings (such as the lighthouse, now known as St Catherine's Oratory). These structures display a 'rude' vigour (101) and suggest a noble morality more in keeping with the 'rude and majestic' coastal scenery (76). Englefield's antiquarianism is a moral discourse to the extent that his encounter with ruins—both geological and architectural—leads him to lament a perceived moral and social decline. He makes a contrast, for example, between the 'venerable inhabitant' of this early Christian lighthouse and modern lighthouse keepers motivated by 'love of idleness' in place of 'sublime benevolence' (95). Webster, however, is less apt to associate ruin with a narrative of decline because it has a primarily geological significance for him. His correspondingly greater emphasis on natural scenery—and thus in a loose sense on public land—also suggests a weakening of the association between the picturesque and landed property. In both cases, however, the investment in ruins is driven by specific intellectual pursuits more than by sentimental nationalism.

Englefield does have his geological moments, and it would be a mistake to overdraw the distinction between his brand of antiquarianism and Webster's. Here too, as we have seen in earlier cases, antiquarianism provided a space for the explicit negotiation, as well as the mediation, of class difference. The epistolary form used by Webster, a basic form for scientific communication since the beginning of the *Philosophical Transactions*, also tended to promote a common language for correspondents. When Englefield admires the 'naked and stupendous ruin' presented by the south front of St Catherine's Hill (11), he is merely setting a scene. But when he notes that Hurst Castle (opposite the northwest shoulder of the island) is set 'on the brow of a submarine cliff' (14), he is also making a significant observation about the

[35] Evans, *History of the Society of Antiquaries* 207–14; cf. Ch. 6 herein.

contours of the Channel. In at least one case his attention to scenery and topography leads to an original geological observation and a credible hypothesis. Reflecting on 'the shattered state of the flints in the chalk' (30), Englefield presents evidence that they could not have been deposited alternately with the chalk and then concludes that they must have been altered by the same event that 'subverted the strata of chalk' as a whole. This problem is the subject of his two brief geological articles, published separately after his initial observations on the island were completed.[36] In the *Description* he illustrates this hypothesis in the best antiquarian fashion by comparing this hypothetical 'immense concussion' to the impact on a wooden pier of a 'vast stone' landed at St Petersburg to form the base for a 'colossal statue' of Peter the Great (33). Both Englefield and Webster, in their different ways, perceived that geological monuments too—however slowly—were being gradually 'effaced from the earth', like the shallow graves of the 1782 shipwreck victims (60).

7.6 GEOLOGICAL CONCLUSIONS

While the elegiac mode is latent rather than manifest throughout the *Description of the Isle of Wight*, the motif of ruin must be central to any understanding of how this book became the forum for Webster's path-breaking geological investigations. Many of Webster's descriptions are initially ambivalent, and could apply equally to geological or antiquarian research: 'I climbed, therefore, over the ruins'—to quote his narrative of the Undercliff again—'and searched, among the fallen masses, for some specimens of what you had seen' (133). Englefield's antiquarian vocabulary provides the medium required by Webster to express the historicity of these geological ruins. It facilitates the reconstruction of stratigraphic sequence for which these letters, sections, and his 1814 essay on the chalk are so justly celebrated.

Webster, however, with his strong background in applied science, concluded his portion of the book by distinguishing geology (as practical) from antiquarian and aesthetic reverie. His penultimate letter differs in tone and content from the others: it is long and boldly speculative, but then seeks to justify its theoretical reach—and perhaps also to question the values of the leisured elite that dominate Englefield's chapters and colour many of his own observations—by underscoring the utility of his model of an 'imaginary vault' or arch (the Isle of Wight Basin) produced by 'wonderful revolutions' (205, 215):

36 Englefield, 'Observations on Some Remarkable Strata' and 'Additional Observations'.

> Now that they are known to be the remnants of a great whole, the original of which can be traced only by its ruins, the subject appears in a more interesting point of view: every detached fact, though by itself of little value, becomes important, as forming a link in the great chain: and Geology, instead of being a collection of the dreams of philosophers, affords among its other important purposes, a basis upon which may be founded the improvement of Agriculture, and hence becomes a science of the highest practical utility. (225)

Though agricultural improvement was certainly important to gentleman landowners, Englefield follows picturesque convention in objecting on aesthetic grounds to agricultural land use (e.g. 58), and his part of the book omits this connection between geology and agriculture.[37]

Webster is optimistic in this letter for good reason. He was summing up two very substantial summers of fieldwork on the island (1811) and the opposite coast of Dorset (1812) and writing from London, where he had had significant time to elaborate the hypothesis of a regional fold structure (the 'arch or vault' depicted in Figure 7.1) and test its merit by reconstructing the sequence of strata across a much larger region. He wrote one final, unexpected letter from the island the next winter (February 1813) explaining to Englefield that he had returned once more—evidently at his own cost—to investigate more concretely the relationships between the Isle of Wight Basin and the Paris Basin described by Cuvier and Alexandre Brongniart (226).[38] Webster had evidently read Cuvier and Brongniart's memoir before he began his work for Englefield, but according this letter, he had now seen for the first time (winter 1812–13) the specimens that Brongniart sent to London to illustrate the memoir. He was evidently embarking on the more extended fieldwork along the coast that would lead to his 1814 essay 'On the freshwater formations'. This last letter contains the kernel of that paper's argument and includes a detailed table showing 'The Order of the Upper Strata' (235–37). Here Webster also fleshes out into a narrative the hypothetical abstract of deposition and 'denudation' included in the previous letter (223–24). His research was beginning to bear geological fruit, instilling a confidence also apparent in the 'Explanation of the Plates', composed last, as when he argues the originality and importance of his observations at Handfast Point in Dorset (Figure 7.4): 'no one had yet noticed

[37] Here Englefield is most likely following William Gilpin, the father of picturesque theory, who dismissed marks of 'cultivation' in the landscape as 'disgusting' in *Observations on Several Parts of England* I.6–7.

[38] In March 1822 Webster wrote to T. R. Underwood concerning the cost of his return trips to the island, beginning in 1813 and continuing periodically at least to the early 1820s when his views were under attack by George Sowerby and others: 'I have expended no less than 200 pounds out of my own pocket' in 'my Isle of Wight affair'. A letter to Brongniart, dated November of the same year, affirms that Brongniart and Cuvier's memoir on the Paris Basin originally inspired this work. Both letters are printed in Challinor, 'Some Correspondence [II]' 153–55, 168–70.

strata of chalk *quite* vertical', he writes, and this unparalleled phenomenon seems to demand an entirely new order of explanation (xvii).[39]

Although Webster's correspondence, much of it published from manuscript by John Challinor, abundantly demonstrates the respect that Webster accrued from geologists such as W. H. Fitton and William Conybeare, it also tells a more melancholy story. As keeper of the museum, librarian, house secretary, and de facto editor of the Geological Society's *Transactions*, Webster had little opportunity to pursue additional fieldwork or publication.[40] Many of his letters between 1812 and 1826 (when he resigned) express his frustration with these time constraints, his inadequate salary, and what he perceived to be his subordinate position in the Geological Society. As Wendy Kirk has put it, 'his time at the Geological Society was not a particularly happy one'. In Webster's own words from an 1822 letter to T. R. Underwood, 'I do not think meanly of Englishmen generally, but I have got among a bad lot'.[41] New evidence indicates that he prepared and attempted to publish his own book on the geology of the Isle of Wight at this stage. An uncatalogued publisher's prospectus recently discovered by Hugh Torrens in the British Library, dated 1827, attempts to promote the publication of this book, but there is no further record of its existence. In this *Proposal for Publishing (by Subscription) A Description of the Isle of Wight* Webster describes himself as a 'professional geologist' and emphasizes both his priority as discoverer of the island's tertiary formations and his qualifications as an educator: the volume, he says, would be aimed primarily at the 'young geologist' and priced at no more than ten shillings.[42] Webster lectured on geology until his death in 1844, but evidently continued to struggle financially even after he was named Professor of Geology at the University of London (now UCL) in 1841.

In 1822 he wrote to Adam Sedgwick gracefully but somewhat bitterly pleading for the respect that he felt eluded him as 'a professional man dependent upon only his own exertions for the means of existence'.[43] After the heady days of Webster's Isle of Wight research, being a professional geologist turned out to mean administrative work, editing, and lecturing for a return of not much more than £100 a year. This was Webster's salary at the Geological Society, and although his initial contract only called for a three-day work week, all sources indicate that his manifold duties there increasingly took

[39] A reference to the 'upper freshwater formation' (*Description* xv) and a correlation with the Paris Basin (xi) in the 'Explanation of the Plates' prove that it was written after the 1813 letter.

[40] Challinor, 'Some Correspondence [I]' 175.

[41] Kirk, 'Thomas Webster (1772–1844)' 311; Challinor, 'Some Correspondence [II]' 155.

[42] *Proposal* 2. Thanks to Hugh Torrens for sharing this discovery. Webster's letters make it clear that he was occupied for some years with defending his views of the Isle of Wight and collecting more evidence. Presumably he intended to include this material in the book, which must be more or less identical with the work in progress alluded to in his letters.

[43] Challinor, 'Some Correspondence [II]' 148.

up all his time. According to one recent authority, the minimum weekly income of a gentleman in the early nineteenth century was £5, and the 'large constituency of potential readers' in the next-lowest bracket of 50–100s. (£125–250 p.a.) could afford to buy books only occasionally.[44] Based on what we know about his income, it seems that Webster would have entered the ranks of the book-buying public only in a very good year. In 1841, when Fitton testified that Webster's works '[were] referred to by all the geologists in Europe', Webster applied for and received a government pension of £50 per annum. Only during his most successful year at the University of London, based on figures provided by Kirk, would he have equalled his combined income from Geological Society work and occasional lecturing.[45] Yet Webster was chosen over other candidates by the university because of his professional experience and his bent for applied science: he was clearly the best man to teach geology to future engineers.

In light of these professional ambitions, it is thoroughly ironic that a stint of old-fashioned aristocratic patronage in 1811–12 allowed Webster to do his only substantial research. As a largely self-taught geologist with a professional background in art and architecture, he was able in this context to adapt the languages of antiquarianism and the picturesque to scientific ends and set new standards for an emerging geology. This accomplishment would have been impossible at a later stage when geologists' curricula became more standardized and their social status more secure. For the same reasons, however, Webster's career in geology never fulfilled the promise of his pioneering research on the Isle of Wight and the chalk formation. Even in 1827, some fifty-four years after Stanfield Parkinson celebrated his brother's 'profession of natural history', the word still carries the sense of 'vocation' rather than remunerative employment. Knowledge work of the kind taken up later by researchers in universities was still very far from being a 'profession' as law and medicine were. William Buckland's more impressive career as Reader in Geology at Oxford (1819–25) comes closer to the twentieth-century paradigm of professional prestige.[46] This contrast, however, also illustrates more fundamental incongruities of status and intellectual achievement that persist beyond historical transformations of intellectual life.

[44] Edwards, 'Thomas Webster (circa 1772–1844)' 472n51; St Clair, *Reading Nation* 193–5.

[45] Kirk, 'Thomas Webster (1772–1844)' 313, 325n. 9.

[46] Buckland was also (from 1824) President of the Geological Society and probably the main obstacle to Webster's salary negotiations in 1826, the breakdown of which led to his resignation. In contrast to the classically educated Oxford clergyman Buckland, Webster's education at Aberdeen was limited and broken off early for financial reasons.

Conclusion

Thomas Chatterton coyly questioned the authenticity of ancient heroes when he 'defied' Homer and Virgil to 'produce' a hero comparable to Sir Charles Baldwin, a local celebrity in fifteenth-century Bristol. Chatterton's challenge appears in one of several footnotes to his *Bristowe Tragedie* (1768), a long narrative in ballad stanzas 'supposed to have been written' by a monk named Thomas Rowley. The last of these notes reminds Chatterton's antiquarian readers that their manuscript collections too may prove on closer inspection to hold legitimate antiquities 'of our native country'. The proliferation of antiquities has an especially problematic sense in Chatterton's case, since he multiplied authenticating documents and regularly offered 'copys of several ancient poems' to James Dodsley and other publishers. As Margaret Russett observes, with Chatterton, 'the ancient is actually modern; the youth is ancient; the copy is the original; the original is the copy; the author is a fictional character; the editor is the author'.[1]

William Smith, born just a year before Chatterton's death, reclaimed antiquarianism for more scientific purposes: 'the antiquities of the earth', in Smith's *Stratigraphical System of Organized Fossils*, document a different order of time and require no transcription or translation, being composed in 'characters on which all must agree'. Nonetheless, like Chatterton, Smith found his antiquities in unsuspected provincial archives—coal canal excavations, in this case—and claimed them for the nation. The analogy between fossils and artefacts remains a slippery one because it comes from an ancient text itself, Pliny's *Natural History*. Not long before Smith, John Whitehurst used the same analogy to argue that human history should be backdated to match the expanding antiquity of the earth—a solution to the problem of deep time that proved less popular with modern geology.[2]

John Carter's antiquarian science of architecture shares something both of the solidity of the strata and of the zeal of invented literary history. The middle term is excavation, a source of evidence cited frequently by Carter, along with

[1] Chatterton, *Complete Works* I. 17, 20, 157 (cp. 51–56, 171–72); Russett, *Fictions and Fakes* 64–5.
[2] Smith, *Stratigraphical System* ix–x; Whitehurst, *Inquiry* 272.

his own professional qualifications, in his public defence of the nation's antiquities. One of among the more prolific writers treated in this study, Carter also made a significant lexical contribution to the proliferation of antiquities simply by reclaiming the epithet 'ancient'. The word occurs regularly over twenty years in his monthly *Gentleman's Magazine* column as an epithet for Gothic architecture—'Gothic' being a misnomer, in Carter's view—as well as in the titles of his other works, such as *Specimens of the Ancient Sculpture and Painting, Now Remaining in this Kingdom* (1780–94).[3] His column in the *Gentleman's*, published under the pseudonym 'An Architect', allowed him to cultivate the fiery populist rhetoric that may have been his strongest contribution toward establishing a preservation movement. One aspect of the recovery of ancient British architecture that Carter clearly relished was the need to repel architectural classicism as an 'invader', a significant gesture in the late 1790s. His dismissal of 'armless Venuses' occurs in the middle of a diatribe that helped to stop the aggressive programme of architectural modernization in Durham Cathedral. He complains of statues of kings and bishops having been 'thrown down' from the east front of the cathedral and notices others 'lying half buried in the ground . . . Is it thus with your noseless Mark Antonys and Cleopatras, your armless Venuses, your legless Herculeses ? . . . Do these collections of 'marbles' lie a rubbish show? No; gilded domes and jewelled cabinets shroud them round; unlimited adoration is their meed'. The native 'grit-stone' sculptures of Edward, Eleanora, and other British worthies are simply not as rare as the classical marbles, but Carter makes this abundance a point in their favour, as it contributes greatly toward the proliferation of antiquities.[4]

Carter and other knowledge workers often had to reckon more fully with the class conflict staged here as an assault on connoisseurs and classicism. Class and status difference mattered tremendously for these practical scholars, less as an openly contested issue than as an incentive to elevate the status of their work from menial to intellectual labour. Smith, a surveyor by profession, failed to achieve any acknowledgment, much less membership, from the Geological Society until very late in life. Chatterton, though humbly born, might in time have achieved some recognition as an antiquary, and the 'Rowley' poems found a scholarly champion after Chatterton's death in Dean Jeremiah Milles, then president of the Society of Antiquaries. Whitehurst, a clockmaker by trade, was recognized with a fellowship in the Royal

[3] For one characteristic example, see Carter, *Pursuits*, rpt. *Gentleman's Magazine Library* X.88–9. The first influential use of the word 'ancient' in English discussions of medieval antiquities is probably that of Batty and Thomas Langley, *Ancient Architecture* (1742).

[4] *Gentleman's Magazine Library* X.89, XI.180–1. Robert Browning's famous bishop is less sure about the virtues of 'grit-stone', though the word provides a curious English trace in the otherwise Italian context of 'The Bishop Orders His Tomb at St. Praxed's Church' (line 116).

Society after the publication of his book, but he had powerful support from fellow members of the Birmingham Lunar Society.

For the knowledge workers featured in this study, who relied on patronage to some degree, the obstacles were more formidable. Schnebbelie died at thirty-one from 'too intense an application to his professional engagements', or what might be construed as an effort to overcome structural obstacles to professional advancement.[5] These words from Schnebbelie's obituary could also be applied to Sydney Parkinson, another scientifically minded young draftsman who worked for two and a half years under even more dangerous conditions before succumbing to dysentery on the *Endeavour*'s homeward voyage. His brother and heir, Stanfield, though often dismissed as paranoid, nonetheless rightly anticipated the prestige of Sydney's collection of South Seas artefacts, a class of objects increasingly in demand under the guise of modern antiquities. Stanfield—perhaps with some help from his ghostwriter William Kenrick—advanced claims to independent expertise and professional status on Sydney's behalf that are comparable to those of Schnebbelie and Carter. Banks, he insisted, was an employer and not a patron. Stanfield was also canny enough to note Sydney's collection of artefacts as a legitimating adjunct of what he called his brother's 'profession of natural history'.[6]

The empirical study of antiquities developed in tandem with natural history to open new domains of non-classical antiquity. Rediscovered British antiquities and newly discovered South Seas artefacts appealed to readers and spectators in the self-consciously imperial centre of Britain for some of the same reasons. According to John Reinhold Forster's representative description of Pacific artefacts, they were relics of the 'ancient systems' of arts and manufactures established in these tropical islands, 'originally the seat of the human race'.[7] Even on the classic ground of southern Italy, natural history and antiquities became increasingly independent of classical scholarship, as I argued in the middle section of this book. But the classicizing paradigm, a predisciplinary factor, remained more powerful there for obvious reasons. The proliferation of antiquities in the form of artefacts appeared to be the work of nature, in particular Vesuvius, which alternately claimed and then relinquished whole cities. Hamilton, the self-appointed curator of both, assembled many fieldworkers and artists in his pursuit of monumental publications in Naples, like Banks and Gough in London. An important difference is that one of Hamilton's knowledge workers, d'Hancarville, really did achieve the status of a widely recognized expert, more or less at Hamilton's expense. Given enough time, Parkinson and Schnebbelie too might have achieved considerable recognition.

[5] 'Schnebbelie' (Obituary) 189–90.

[6] *Journal of a Voyage to the South Seas* vii.

[7] J. R. Forster, *Observations* 196, 342.

History, however, has neglected the contributions of these knowledge workers, to dwell once more on my deliberately anachronistic name for them.[8] To recover their work is not to diminish the contributions of their patrons, whom I have represented as collaborators engaged in intellectual exchange as much as patronage or exploitation. Banks, Hamilton, Gough, and Englefield certainly deserve their place in the history of ideas, but the privileged or even exclusive association of their names with certain intellectual domains is an artefact of history, or even of the professionalization of history. The standard Anglophone narrative of the rise of global natural history relies too heavily on Banks, the standard narrative of early volcanology and vase collecting relies too heavily on Hamilton, and so on, because of the norm of individual intellectual achievement put in place by nineteenth-century historicism, which in turn mirrored the norm of the professionalized academic disciplines.

The story of these knowledge workers is not a prehistory of professionalism in this sense. Rather, the use of 'profession' and its derivatives in period narratives about their careers stands out precisely because it carries a sense largely different from our own. It is this historical distance that makes their knowledge work visible for what it is—innovative, precariously situated, deeply felt, often autodidactic intellectual work performed in the absence of anything like the status or stability associated with intellectual work in the era of the research university.[9] As Mary Poovey points out in her reading of Adam Smith, 'the disinterestedness of the elite, individualized spokesman for the "public" both depends on and produces another subject position', whose 'undifferentiated' representatives form what Smith calls the 'great body of the people'. Just as this 'great body' can appear a 'rudimentary version of... the "working class"', the class position created by the division of intellectual labour may be recognized retrospectively as that of the knowledge worker.[10]

Decoupling the production of knowledge from high-status individuals also means decoupling it from a narrative of progress. Progress is inscribed not only in the more-or-less exploded 'Whiggish' paradigm of the history of science, but also in the Foucauldian point of view that makes disciplinarity itself the touchstone of modernity. I have argued that natural history, arguably the older and more stable discipline, helped to give a recognizable shape to antiquarianism. Prehistory is one new disciplinary domain made visible by this intersection: generated by natural history and classical scholarship, as well

[8] Although Parkinson's contribution to natural history has been widely acknowledged, he is still dismissed as 'a drawer of flies and flowers rather than an accomplished artist' by Nicholas Thomas, *Cook* 191.

[9] Ironically, that status and stability are now increasingly being denied to intellectual workers at many universities as permanent faculty positions are replaced by sessional and part-time instructors.

[10] Poovey, 'Social Constitution of "Class"' 44–5.

as by the history of civil society, prehistory eventually became the domain of new disciplines such as geology and archaeology (where it became the subject of a strongly defined sub-discipline). This admittedly schematic narrative presents a cyclical and materialist model (disciplines generating domains generating more disciplines) that has more affinities with Andrew Abbott's synchronic *Chaos of Disciplines* than with diachronic models such as Foucault's historical turn or Clifford Siskin's *The Work of Writing*. Luisa Calè, Adriana Craciun, and Luciana Martins have recently proposed 'predisciplinarity' as a rubric for scholarship seeking to avoid 'teleological short circuits' and attend more closely to the 'unruly dynamism' of 'predisciplinary encounters', with their 'varying degrees of agency'.[11]

In this study I have sought to credit antiquarianism and natural history as disciplines in their own right, as cognate sciences of antiquity. I have focused on d'Hancarville particularly because his synthesis mobilizes concepts from scientific voyaging to address the problem of human origins, including the origin of the arts, a strategy that I would describe as a prehistoric turn distinct from the historical turn more familiar from histories of modern knowledge. To put it in the abstract terms suggested by 'predisciplinarity', this book sought to redefine eighteenth-century antiquarianism through its intimate connection with natural history and to show the implications of this connection for the study of European, colonial, and Greco-Roman antiquities. In concrete terms this means, in part, recovering the writing and fieldwork of the knowledge workers who left increasingly visible traces on the knowledge projects of the late Enlightenment, even in the absence of legitimating disciplinary frameworks—or rather, in spite of the now unfamiliar disciplinary structures that enabled their work.

The social function of classical education as a carrier of class privilege partly explains the exclusive association between antiquity and the classics that persists even today—making it necessary to specify 'human antiquity' or 'Hebrew antiquity' (for example) to mark off terrain separate from Greco-Roman antiquity. Commenting on the prestige of the classics, Richard Hamblyn has argued that volcanology is one science that emerged only because a new class of inquirers achieved a mode of observation uncoloured by the 'time-worn classical schedules of aristocratic travel'.[12] This group includes several continental naturalists as well as Whitehurst, who correctly observed evidence of very ancient volcanism without ever leaving his native Derbyshire. Fabris, as

[11] Siskin derives the discipline, 'in its distinctly modern sense', so consequentially from the rise of print culture that the premodern senses of the word are all but erased, even as he acknowledges its classical and pedagogical roots (*Work of Writing* 20). If disciplinarity itself is a 'new classification system', 'constituted' by writing, then the 'distinctly modern form' of disciplines is effectively conflated with disciplinarity itself, the discipline 'as opposed to other forms of knowledge' (57). On 'predisciplinarity', see, again, Calè et al., 'Romantic Disorder'.

[12] 'Private Cabinets and Popular Geology' 195.

the creator of the most naturalistic visual documents of volcanism then existing, might also be included in this category, since he does not emphasize classical landscape elements in most of his plates to *Campi Phlegraei* and leaves the scholarly exposition to Hamilton. But Fabris does render natural history in the context of the living Neapolitan population, including Hamilton's indefatigable guide, Bartolomeo Pumo. This association between nature and labouring-class customs and manners—also known as 'popular antiquities'—maps onto the ethnographic turn of natural history, still present in Webster's treatment of the Isle of Wight. Just as present evidence provokes inferences about past volcanic events, so too the noble and ignoble savages of Naples seem to invite comparison with their classical ancestors.

In *Campi Phlegraei* readers are left to make their own connections between the classicizing text and the ethnographic illustrations. D'Hancarville makes the connection explicit and in doing so finds a powerful instrument for the proliferation of antiquities. He multiplies the times and spaces of antiquity somewhat indiscriminately, so the analogy between his cultural stratigraphy and Hamilton's reconstruction of prehistoric volcanism remains merely suggestive. But d'Hancarville's opportunistic use of his classical education also serves as a caution against associating classicism too exclusively or reductively with social privilege.

Hamilton's vases themselves, though at times manipulated in the engravings, deliver a more solid performance as evidence than do Chatterton's manuscripts on ancient heroes such as Sir Charles Baldwin. Other artefacts, including engraved gems, bronzes, reliefs, and architectural fragments, proliferate in d'Hancarville's text through ornamental engravings; antiquities also proliferate in virtual form, as absent originals that stimulate inspired conjecture, more comparable to James Macpherson's phantasmatic Ossian manuscripts. But if the vase paintings are the oldest available—'the most ancient pieces that can exist', as d'Hancarville says at one point—the culture they document lives on in modern Naples. When d'Hancarville descends from high theory to gloss the enigmatic vase paintings, he displays the situated knowledge of one immersed in Neapolitan popular antiquities, as well as classical antiquity, during the six years that he was involved in publishing and archaeological fieldwork there.[13] D'Hancarville cites contemporary pottery, musical instruments, dancing, hand gestures, and religious rites to explain the ancient vase paintings. This method promotes the romantic notion that classical form is perpetuated through folk memory rather than high culture. This ethnographic junction between natural history

[13] As richly demonstrated by the 2007 special issue of the *Journal of the History of Collections* (19:2), Naples had its own traditions of antiquarianism and natural history, without which Hamilton's and d'Hancarville's work would not have been possible—though they (the latter especially) acknowledged it all too rarely.

and antiquariunism typifies the dynamic contradictions that transformed the study of antiquity. Newly inflected by the anticlassicism of early Gothic revivalists such as Carter and by the ethnographic interests of artists and voyagers such as Fabrís and Parkinson, classical antiquity itself merged with the rapidly expanding domain of antiquities first charted by the knowledge workers of the late eighteenth century.

Bibliography

MS sources

British Library

BL pressmark 1879.b.1. Thomas Webster, *Proposal for Publishing (by Subscription) a Description of the Isle of Wight*, in *Prospectuses of Books, Engravings, Lithographs, &c. collected, with the view, not only of showing the state of the literature, arts, & sciences of the passing day, but as too often illustrative of the vanity of human wishes, etc.*, vol. 4, Dawson Turner Collection.

Royal Society of London

Royal Society MSS 2–9. Padre Antonio Piaggio, *Abbozzi di Fenomeni del Vesuvio con qualche. Riflessioni per la Dissertazione di N. N.*

Society of Antiquaries of London

Antiquarian Society MS 267. *Letters from and to Mr. Jacob Schnebbelie, Draughtsman to the Society of Antiquaries, chiefly addressed to Richard Gough, Esq., and John Nichols Esq., to Which are prefixed Memoirs of Mr. Schnebbelie Mr. Gough & Mr. Nichols.*

Antiquarian Society MS 891/2. Rooke Papers.

Digital Editions and Online Resources

Endeavour Botanical Illustrations. National History Museum in Kensington. <http://www.nhm.ac.uk/jdsml/nature-online/endeavour-botanical/>

South Seas: Voyaging and Cross-Cultural Encounters in the Pacific 1760–1800. National Library of Australia. <http://southseas.nla.gov.au/>

Print Sources—Primary

'The Antiquaries Museum', *Monthly Register of Literature*, 1 (1792), 282.

'An Antiquary', *Gentleman's Magazine*, 73 (1803), 105–6.

—— *Gentleman's Magazine*, 73 (1803), 316.

Albinus, Bernhard Siegfried, *Albinus on Anatomy* ed. Robert Beverly Hale and Terence Coyle (New York: Dover, 1988).

Banks, Joseph, 'Copy of an Original Manuscript Entitled "a Breviate Touching the Order and Governmente of a Nobleman's House," &c.', *Archaeologia*, 13 (1800), 315–89.

—— 'A Speech Delivered to the Royal Society', *Philosophical Transactions of the Royal Society of London*, 71 (1781), 1–7.

Beaglehole, J. C. (ed.), *The Endeavour Journal of Joseph Banks, 1768–1771*, 2 vols (Sydney: Trustees of the Public Library of New South Wales in Association with Angus and Robertson, 1962).

—— (ed.), *The Journals of Captain James Cook on his Voyages of Discovery*, 3 vols (London: Hakluyt Society, 1955).

Bertuch, J. F., *Bertuch's Bilderbuch für Kinder*, 12 vols (Weimar: Verlag des Landes-Industrie-Comptoirs, 1798–1830).

Blake, William, *The Complete Poetry and Prose of William Blake*, ed. David Erdman (New York: Anchor Books, 1988).

Bohls, Elizabeth A. and Ian Duncan (eds), *Travel Writing 1700–1830: An Anthology* (Oxford: Oxford University Press, 2005).

Boydell, John, *A Catalogue of the Pictures in the Shakespeare Gallery* (London, 1789).

Carter, John, *Architectural Antiquities*, ed. G. L. Gomme, vols 10–11 of *The Gentleman's Magazine Library* (London: E. Stock, 1890–1891). Rpt. of *Pursuits of Architectural Innovation*, 1789–1817.

—— 'Durham Cathedral', *Gentleman's Magazine*, 72 (1802), 228–31, 399–402.

—— [J. C.], 'Publication of Cathedrals by the Antiquarian Society', *Gentleman's Magazine*, 73 (1803), 106–7.

—— 'Pursuits of Architectural Innovation', No. II. *Gentleman's Magazine*, 68 (1798), 824–5.

—— 'Pursuits of Architectural Innovation', No. XXXV. *Gentleman's Magazine*, 71 (1801), 413–18.

—— 'Pursuits of Architectural Innovation', No. XXXVI. *Gentleman's Magazine*, 71 (1801), 524–8.

—— 'Pursuits of Architectural Innovation', No. LXIV. *Gentleman's Magazine*, 73 (1803), 735–7.

—— *Specimens of the Ancient Sculpture and Painting, Now Remaining in this Kingdom, from the Earliest Period to the Reign of Henry ye VIII*, 2 vols (London: John Carter, 1780–94).

Challinor, John, 'Some Correspondence of Thomas Webster, Geologist (1773–1844)—I', *Annals of Science*, 17 (1961), 175–95.

—— 'Some Correspondence of Thomas Webster, Geologist (1773–1844)—II', *Annals of Science*, 18 (1961), 147–75.

Chatterton, Thomas, *The Complete Works of Thomas Chatterton*, ed. Donald S. Taylor and Benjamin B. Hoover, 2 vols (Oxford: Clarendon Press, 1971).

Christie, James, *A Disquisition upon Etruscan Vases; Displaying their Probable Connection with the Shews at Eleusis, and the Chinese Feast of Lanterns* (London: Printed by W. Bulmer for T. Becket, 1806).

Cicero, Marcus Tullius, *De re publica, De legibus, with an English translation by Clinton Walker Keyes* (Cambridge: Harvard University Press, 1928).

Cook, James, *A Voyage Towards the South Pole, and Round the World. Performed in His Majesty's Ships the Resolution and Adventure, in the Years, 1772, 1773, 1774, and 1775* (London: Printed for W. Strahan & T. Cadell, 1777).

Davy, Humphry, *Humphry Davy on Geology: The 1805 Lectures for the General Audience*, ed. Robert Siegfried and Robert H. Dott (Madison: University of Wisconsin Press, 1980).

Delany, Mary, *The Autobiography and Correspondence of Mary Granville, Mrs. Delany*, 6 vols (New York: AMS Press, 1974).

Dionysius of Halicarnassus, *Roman Antiquities*, trans. Earnest Cary, 7 vols (Cambridge: Loeb-Harvard University Press, 1937–50).

Edwards, Nicholas, 'Some Correspondence of Thomas Webster (circa 1772–1844), Concerning the Royal Institution', *Annals of Science*, 28 (1972), 43–60.

Englefield, Henry C., 'Additional Observations on some Remarkable Strata of Flint in the Isle of Wight', *Transactions of the Linnean Society*, 6 (1802), 303–8.

—— 'Observations on some remarkable Strata of Flint in a Chalk-pit in the Isle of Wight', *Transactions of the Linnean Society*, 6 (1802), 103–9.

Englefield, Henry C. and Thomas Webster, *A Description of the Principal Picturesque Beauties, Antiquities, and Geological Phaenomena, of the Isle of Wight* (London: Printed by William Bulmer for Payne and Foss, 1816).

Fabris, Pietro, *Raccolta di varii vestimenti ed arti del Regno di Napoli* (1773) ed. Franco Mancini (Napoli: Guida, 1985).

Ferguson, Adam, *An Essay on the History of Civil Society*, ed. Fania Oz-Salzberger (Cambridge: Cambridge University Press, 1995).

Forster, George, *A Voyage Round the World*, ed. Nicholas Thomas and Oliver Berghof, 2 vols (Honolulu: University of Hawaii Press, 2000).

Forster, John Reinhold, *Observations Made During a Voyage Round the World*, ed. Nicholas Thomas, Harriet Guest, and Michael Dettelbach (Honolulu: University of Hawaii Press, 1996).

Fothergill, John, Preface to Sydney Parkinson, *A Journal of a Voyage to the South Seas in His Majesty's Ship, the Endeavour* (London: Caliban, 1984), 1–18.

Gilpin, William, *Observations, on Several Parts of England, Particularly the Mountains and Lakes of Cumberland and Westmoreland, Relative Chiefly to Picturesque Beauty*, 2 vols (3rd edn, London: Printed for T. Cadell and W. Davies, 1808).

—— *Three Essays* (London: Printed for R. Blamire, 1792).

Goethe, Johann Wolfgang von, *Italian Journey*, trans. W. H. Auden and Elizabeth Mayer (San Francisco: North Point, 1982).

Gough, Richard, Introduction, *Archaeologia*, 1 (1770), i–xxxix.

—— *Sepulchral Monuments in Great Britain. Applied to Illustrate the History of Families, Manners, Habits, and Arts, and the Different Periods from the Norman Conquest to the Seventeenth Century*, 2 vols in 5 (London: Printed by J. Nichols, for the author, 1786–96 [1799]).

Guasco, Octavien de, *De l'usage des Statues chez les Anciens. Essai historique* (Brussels: J. L. de Boubers, 1768).

Hamilton, William, 'Account of the Discoveries at Pompeii', *Archaeologia*, 4 (1777), 160–75.

—— 'An Account of an Eruption of Mount Vesuvius', *Philosophical Transactions of the Royal Society of London*, 70 (1780), 42–84.

—— 'An Account of the Late Eruption of Mount Vesuvius', *Philosophical Transactions of the Royal Society of London*, 85 (1795), 73–116.

—— *Campi Phlegraei: Observations on the Volcanos of the Two Sicilies*, 2 vols (Naples: Pietro Fabris, 1776).

—— *Collection of Engravings from Ancient Vases*, 4 vols (Naples: Tischbein, 1791–95 [1793–1803]).

—— *Observations on Mount Vesuvius, Mount Etna, and Other Volcanos* (London: Printed for Thomas Cadell, 1772).

—— 'An account of the Earthquakes which Happened in Italy', *Philosophical Transactions of the Royal Society of London*, 73 (1783), 169–208.

—— *Remarks upon the Nature of the Soil of Naples, and Its Neighborhood* (London: W. Bowyer and J. Nichols, 1771).

—— 'Some Particulars of the Present State of Mount Vesuvius; With the Account of a Journey into the Province of Abruzzo, and a Voyage to the Island of Ponza', *Philosophical Transactions of the Royal Society of London*, 76 (1786), 365–81.

—— *Supplement to the Campi Phlegraei: Being an Account of the Great Eruption of Mount Vesuvius in the month of August 177* (Naples: Pietro Fabris, 1779).

Hancarville, Pierre de [P. F. Hugues], *Collection of Etruscan, Greek, and Roman Antiquities from the Cabinet of Sir William Hamilton*, 4 vols (Naples: F. Morelli, 1766–1767 [1768–1776]).

—— *Recherches sur l'origine, l'esprit et les progrès des arts de la Grèce* (1785), ed. and intr. Burton Feldman, 3 vols (New York: Garland, 1984).

Harriot, Thomas, *A Briefe and True Report of the New Found Land of Virginia*, illus. Theodor de Bry, 1590 (New York: Dover Publications, 1972).

Hawkesworth, John, et al., *An Account of the Voyages Undertaken by the Order of His Present Majesty for Making Discoveries in the Southern Hemisphere, and Successively Performed by Commodore Byron, Captain Wallis, Captain Carteret, and Captain Cook*, 3 vols (London: Printed for W. Strahan and T. Cadell, 1773).

Henderson, W. A., *The Housekeeper's Instructor* (London: Printed and sold by J. Stratford, 1805).

'Hints to the Society of Antiquaries', *Gentleman's Magazine*, 72 (1802), 1181–3.

Hogarth, William, *The Analysis of Beauty* (1753), ed. Ronald Paulson (New Haven: Yale University Press, 1997).

Horace, *Satires, Epistles and Ars Poetica*, trans. H. Rushton Fairclough (Cambridge MA: Loeb-Harvard University Press, 1966).

Hutton, James, 'Theory of the Earth', *Transactions of the Royal Society of Edinburgh*, 1:2 (1788), 209–304.

'Indagator', *Gentleman's Magazine*, 57 (1787), 661–4.

Jameson, Robert (ed. and trans), *Essay on the Theory of the Earth by M. Cuvier* (4th edn, Edinburgh: Blackwood, 1822).

Jorio, Andrea de, *Gesture in Naples and Gesture in Classical Antiquity: A Translation of* La mimica degli antichi, trans. Adam Kendon (Bloomington: Indiana University Press, 2000).

King, Edward, *A Speech Delivered by Edward King* (London: Printed by J. Nichols, 1784).

Knight, Richard Payne, *An Account of the Remains of the Worship of Priapus, Lately Existing at Isernia, in the Kingdom of Naples* (London: Printed by T. Spilsbury, 1786).

Lafitau, Joseph François, *Customs of the American Indians Compared with the Customs of Primitive Time*, ed. and trans. William N. Fenton and Elizabeth L. Moore, 2 vols (1724) (Toronto: Champlain Society, 1974–1977).

Langley, Batty and Thomas Langley, *Ancient Architecture, Restored, and Improved* (London, 1742).

La Pérouse, Jean-François, Comte de, *Voyage de La Pérouse autour du monde*, ed. M. L. A. Milet-Mureau, 4 vols (Paris: Imprimierie de la Republique, 1797).

Laugier, Marc-Antoine, *An Essay on Architecture* (1753), trans. Wolfgang and Anni Herrman (Los Angeles: Hennessey and Ingalls, 1977).

Le Grand d'Aussy, Philibert, 'Mémoire sur les anciennes sépultures nationales . . . et sur un projet de fouilles à faire dans nos départemens', *Mémoires de l'Institut National des Sciences et Arts* [Sciences morales et politiques], 2 (1799), 411–680.

Lessing, G. E., *Laocoön: An Essay on the Limits of Painting and Poetry*, trans. E. A. McCormick (Baltimore: Johns Hopkins University Press, 1984).

Lyell, Charles, *Principles of Geology*, 3 vols (London: John Murray, 1830–33).

Macpherson, James, *The Poems of Ossian and Related Works*, ed. Howard Gaskill (Edinburgh: Edinburgh University Press, 1996).

Millin, Aubin-Louis, *Antiquités Nationales*, 5 vols (Paris: Imprimerie de la République, 1790–99).

Moore, James, *Monastic Remains and Ancient Castles in England and Wales* (London: Printed by George Stafford, and sold by Mr Schnebbelie, 1791).

Morrison, Alfred, *The Collection of Autograph Letters and Historical Documents Formed by Alfred Morrison*, 2nd ser. [1882–1893] ([London]: Printed for Private Circulation, 1893).

[Moses, Henry], *Vases from the Collection of Sir Henry Englefield, Bart., Drawn and Engraved by H. Moses* (London: Printed for Rodwell and Martin, [1820]).

Nichols, John, *The History and Antiquities of the County of Leicester*, 4 vols (London: John Nichols, 1795–1815).

—— *Literary Anecdotes of the Eighteenth Century: Comprising Biographical Memoirs of William Bowyer, Printer, F.S.A., and Many of His Learned Friends; an Incidental View of the Progress and Advancement of Literature in This Kingdom During the Last Century; and Biographical Anecdotes of a Considerable Number of Eminent Writers and Ingenious Artists*, 9 vols (London: John Nichols, 1812–16).

—— and John Bowyer Nichols (eds), *Illustrations of the Literary History of the Eighteenth Century: Consisting of Authentic Memoirs and Original Letters of Eminent Persons; and Intended as a Sequel to the Literary Anecdotes*, 8 vols (London: Nichols & Bentley, 1817–58).

Nietzsche, Friedrich, *On the Advantage and Disadvantage of History for Life*, trans. Peter Preuss (Indianapolis: Hackett, 1980).

Owen, John, *A Compleat and Impartial History of the Ancient Britons from the Earliest Account of Time to the End of the Reign of King Henry VIII* (London: Printed for the Author, 1743).

Parkinson, Sydney and Stanfield Parkinson, *A Journal of a Voyage to the South Seas, in His Majesty's Ship, the* Endeavour (Adelaide: Libraries Board of South Australia, 1972).

Pausanias, *Description of Greece*, trans. W. H. S. Jones, 5 vols (Cambridge, MA: Loeb-Harvard University Press, 1918–35).

Pennant, Thomas, *The Literary Life of the Late Thomas Pennant, Esq., Written by Himself* (London: B. and J. White, and R. Faulder, 1793).

Percy, Thomas, *Reliques of Ancient English Poetry*, 3 vols (London: Printed for J. Dodsley, 1765).

Pinkerton, John ['Philistor'], 'Letters to the People of Great Britain (I)', *Gentleman's Magazine*, 58 (1788), 125–7.

Pliny, *Natural History*, trans. H. Rackham, 10 vols (Cambridge, MA: Loeb-Harvard University Press, 1938–63).

Plutarch, *Moralia*, trans. Frank Cole Babbitt, 16 vols (Cambridge, MA: Loeb-Harvard University Press, 1927–69).

'Preservation of Life from Shipwreck', *The Saturday Magazine*, 4 (1834), 92–3.

Pope, Alexander, 'An Essay on Criticism', *The Poems of Alexander Pope*, ed. John Butt (New Haven: Yale University Press, 1963), 143–68.

Rawdon, Elizabeth, Countess of Moira, 'Particulars Relative to a Human Skeleton', *Archaeologia*, 7 (1785), 90–110.

Rehberg, Friedrich, *Drawings Faithfully Copied from Nature at Naples* (Rome, 1794).

Rev. of *A Journal of a Voyage to the South Seas, in His Majesty's Ship, the* Endeavour. *Critical Review*, 37 (1774), 1–10.

Rev. of *A Journal of a Voyage to the South Seas, in His Majesty's Ship, the* Endeavour. *Gentleman's Magazine*, 55 (1785), 52.

Rev. of *A Journal of a Voyage to the South Seas, in His Majesty's Ship, the* Endeavour. *Monthly Review*, 50 (1774), 410–11.

Reynolds, Joshua, *Discourses on Art*, ed. Robert R. Wark (New Haven: Yale University Press, 1975).

Richardson, George, *A New Collection of Chimney Pieces Ornamented in the Style of Etruscan, Greek, and Roman Architecture* (London: Printed for the Author, 1781).

Rix, Helmut, *Etruskische Texte: editio minor* (Tübingen: G. Narr, 1991).

Robertson, William, *The History of Scotland During the Reigns of Queen Mary and of King James VI*, 2 vols (London: Printed for A. Millar, 1759).

Rousseau, Jean-Jacques, 'Discourse on the Origin and the Foundations of Inequality Among Mankind', *The Social Contract and The First and Second Discourses*, ed. Susan Dunn (New Haven: Yale University Press, 2002), 69–148.

Schiller, Friedrich, 'On Naïve and Sentimental Poetry', *Essays*, ed. Walter Hinderer and Daniel O. Dahlstrom (New York: Continuum, 1993), 179–260.

Schlegel, Friedrich, *Kritische und theoretische Schriften*, ed. Andreas Huyssen (Stuttgart: Reclam, 1978).

'Schnebbelie, Jacob', Obituary. *Gentleman's Magazine*, 62 (1792), 189–90.

Schnebbelie, Jacob, *The Antiquaries Museum, Illustrating the Ancient Architecture, Painting, and Sculpture of Great Britain, from the Time of the Saxons to The . . . Reign of King James I* (London: Printed by J. Nichols, 1791–1800).

Scott, Walter, *The Antiquary*, ed. Nicola J. Watson (Oxford: Oxford University Press, 2002).

'Sir William Hamilton,' Obituary. *Gentleman's Magazine*, 73 (1803), 390.

Smith, Adam, *The Early Writings of Adam Smith*, ed. J. Ralph Lindgren (New York: A. M. Kelley, 1967).

Smith, Charlotte, *Emmeline: The Orphan of the Castle* (1788), ed. Loraine Fletcher. (Peterborough, Ont.: Broadview, 2003).

Smith, James Edward, 'Discourse on the Rise and Progress of Natural History,' *Tracts Relating to Natural History* (London: Printed for the Author by J. Davis, 1798), 49–162.

Smith, William, *Strata Identified by Organized Fossils* (London: W. Arding, 1816).

—— *Stratigraphical System of Organized Fossils* (London: Printed for E. Williams, 1817).

Society of Antiquaries of London, *Queries Proposed to Gentlemen in the Several Parts of Great Britain, in Hope of Obtaining from Their Answers a Better Knowledge of Its Antiquities and Natural History* (London, 1754).

—— *Vetusta Monumenta*, Vol. 2 (London: SAL, 1789).

—— *Vetusta Monumenta*, Vol. 3 (London: SAL, 1796).

Stothard, C. A., *The Monumental Effigies of Great Britain* (London: J. McCreery, 1817).

Tischbein, Wilhelm, *Aus meinem Leben*, ed. Lothar Brieger (Berlin: Propyläen-Verlag, 1922).

Vico, Giambattista, *The New Science of Giambattista Vico* (1744). rev. edn, trans. Thomas Goddard Bergin and Max Harold Fisch (Ithaca: Cornell University Press, 1968).

Virgil, *The Aeneid*, trans. Robert Fitzgerald (New York: Random House, 1983).

Vitruvius, *On Architecture*, trans. Frank Granger, 2 vols (Cambridge, MA: Loeb-Harvard University Press, 1970).

—— *The Ten Books on Architecture*, trans. Morris Hicky Morgan (New York: Dover, 1960).

Walpole, Horace, *Walpoliana*, 2 vols (London: Printed for R. Phillips, [1799]).

—— *The Yale Edition of Horace Walpole's Correspondence*, ed. W. S. Lewis (New Haven: Yale University Press, 1937–).

Warton, Thomas, *The History of English Poetry*, 4 vols (London: Printed for J. Dodsley, 1774–81).

—— *Observations on the Fairy Queen of Spenser*, 2 vols (2nd edn, London: Printed for R. and J. Dodsley, 1762).

Webster, Thomas, 'On Some New Varieties of Fossil Alcyonia', *Transactions of the Geological Society of London*, 2 (1814), 377–87.

—— 'On the freshwater formations in the Isle of Wight, with some observations, on the strata over the chalk in the south-east part of England', *Transactions of the Geological Society of London*, 2 (1814), 161–254.

Wedgwood, Josiah, *Account of the Barberini, Now Portland Vase* ([London], [1788]).

Whitehurst, John, *Inquiry into the Original State and Formation of the Earth* (1786) (2nd edn, New York: Arno Press, 1978).

Winckelmann, Johann Joachim, *Briefe*, ed. Walther Rehm, 4 vols (Berlin: W. de Gruyter, 1952–57).

—— *History of the Art of Antiquity*, trans. Harris Francis Mallgrave (Los Angeles: Getty Research Institute, 2006).

—— *Reflections on the Imitation of Greek Works in Painting and Sculpture* (1755), trans. Elfriede Heyer and Roger C. Norton (La Salle, Ill: Open Court, 1987).

Wolcot, John ['Peter Pindar'], *A Benevolent Epistle to Sylvanus Urban, Esq.* (London: G. Kearsley, 1790).

Wraxall, Nathaniel William, *Historical Memoirs of My Own Time*, 2 vols (2nd edn, London: Printed for T. Cadell and W. Davies, 1815).

Print Sources—Secondary

Abbott, Andrew, *Chaos of Disciplines* (Chicago: University of Chicago Press, 2001).

Adams, Brian, *The Flowering of the Pacific: Being an Account of Joseph Banks' Travels in the South Seas and the Story of his* Florilegium (London: Natural History Museum, 1986).

Amar, A. D., *Managing Knowledge Workers* (Westport, CT: Quorum, 2001).

Amyx, Darrell A., *Corinthian Vase-Painting of the Archaic Period*, 3 vols (Berkeley: University of California Press, 1988).

'Antiquity.' Def. II. 4. b. *Oxford English Dictionary* (online edn, Oxford University Press, 2011).

Arnold, Dana and Stephen Bending (eds), *Tracing Architecture: The Aesthetics of Antiquarianism* (Oxford: Blackwell, 2003).

Aubin, Robert A., 'Grottoes, Geology, and the Gothic Revival', *Studies in Philology*, 31 (1934), 408–16.

Banks, R. E. R., et al. (eds), *Sir Joseph Banks: A Global Perspective* (Kew: Royal Botanic Gardens, 1994).

Barrell, John, *The Idea of Landscape and the Sense of Place, 1730–1840: An Approach to the Poetry of John Clare* (Cambridge: Cambridge University Press, 1972).

Bending, Stephen, 'Every Man is Naturally an Antiquarian: Francis Grose and Polite Antiquities' in Arnold and Bending, *Tracing Architecture*, 100–10.

Benjamin, Walter, *Illuminations*, ed. Hannah Arendt (New York: Schocken, 1969).

Bermingham, Ann, *Landscape and Ideology: The English Rustic Tradition 1740–1860* (Berkeley: University of California Press, 1986).

Bewell, Alan, 'Romanticism and Colonial Natural History', *Studies in Romanticism*, 43:1 (2004), 5–34.

Blunt, Wilfred, 'Sydney Parkinson and His Fellow Artists', in Carr, *Sydney Parkinson*, 14–45.

Briggs, C. Stephen, 'Prehistory in the Nineteenth Century', in Pearce, *Visions of Antiquity*, 226–65.

Brylowe, Thora, 'Of Gothic Architects and Grecian Rods: William Blake, Antiquarianism, and the History of Art', *Romanticism*, 18:1 (2012), 89–104.

—— 'Two Kinds of Collections: Sir William Hamilton's Vases, Real and Represented', *Eighteenth-Century Life*, 32 (2008), 23–52.

Buchanan, Alexandrina, 'Science and Sensibility: Architectural Antiquarianism in the Early Nineteenth Century', in Myrone and Peltz, *Producing the Past*, 169–86.

Butler, Marilyn, 'Antiquarianism (Popular)', in *An Oxford Companion to the Romantic Age: British Culture 1776–1832*, ed. Iain McCalman (Oxford: Oxford University Press, 1999), 328–37.

Calkin, J. B., ' "Kimmeridge Coal-Money": the Romano-British Shale Armlet Industry', *Proceedings of the Dorset Natural History and Archaeological Society*, 75 (1955), 45–71.

Calè, Luisa and Adriana Craciun, 'The Disorder of Things', *Eighteenth-Century Studies*, 45 (2011), 1–13.

—— —— and Luciana Martins, 'Romantic Disorder: Predisciplinarity and the Divisions of Knowledge, 1750-1850', Birkbeck, University of London. <http://www.bbk.ac.uk/18/news/past-events/romantic-disorder-predisciplinarity-and-the-divisions-of-knowledge-1750–1850>

Carr, D. J. (ed.), *Sydney Parkinson: Artist of Cook's Endeavour Voyage* (London: Natural History Museum in Association with Croom Helm, 1983).

Carter, Harold B., *Sir Joseph Banks (1743–1820): A Guide to Biographical and Bibliographical Sources* (Winchester: St Paul's Bibliographies in Association with Natural History Museum, 1987).

Ceserani, Giovanna, 'The Antiquary Alessio Simmaco Mazzocchi: Oriental Origins and the Rediscovery of Magna Graecia in Eighteenth-Century Naples', *Journal of the History of Collections*, 19 (2007), 249–59.

Coe, Kathryn, *The Ancestress Hypothesis: Visual Art as Adaptation* (New Brunswick, NJ: Rutgers University Press, 2003).

Coltman, Viccy, *Fabricating the Antique: Neoclassicism in Britain, 1760–1800* (Chicago: University of Chicago Press, 2006).

—— 'Sir William Hamilton's Vase Publications (1766–1776): A Case Study in the Reproduction and Dissemination of Antiquity', *Journal of Design History*, 14:1 (2001), 1–16.

Constantine, David, *Fields of Fire: A Life of Sir William Hamilton* (London: Weidenfeld and Nicolson, 2001).

—— 'Winckelmann and Sir William Hamilton', *Oxford German Studies*, 22 (1993), 55–83.

Coote, Jeremy, 'An Interim Report on a Previously Unknown Collection from Cook's First Voyage', *Journal of Museum Ethnography*, 16 (2004), 111–21.

—— 'Joseph Banks's Forty Brass Patus', *Journal of Museum Ethnography*, 20 (2008), 49–68.

—— Peter Gathercole, and Nicolette Meister, '"Curiosities Sent to Oxford": The Original Documentation of the Forster Collection at the Pitt Rivers Museum', *Journal of the History of Collections*, 12:2 (2000), 177–92.

[Coote, Jeremy], *Curiosities from the* Endeavour: *A Forgotten Collection - Pacific Artefacts Given by Joseph Banks to Christ Church, Oxford after the First Voyage* (Whitby: Captain Cook Memorial Museum, 2004).

Cox, J. Charles and R. M. Serjeantson, *A History of the Church of the Holy Sepulchre, Northampton* (Northampton: William Mark, 1897).

Cox, Jeffrey N., *Poetry and Politics in the Cockney School: Keats, Shelley, Hunt and Their Circle* (Cambridge: Cambridge University Press, 1998).

Craciun, Adriana, 'What Is an Explorer?', *Eighteenth-Century Studies*, 45 (2011), 29–51.

Crook, J. Mordaunt, *John Carter and the Mind of the Gothic Revival* (London: Society of Antiquaries, 1995).

Cunningham, Andrew and Nicholas Jardine (eds), *Romanticism and the Sciences* (Cambridge: Cambridge University Press, 1990).

Daston, Lorraine J. (ed.), *Biographies of Scientific Objects* (Chicago: University of Chicago Press, 2000).

—— (ed.), *Things That Talk: Object Lessons from Art and Science* (New York: Zone Books, 2004).

—— and Katharine Park, *Wonders and the Order of Nature* (New York: Zone Books, 1998).

Dean, Dennis R., *Romantic Landscapes: Geology and Its Cultural Influence in Britain, 1765–1835* (Ann Arbor: Scholars' Facsimiles and Reprints, 2007).

—— 'The Word "Geology" ', *Annals of Science*, 36 (1979), 35–43.

Décultot, Elisabeth, 'Comment l'art est-il venu aux Grecs? Winckelmann face à Shaftesbury, Caylus, et Herder', *Klassizismen und Kosmopolitismus: Programm oder Problem? Austausch in Kunst und Kunsttheorie im 18. Jahrhundert*, ed. P. Griener and K. Imesch (Zürich: Schweizerisches Institut für Kunstwissenschaft, 2004), 45–58.

Derrida, Jacques, *Of Grammatology* (1967), trans. Gayatri Spivak (Baltimore: Johns Hopkins University Press, 1976).

Durrans, Brian, 'Ancient Pacific Voyaging', in Mitchell, *Captain Cook and the South Pacific*, 137–66.

Edwards, Nicholas, 'Thomas Webster (*Circa* 1772–1844)', *Journal of the Society for the Bibliography of Natural History*, 5 (1971), 468–73.

—— 'Webster, Thomas (1772–1844)', *Oxford Dictionary of National Biography* (online edn, Oxford: Oxford University Press, 2004).

Engelstein, Stefani, *Anxious Anatomy: The Conception of the Human Form in Literary and Naturalist Discourse* (Albany: SUNY Press, 2008).

Evans, Joan, *A History of the Society of Antiquaries* (London: Society of Antiquaries, 1956).

Findlen, Paula, *Possessing Nature: Museums, Collecting, and Scientific Culture in Early Modern Italy* (Berkeley: University of California Press, 1994).

Foucault, Michel, *The Order of Things. An Archaeology of the Human Sciences* (New York: Random House, 1970).

Freeman, Michael J., *Victorians and the Prehistoric: Tracks to a Lost World* (New Haven: Yale University Press, 2004).

Fulford, Tim, Debbie Lee, and Peter J. Kitson, *Literature, Science, and Exploration in the Romantic Era: Bodies of Knowledge* (Cambridge: Cambridge University Press, 2004).

Gaimster, David, Sarah McCarthy, and Bernard Nurse (eds), *Making History: Antiquaries in Britain, 1707–2007* (London: Royal Academy, 2007).

Gascoigne, John, *Joseph Banks and the English Enlightenment: Useful Knowledge and Polite Culture* (Cambridge: Cambridge University Press, 1994).

—— *Science in the Service of Empire: Joseph Banks, the British State and the Uses of Science in the Age of Revolution* (Cambridge: Cambridge University Press, 1998).

Gates, Barbara T. and Anne Shteir (eds), *Natural Eloquence: Women Reinscribe Science* (Madison: University of Wisconsin Press, 1997).

Gell, Alfred, *Art and Agency: An Anthropological Theory* (Oxford: Oxford University Press, 1998).

Gisler-Huwiler, M., et al. (eds), *The Complete Collection of Antiquities from the Cabinet of Sir William Hamilton* (Köln: Taschen, 2004).

Goloubew, Victor, 'Les tambours magiques en Mongolie', *Bulletin de l'Ecole française d'Extrême-Orient,* 23 (1923), 407–9.

Gramsci, Antonio, *Selections from the Prison Notebooks of Antonio Gramsci,* ed. Quintin Hoare and Geoffrey Nowell Smith (New York: International, 1971).

Greenberg, Daniel S., *Science for Sale: The Perils, Rewards, and Delusions of Campus Capitalism* (Chicago: University of Chicago Press, 2007).

Griener, Pascal, *Le Antichità Etrusche Greche e Romane 1766–1776 di Pierre Hugues d'Hancarville. La pubblicazione delle ceramiche antiche della prima collezione Hamilton* (Rome: Edizione dell'Elefante, 1992).

Hall, Marie Boas, *All Scientists Now: The Royal Society in the Nineteenth Century* (New York: Cambridge University Press, 1984).

Hamblyn, Richard, *Landscape and the Contours of Knowledge: The Literature of Travel and the Sciences of the Earth in Eighteenth-Century Britain* (Diss: Cambridge University Press, 1994).

—— 'Private Cabinets and Popular Geology: The British Audiences for Volcanoes in the Eighteenth Century', in Chloe Chard and Helen Langdon (eds), *Transports: Travel, Pleasure, and Imaginative Geography, 1600–1830* (New Haven: Yale University Press, 1996), 179–205.

Harding, Sandra, *Sciences from Below: Feminisms, Postcolonialities, and Modernities* (Durham, NC: Duke University Press, 2008).

Haskell, Francis, 'The Baron d'Hancarville: An Adventurer and Art Historian in Eighteenth-Century Europe', in *Past and Present in Art and Taste* (New Haven: Yale University Press, 1987), 30–45.

—— and Nicholas Penny, *Taste and the Antique: The Lure of Classical Sculpture, 1500–1900* (New Haven: Yale University Press, 1981).

Hauser-Schäublin, Brigitte and Gundolf Krüger (eds), *James Cook: Gifts and Treasures from the South Seas* (Munich: Prestel, 1998).

Heringman, Noah, 'Deep Time in the South Pacific: Scientific Voyaging and the Ancient/Primitive Analogy', in Joel Falfak (ed.), *Marking Time: Romanticism and Evolution* (Toronto: University of Toronto Press, 2013).

—— 'Natural History in the Romantic Period', *A Concise Companion to the Romantic Age,* ed. Jon Klancher (Oxford: Blackwell, 2009), 141–67.

—— 'Peter Pindar, Joseph Banks, and the Case against Natural History', *The Wordsworth Circle,* 35 (2004), 21–30.

—— (ed.), *Romantic Science: The Literary Forms of Natural History* (Albany: SUNY Press, 2003).

—— 'Stones So Wonderous Cheap', *Studies in Romanticism,* 37:1 (1998), 43–62.

—— 'The Style of Natural Catastrophes', *Huntington Library Quarterly,* 66 (2003), 97–133.

—— ' "Very vain is Science proudest boast": the resistance to geological theory in early nineteenth-century England', in Gary D. Rosenberg (ed.), *The Revolution in Geology from the Renaissance to the Enlightenment* (Boulder: Geological Society of America, 2009), 235–45.

Herries Davies, Gordon L., *Whatever Is under the Earth: The Geological Society of London, 1807–2007* (London: Geological Society, 2007).

Holmes, Maurice, *Captain James Cook, A Bibliographical Excursion* (New York: B. Franklin, 1968).

Hoock, Holger, *Empires of the Imagination: Politics, War and the Arts in the British World, 1750–1850* (London: Profile Books, 2010).

Imbruglia, Girolamo (ed.), *Naples in the Eighteenth Century: The Birth and Death of a Nation-State* (Cambridge: Cambridge University Press, 2000).

Ironmonger, Edward, 'The Royal Institution and the Teaching of Science in the 19th Century', *Proceedings of the Royal Institution*, 37 (1958), 139–58.

Janowitz, Anne F., *England's Ruins: Poetic Purpose and the National Landscape* (Cambridge, MA: Blackwell, 1990).

Jardine, Nicholas, 'Whigs and Stories: Herbert Butterfield and the Historiography of Science', *History of Science*, 41 (2003), 125–40.

—— James A. Secord, and Emma C. Spary (eds), *Cultures of Natural History* (Cambridge: Cambridge University Press, 1996).

Jasanoff, Maya, *Edge of Empire: Lives, Culture, and Conquest in the East, 1750–1850* (New York: Knopf, 2005).

Jenkins, Ian, ' "Contemporary Minds": Sir William Hamilton's Affair with Antiquity', in Jenkins and Sloan, *Vases and Volcanoes*, 40–64.

—— and Kim Sloan (eds), *Vases & Volcanoes: Sir William Hamilton and His Collection* (London: British Museum, 1996).

Joppien, Rüdiger and Bernard Smith, *The Art of Captain Cook's Voyages*, 3 vols (New Haven: Yale University Press, 1985–88).

Justi, Carl, *Winckelmann und seine Zeitgenossen* (1872), 3 vols, ed. Walther Rehm (Köln: Phaidon, 1956).

Kaeppler, Adrienne, *'Artifical Curiosities': Being an Exposition of Native Manufactures Collected on the Three Pacific Voyages of Captain James Cook* (Honolulu: Bishop Museum Press, 1978).

—— 'The Göttingen Collection in an International Context', in Hauser-Schäublin and Krüger, *James Cook: Gifts and Treasures*, 86–93.

—— 'Tracing the History of Hawaiian Cook Voyage Artifacts in the Museum of Mankind', in Mitchell, *Captain Cook and the South Pacific*, 167–98.

Kelly, Jason, *The Society of Dilettanti: Archaeology and Identity in the British Enlightenment* (New Haven: Yale University Press, 2009).

Kirk, Wendy, 'Thomas Webster (1772–1844): First Professor of Geology at University College London', *Archives of Natural History*, 23 (1996), 309–26.

Klancher, Jon (ed.), Romanticism and its Publics: A Forum, *Studies in Romanticism*, 33 (1994).

Kleinman, Daniel, *Impure Cultures: University Biology and the World of Commerce* (Madison: University of Wisconsin, 2003).

Knight, Carlo, 'William Hamilton and the "Art of Going Through Life"', in Jenkins and Sloan, *Vases and Volcanoes*, 11–23.

—— *Hamilton a Napoli: cultura, svaghi, civiltà di una grande capitale europea* (Napoli: Electa, 1990).

—— 'Un Inedito di Padre Piaggio', *Rendiconti della Accademia di Archeologia, Lettere, e Belle Arti*, 62 (1989–90), 59–131.

—— 'Sir William Hamilton's "Campi Phlegraei" and the Artistic Contribution of Peter Fabris', in E. Chaney and N. Ritchie (eds), *Oxford, China, and Italy: Writings in Honour of Sir Harold Acton on His Eightieth Birthday* (London: Thames and Hudson, 1984), 192–208.

Krafft, Maurice, *Volcanoes: Fire from the Earth* (London: Thames and Hudson, 1993).

Kramnick, Jonathan Brody, 'The Making of the English Canon', *PMLA*, 112 (1997), 1087–1101.

Krüger, Gundolf, *'Sportlicher Wettkampf' auf Hawaii: eine Konfiguration und ihr Wandel als Gegenstand ethnohistorischer Forschung* (Göttingen: Edition Herodot, 1986).

—— 'Tahiti and the Society Islands—An Archipelago as Seen in the Light of European Yearnings', in Hauser-Schäublin and Krüger, *James Cook: Gifts and Treasures*, 141–72.

Lamb, Jonathan, 'Circumstances Surrounding the Death of John Hawkesworth', *Eighteenth-Century Life*, 18:3 (1994), 97–113.

—— *Preserving the Self in the South Seas, 1680–1840* (Chicago: University of Chicago Press, 2001).

—— Vanessa Smith, and Nicholas Thomas (eds), *Exploration and Exchange: A South Seas Anthology, 1680–1900* (Chicago: University of Chicago Press, 2000).

Lansdown, Richard, *Strangers in the South Seas: The Idea of the Pacific in Western Thought* (Honolulu: University of Hawaii Press, 2006).

Larson, Magali S., *The Rise of Professionalism: A Sociological Analysis* (Berkeley: University of California Press, 1977).

Latour, Bruno, 'On the Partial Existence of Existing *and* Nonexisting Objects', in Daston, *Biographies of Scientific Objects*, 247–69.

—— *Reassembling the Social: An Introduction to Actor-Network Theory* (Oxford: Oxford University Press, 2005).

Lepenies, Wolf, *Das Ende der Naturgeschichte: Wandel kultureller Selbstverständlichkeiten in den Wissenschaften des 18. und 19. Jahrhundert* (Munich: Hanser, 1976).

Lewis, Michael J., *The Gothic Revival* (London: Thames and Hudson, 2002).

Lissarrague, François and Marcia Reed, 'The Collector's Books', *Journal of the History of Collections*, 9:2 (1997), 275–94.

Little, Stephen and Peter Ruthenberg (eds), *Life in the Pacific of the 1700s: The Cook/Forster Collection of the University of Göttingen*, 3 vols (Honolulu: Honolulu Academy of the Arts, 2006).

Liu, Alan, *The Laws of Cool: Knowledge Work and the Culture of Information* (Chicago: University of Chicago Press, 2004).

Lolla, Maria Grazia, 'Ceci n'est pas un monument: *Vetusta Monumenta* and Antiquarian Aesthetics', in Myrone and Peltz, *Producing the Past*, 15–34.

Lyons, Claire, 'The Neapolitan Context of Hamilton's Antiquities Collection', *Journal of the History of Collections*, 9:2 (1997), 229–39.

Lysaght, Averil, 'Banks's Artists and His *Endeavour* Collections', in Mitchell, *Captain Cook and the South Pacific*, 9–80.

MacKendrick, Paul, *The Mute Stones Speak: The Story of Archaeology in Italy* (2nd edn, New York: Norton, 1983).

McLane, Maureen, *Balladeering, Minstrelsy, and the Making of British Romantic Poetry* (Cambridge: Cambridge University Press, 2008).

Meek, Ronald, *Social Science and the Ignoble Savage* (Cambridge: Cambridge University Press, 1976).

Megill, Kenneth, *Thinking for a Living: The Coming Age of Knowledge Work* (München: K. G. Saur, 2004).

Menely, Tobias, 'Traveling in Place: Gilbert White's Cosmopolitan Parochialism', *Eighteenth-Century Life*, 28 (2004), 46–65.

Miller, David Philip, 'Joseph Banks, Empire, and "Centers of Calculation" in Late Hanoverian London', in Miller and Reill, *Visions of Empire*, 21–37.

—— and Peter Hanns Reill (eds), *Visions of Empire: Voyages, Botany, and Representations of Nature* (Cambridge: Cambridge University Press, 1996).

Mitchell, T. C. (ed.), *Captain Cook and the South Pacific*, The British Museum Yearbook 3 (London: British Museum Publications, 1979).

Momigliano, Arnaldo, *The Classical Foundations of Modern Historiography* (Berkeley: University of California Press, 1990).

Moore, D. T., 'Sir William Hamilton's Volcanology and His Involvement in *Campi Phlegraei*', *Archives of Natural History*, 21 (1994), 169–93.

Myrone, Martin, 'The Society of Antiquaries and the Graphic Arts: George Vertue and His Legacy', in Pearce, *Visions of Antiquity*, 98–121.

—— and Lucy Peltz (eds), *Producing the Past: Aspects of Antiquarian Culture and Practice, 1700–1850* (Aldershot: Ashgate, 1999).

Neill, Michael, 'From the Editor', *Shakespeare Quarterly*, 52:4 (2001), iii–x.

Nevers, Kevin, 'Immovable Objects, Irresistible Forces: The Sublime and the Technological in the Eighteenth Century', *Eighteenth-Century Life*, 19:1 (1995), 18–38.

Nolta, David, 'The Body of the Collector and the Collected Body in William Hamilton's Naples', *Eighteenth-Century Studies*, 31 (1997), 108–14.

Nurse, Bernard, 'Englefield, Sir Henry (1752–1822)', *Oxford Dictionary of National Biography* (online edn, Oxford: Oxford University Press, 2004).

O'Connor, Ralph, 'Facts and Fancies: The Geological Society of London and the Wider Public, 1807–1837', in C. L. E. Lewis and S. J. Knell (eds), *The Making of the Geological Society of London* (London: Geological Society, 2009), 331–40.

—— *The Earth on Show: Fossils and the Poetics of Popular Science, 1802–1856* (Chicago: University of Chicago Press, 2007).

O'Donoghue, F. M. and Rosie Dias, 'Schnebbelie, Jacob (1760–1792)', *Oxford Dictionary of National Biography* (online edn, Oxford: Oxford University Press, 2004).

Pearce, Susan, 'Antiquaries and the Interpretation of Ancient Objects, 1770–1820', in Pearce, *Visions of Antiquity*, 147–72.

—— (ed.), *Visions of Antiquity: The Society of Antiquaries of London, 1707–2007* (London: SAL, 2007).

Peterfreund, Stuart, '"Great Frosts and ... Some Very Hot Summers": Strange Weather, the Last Letters, and the Last Days in Gilbert White's *The Natural History of Selborne*', in Heringman, *Romantic Science*, 85–109.

Piggott, Stuart, *Ruins in a Landscape: Essays in Antiquarianism* (Edinburgh: Edinburgh University Press, 1976).

Poovey, Mary, *A History of the Modern Fact: Problems of Knowledge in the Sciences of Wealth and Society* (Chicago: University of Chicago Press, 1998).

—— 'The Social Constitution of "Class": Toward a History of Classificatory Thinking', in Wai Chee Dimock and Michael T. Gilmore (eds), *Rethinking Class: Literary Studies and Social Formations* (New York: Columbia University Press, 1994), 15–56.

Porter, Roy, *The Making of Geology: Earth Science in Britain 1660–1815* (Cambridge: Cambridge University Press, 1977).

Ramage, Nancy, 'The Initial Letters in Sir William Hamilton's "Collection of Antiquities" ', *Burlington Magazine*, 129 (1987), 446–56.

—— 'The Publication Dates of Sir William Hamilton's Four Volumes', *Ars Ceramica*, 8 (1991), 35–36.

—— 'Sir William Hamilton as Collector, Exporter, and Dealer: The Acquisition and Dispersal of His Collections', *American Journal of Archaeology*, 94:3 (1990), 469–80.

Reeve, Matthew, 'Jacob Schnebbelie, Draughtsman to the Society of Antiquaries (1760–92), and the Politics of Preservation in Late-Eighteenth-Century England', *Transactions of the Ancient Monuments Society*, 51 (2007), 160–89.

—— *Thirteenth-century Wall Painting of Salisbury Cathedral: Art, Liturgy, and Reform* (Suffolk: Boydell and Brewer, 2008).

Rennie, Neil, *Far-Fetched Facts: The Literature of Travel and the Idea of the South Seas* (Oxford: Clarendon Press, 1995).

Rice, Tony, *Voyages of Discovery: Three Centuries of Natural History Exploration* (London: Scriptum Editions, 1999).

Rossi, Paolo, *The Dark Abyss of Time: The History of the Earth & the History of Nations from Hooke to Vico* (1979), trans. Lydia G. Cochrane (Chicago: University of Chicago Press, 1984).

Rousseau, George S., 'The Sorrows of Priapus: Anticlericalism, Homosocial Desire, and Richard Payne Knight', in George S. Rousseau and Roy Porter (eds), *Sexual Underworlds of the Enlightenment* (Chapel Hill: University of North Carolina Press, 1988), 101–53.

Rudwick, Martin J. S., *Bursting the Limits of Time: The Reconstruction of Geohistory in the Age of Revolution* (Chicago: University of Chicago Press, 2005).

—— 'The Emergence of a Visual Language for Geological Science, 1760–1840', *History of Science*, 14:3 (1976), 149–95.

—— *The Great Devonian Controversy: The Shaping of Scientific Knowledge among Gentlemanly Specialists* (Chicago: University of Chicago Press, 1985).

Russett, Margaret, *Fictions and Fakes: Forging Romantic Authenticity, 1760–1845* (Cambridge: Cambridge University Press, 2006).

Ruthenberg, Peter, 'John Webber—Portraits from the Pacific of the 1700s', in Little and Ruthenberg, *Life in the Pacific of the 1700s*, Vol. III, 42–43.

Sachs, Jonathan, *Romantic Antiquity: Rome in the British Imagination, 1789–1832* (Oxford: Oxford University Press, 2010).

St. Clair, William, *The Reading Nation in the Romantic Period* (Cambridge: Cambridge University Press, 2004).

Schnapp, Alain, 'Antiquarian Studies in Naples at the End of the Eighteenth Century. From Comparative Archaeology to Comparative Religion', in Imbruglia, *Naples in the Eighteenth Century*, 154–66.

—— *The Discovery of the Past: The Origins of Archaeology* (New York: Harry N. Abrams, 1997).

—— 'Introduction: Neapolitan Effervescence', *Journal of the History of Collections*, 19 (2007), 161–64.

Shell, Alison, 'Publishing Pompeii: A Study in Cultural Censorships', *Biblion*, 4 (1996), 17–34.

Siegel, Jonah, *Desire and Excess: The Nineteenth-Century Culture of Art* (Princeton: Princeton University Press, 2000).

Simpson, David, *Romanticism, Nationalism, and the Revolt against Theory* (Chicago: University of Chicago Press, 1993).

Siskin, Clifford, *The Work of Writing: Literature and Social Change in Britain, 1700–1830* (Baltimore: Johns Hopkins University Press, 1998).

Sleep, Mark, 'Sir William Hamilton (1730–1803): His Work and Influence in Geology', *Annals of Science* 25:4 (1969), 319–38.

Smiles, Sam, 'Art and Antiquity in the Long Nineteenth Century', in Pearce, *Visions of Antiquity*, 123–45.

—— 'Data, Documentation and Display in Eighteenth-Century Investigations of Exeter Cathedral', in Arnold and Bending, *Tracing Architecture*, 80–99.

—— *Eye Witness: Artists and Visual Documentation in Britain, 1770–1830* (Aldershot: Ashgate, 2000).

—— *The Image of Antiquity: Ancient Britain and The Romantic Imagination* (New Haven: Yale University Press, 1994).

—— and Stephanie Moser (eds), *Envisioning the Past: Archaeology and the Image* (Oxford: Blackwell, 2005).

Smith, Bernard, *European Vision and the South Pacific* (2nd edn, New Haven: Yale University Press, 1985).

Smith, Edward, *The Life of Sir Joseph Banks* (1911) (New York: Arno Press, 1975).

Sontag, Susan, *The Volcano Lover: A Romance* (New York: Doubleday, 1992).

Sox, David, *Quaker Plant Hunters: from North America's early frontier to the South Pacific: John & William Bartram of Philadelphia, then Sydney Parkinson on Captain Cook's* Endeavour *Voyage* (York: Sessions Book Trust, 2004).

Spencer, J. R. H., 'Coastal Profiles and Landscapes', in Carr, *Sydney Parkinson*, 260–74.

Stearn, William T., 'Account of the Botanical Explorations and Prints', in *Captain Cook's Florilegium* (London: Lion and Unicorn Press, 1973).

Sweet, Rosemary, *Antiquaries: The Discovery of the Past in Eighteenth-Century Britain* (London: Hambledon and London, 2004).

—— 'Antiquaries and Antiquities in Eighteenth-Century England', *Eighteenth-Century Studies*, 34:2 (2001), 181–206.

—— 'The Incorporated Society and Its Public Role', in Pearce, *Visions of Antiquity*, 74–97.

Tapsell, Paul, 'Footprints in the sand: Banks's Maori collection, Cook's first voyage 1768–7', in Michelle Hetherington and Howard Morphy (eds), *Discovering Cook's Collections* (Canberra: National Museum of Australia Press, 2009), 92–111.

Thomas, Nicholas, *Cook: The Extraordinary Voyages of Captain James Cook* (New York: Walker & Co., 2003).

—— *Entangled Objects: Exchange, Material Culture, and Colonialism in the Pacific* (Cambridge, MA: Harvard University Press, 1991).

—— *In Oceania: Visions, Artifacts, Histories* (Durham, NC: Duke University Press, 1997).

Tilley, Christopher, *The Materiality of Stone: Explorations in Landscape Phenomenology* (Oxford: Berg, 2004).

Tobin, Beth Fowkes, *Colonizing Nature: The Tropics in British Arts and Letters, 1760–1820* (Philadelphia: University of Pennsylvania Press, 2005).

Torrens, Hugh, 'Arthur Aikin's Mineralogical Survey of Shropshire 1796–1816 and the Contemporary Audience for Geological Publications', *British Journal for the History of Science*, 16:2 (1983), 111–53.

—— 'Mary Anning (1799–1847) of Lyme; "The Greatest Fossilist the World Ever Knew" ', *British Journal for the History of Science*, 28:3 (1995), 257–84.

Toscano, Maria, 'The Figure of the Naturalist-Antiquary in the Kingdom of Naples: Giuseppe Giovene (1753–1837) and His Contemporaries', *Journal of the History of Collections*, 19 (2007), 225–37.

Trendall, A. D., *Red Figure Vases of South Italy and Sicily* (London: Thames and Hudson, 1989).

—— and Alexander Cambitoglou, *The Red-Figured Vases of Apulia*, 2 vols (Oxford: Clarendon Press, 1978).

Trumpener, Katie, *Bardic Nationalism: The Romantic Novel and the British Empire* (Princeton: Princeton University Press, 1997).

Urban, Manfred, 'The Acquisition History of the Göttingen Collection', in Hauser-Schäublin and Krüger, *James Cook: Gifts and Treasures*, 56–85.

Vickers, Michael, 'Hamilton, Geology, Stone Vases, and Taste', *Journal of the History of Collections*, 9:2 (1997), 263–74.

—— 'Imaginary Etruscans: Changing Perceptions of Etruria since the Fifteenth Century', *Hephaistos*, 7–8 (1985–86), 153–68.

—— 'Value and Simplicity: A Historical Case' in Michael Vickers and David Gill (eds), *Artful Crafts: Ancient Greek Silverware and Pottery* (Oxford: Oxford University Press, 1994), 1–32.

Weissert, Caecilie, 'Kunsttheorie versus Wissenschaft. D'Hancarville's Kritik and den bemalten antiken Vasen', in Maria Petras (ed.), *Hancarville und die Hamiltonsche Vasensammlung* (Stendal: Winckelmann-Gesellschaft, 2005), 7–20.

Williams, Glyndwr, 'Tupaia: Polynesian Warrior, Navigator, High Priest—and Artist', in Felicity Nussbaum (ed.), *The Global Eighteenth Century* (Baltimore: Johns Hopkins University Press, 2003), 38–51.

Wood, Karen, 'Making and Circulating Knowledge through Sir William Hamilton's *Campi Phlegraei*', *British Journal for the History of Science*, 39 (2006), 67–96.

Worrall, David, 'Blake's *Jerusalem* and the Visionary History of Britain', *Studies in Romanticism*, 16:2 (1977), 189–216.

Zeidler, Wolfgang, 'The Drawings of Hyperiid Amphipods ... by Sydney Parkinson', *Archives of Natural History*, 22:2 (1995), 267–81.

Index